Mechanical Behavior of Shape Memory Alloys: 2022

Mechanical Behavior of Shape Memory Alloys: 2022

Editor

Salvatore Saputo

MDPI • Basel • Beijing • Wuhan • Barcelona • Belgrade • Manchester • Tokyo • Cluj • Tianjin

Editor
Salvatore Saputo
Politecnico di Torino
Torino
Italy

Editorial Office
MDPI
St. Alban-Anlage 66
4052 Basel, Switzerland

This is a reprint of articles from the Special Issue published online in the open access journal *Materials* (ISSN 1996-1944) (available at: https://www.mdpi.com/journal/materials/special_issues/alloys_2021).

For citation purposes, cite each article independently as indicated on the article page online and as indicated below:

LastName, A.A.; LastName, B.B.; LastName, C.C. Article Title. *Journal Name* **Year**, *Volume Number*, Page Range.

ISBN 978-3-0365-8048-7 (Hbk)
ISBN 978-3-0365-8049-4 (PDF)

Contents

About the Editor

Salvatore Saputo

Salvatore Saputo is a mechanical engineer currently working as an assistant professor at Politecnico di Torino. His main research activities involve the mechanical behaviour of composite aeronautical structures under low/high velocity impact, and the numerical and experimental investigation regarding the thermo-mechanical properties of Ni–Ti shape memory alloy used to realize powerful miniaturized actuators. He got his BSc with an activity concerning the effects that the shape relationships of a bottom-heated canal have on a controlled airflow in mixed convention. Then, he gained his MSc in December 2010 at Seconda Università Degli Studi di Napoli discussing a thesis about the optimization of skins and core of a trunk in unidirectional composite fibre material with a sandwich structure. Posteriorly, in April 2016, he obtained his Ph.D. at the University of Campania "Luigi Vanvitelli" with a thesis concerning the development of numerical procedures for intralaminar damage evaluation in composite material induced by low-velocity impacts.

Article

Structure and Mechanical Properties of the NiTi Wire Joined by Laser Welding

Tomasz Goryczka [1,*], Karol Gryń [2], Adrian Barylski [1] and Barbara Szaraniec [2]

[1] Institute of Materials Engineering, University of Silesia in Katowice, 75 Pułku Piechoty 1A, 41-500 Chorzów, Poland; adrian.barylski@us.edu.pl

[2] Faculty of Materials Sciences and Ceramics, Department of Biomaterials and Composites, AGH University of Science and Technology, al. Mickiewicza 30, 30-059 Kraków, Poland; kgryn@agh.edu.pl (K.G.); szaran@agh.edu.pl (B.S.)

* Correspondence: tomasz.goryczka@us.edu.pl; Tel.: +48-32-3497-519

Abstract: Joining wires made of NiTi alloys with shape memory effect and pseudoelasticity causes many technical and structural problems. They result from unwanted phase interactions that occur in high temperatures and negatively affect the characteristics of these materials. Such obstacles are challenging in terms of welding. Hence, an attempt was made to join NiTi wires via an economical and reliable basic laser welding technique which does not require complicated equipment and gas protection. The parameters such as spot diameter and pulse time were constant and only the laser power, calculated as a percentage of the total power, was optimized. The wires were parallelly connected with overlapping seam welds 10 mm long. The welds were examined regarding their microstructure, chemical and phase composition, reversible martensitic transformation, microhardness, and pseudoelasticity. The obtained results showed that the joint was completed at the 12–14% power. The weld revealed good quality with no voids or pores. As the laser power increased, the microhardness rose from 282 (for 4%) to 321 (for 14%). The joint withstood the stress-inducing reversible martensitic transformation. As the transformation was repeated cyclically, the stress value decreased from 587 MPa (initial wire) to 507 MPa (for the 14% power welded wire).

Keywords: NiTi shape memory alloy; welding; microhardness; pseudoelasticity

Citation: Goryczka, T.; Gryń, K.; Barylski, A.; Szaraniec, B. Structure and Mechanical Properties of the NiTi Wire Joined by Laser Welding. *Materials* **2023**, *16*, 2543. https://doi.org/10.3390/ma16072543

Academic Editor: Dezső Beke

Received: 28 February 2023
Revised: 20 March 2023
Accepted: 21 March 2023
Published: 23 March 2023

1. Introduction

Due to their shape memory phenomena, NiTi alloys are known for numerous practical applications [1]. Although almost 50 at.% of content is taken by nickel, the alloys reveal good corrosion resistance. Therefore, they are commonly used for medical implants as well as elements of medical instruments [2–5]. Modern medicine is consistently striving to use less-invasive treatments. For instance, implants are introduced with guide wires through veins, arteries, or urinary ducts which do not exceed a few millimeters in diameter [6,7]. Such procedures require the implant's size and volume to match the available diameter of the supply canal. That is why it is essential to reduce the size of NiTi implant elements made of wires, rods, and tapes measuring below 1 mm in diameter. A good example are stents made of wires or flat bars whose forming elements range from several hundred micrometers to 0.8 mm [8–10].

In order to create a complex, three-dimensional structure of an implant, an appropriate method of forming and connecting elements or their fragments is necessary. There are several techniques of joining NiTi alloys, including: plasma welding [11], vacuum brazing [12], resistance welding [13], tungsten inert gas [14], friction welding [15,16], microplasma arc welding [15], capacitor discharge welding [15], explosive welding [16], ultrasonic welding [16], adhesives joining [16], and laser welding [15–21]. However, most of these methods are used to join sheets, tubes, tapes, or strips. For thin wires, the laser welding is an efficient, economical, and uncomplicated technique that can join particularly thin wires applied in medical procedures [20,22]. Thanks to this method, one can control the size of the welded area on a micro-scale. Moreover, it is unnecessary to use fluxes, solvents, or other additives. Therefore, the material is free from cytotoxic substances. In addition, at high temperature locally generated by the laser beam, the welding process supports the material's sterilization.

Not much literature is devoted to joining thin wires made of NiTi alloy of less than 0.5 mm in diameter. The papers reported the possibility of butt welding of NiTi wires of 100 μm whose strain was 3% at a tension of 500 MPa [17]. In work [20], the welding parameters for a 0.5 mm wire were optimized by selecting the power range of 54 W–72 W. The welding time was relatively long (85–115 ms), but the authors did not analyze the influence of welding conditions on pseudoelasticity. Chan et al. [23] showed the cycling repetition of pseudoelasticity for butt welded 0.5 mm wires whose maximum strain was 4%. Another method of joining wires with a stitch parallel to the wire axis was proposed in [19]. Two NiTi wires of 0.44 mm, intended for cardiac stents, were connected parallelly. The microstructure, thermal transformation behavior, and the strength of the obtained weld were analyzed, yet, again, no attempt was made to assess their pseudoelasticity.

Therefore, in the presented work, we supplemented the reported results with the study of cyclically generated pseudoelasticity in the parallelly connected thin NiTi wires. We also revealed how the laser welding parameters, performed with a low-power device, influenced the weld microstructure, related to the thermal reversibility of martensitic transformation and microhardness. The tests were conducted on the elements made of 0.4 mm wires intended for cardiac surgeries.

2. Materials and Methods

The studies were carried out on a commercially available NiTi wire with a diameter of 0.4 mm (Figure 1a). Two 45 mm long wires were bonded parallelly with the 10 mm overlapping seam weld (Figure 1b). The welding was performed on a desktop jewelry welding machine (PixoLaser OPT-JW100, Shenzhen City, China). In order to investigate the influence of laser welding on the material's characteristics, only one parameter was customized—the power of the laser beam (Curr%). For convenience, it was calculated as a percentage of the total laser power (100 W). The Curr equal to 4, 8, 10, 12, and 14% was applied to obtain the overlapping seam weld. The other welding parameters were constant: the laser beam's cross-section was circular with the 0.1 mm spot diameter and the 0.4 ms pulse duration.

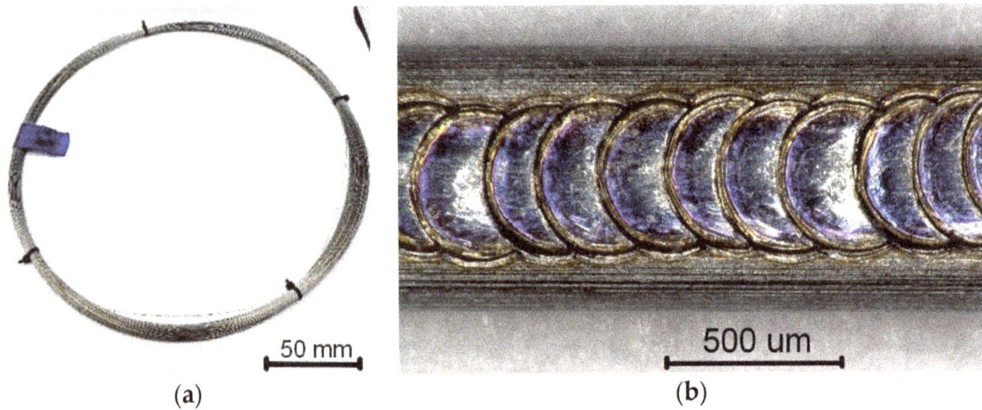

Figure 1. As-delivered NiTi wire (**a**) and an example of the overlapping seam weld (**b**).

The welds' microstructure was observed on their cross-sections using the digital microscope VHX-900F (KEYENCE, Osaka, Japan) and the scanning electron microscope (SEM) JSM 4680 (JEOL, Tokyo, Japan). The SEM was operated at 20 kV and equipped with an X-ray energy dispersive spectrometer (EDS). The 3 mm long samples were embedded in graphite and polished with sandpaper. The final polishing was carried out with polishing pastes of 1µm gradation. The surface was etched in a H_2O:HNO_3:HF solution in the 10:5:1 ratio.

The thermal behavior of the martensitic transformation was studied using the Mettler Toledo DSC 1 (Greifensee, Switzerland) differential scanning calorimeter (DSC).

The structural examination was performed via X-ray diffraction patterns measured with the X'Pert-PRO diffractometer (PANalytical, Almelo, the Netherlands) with $CuK\alpha_{(1\ and\ 2)}$ radiation. The measurements were taken at room temperature using the Bragg–Brentano geometry at the step-scan mode in an angular 2θ range: 10–140°. The ICCD-PDF4 (International Centre for Diffraction Data) database was used for phase identification.

The micromechanical tests were performed with the Micro Combi Tester—MCT3 device (Anton Paar, Corcelles-Cormondrèche, Switzerland). The measurements complied with the ISO 14577 standard. The Berkovich diamond indenter (B-V 83) was used with the maximum load of 250 mN and a load/time of 30 s. The HV_{IT} hardness (hardness determined during indentation) was determined by the Oliver–Pharr method [24,25].

The tensile test to determine the shear strength of the weld was performed on the universal testing machine Zwick 1435 Retroline (Zwick Roell, Ulm, Germany) equipped with the 5 kN load cell. The measurements were carried out at room temperature, following the ASTM F2516 standard due to the practical application of welded wires. The diagram of the sample mounting and the acting force is shown in Figure 2.

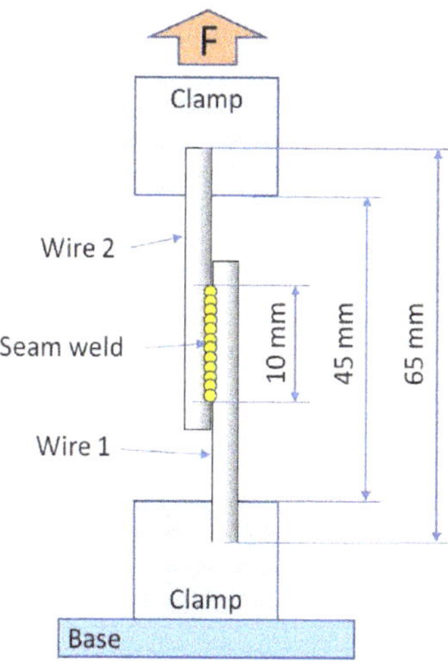

Figure 2. Scheme of mounting the welded wires in the tensile test holders.

3. Results and Discussion

3.1. The Initial State of the Wire

In order to characterize the wire before welding, the detailed analyses of the microstructure and chemical composition were carried out on transverse and longitudinal cross-sections. An example of microscopic observations is shown in Figure 3. The images revealed the compact structure of the wire, free of discontinuities, pores, or micro/macro cracks. However, there were numerous particles ranging from several tens of nanometers to several micrometers in size (Figure 3). The average chemical composition of the matrix oscillated around the equiatomic one. A slight nickel predominance with the differences in tenths of a percent was found, practically the same in both cross-sections (Table 1). The chemical composition analysis of the particles showed that they contained almost twice as much titanium as nickel—which is characteristic of the Ti_2Ni equilibrium phase (Figure 4).

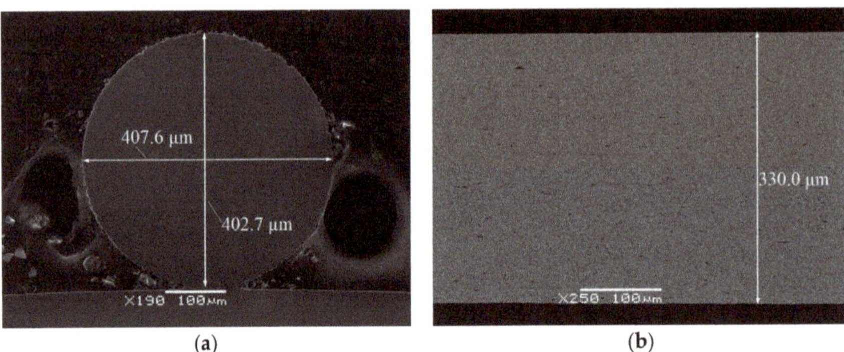

(a) (b)

Figure 3. Images of the wire's cross-sections: (**a**) transverse (SEM-SE); (**b**) longitudinal (SEM-BS).

Table 1. Average chemical composition (at.%) determined at the wires' cross-section.

Alloying Element	Perpendicular	Longitudinal
Ti	49.7 ± 0.4	49.7 ± 0.4
Ni	50.3 ± 0.4	50.3 ± 0.4

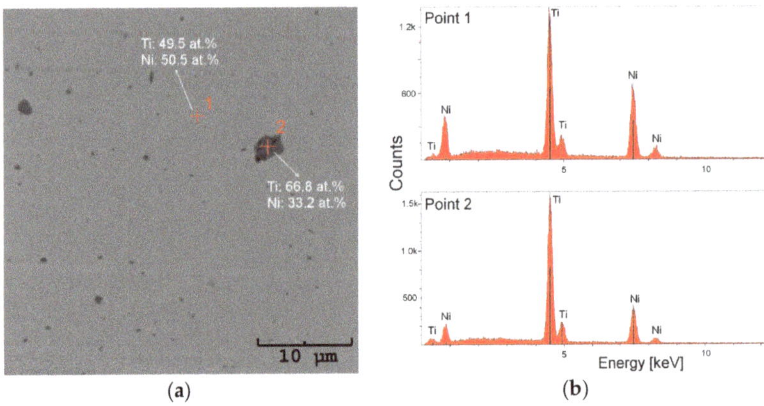

(a) (b)

Figure 4. SEM image of the transverse wire's cross-sections with Ti_2Ni particles (**a**) as well as the measured EDS spectra (**b**).

The phase analysis of the measured diffraction pattern showed that the matrix consisted of the β-NiTi intermetallic phase. Depending on the temperature, this phase can reveal a high-temperature B2-type or a low-temperature structure with the B19′ monoclinic martensite. In our study, the wire's distribution of diffraction peaks and their intensity corresponded only to the B2 structure (ICDD card no. 65-0917) (Figure 5). The parent phase present at room temperature met the requirements for pseudoelasticity [26].

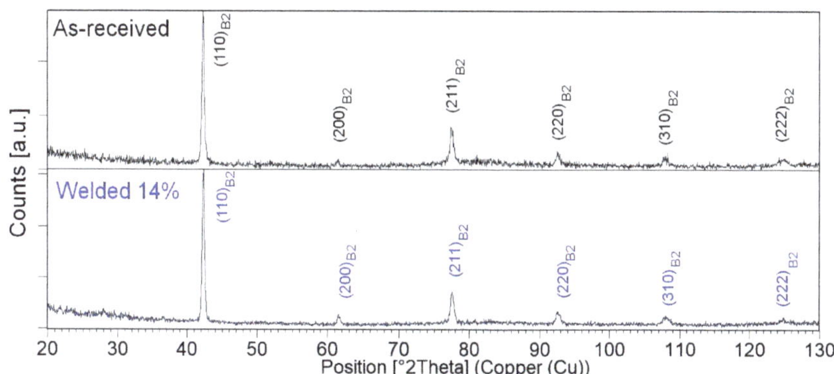

Figure 5. X-ray diffraction patterns measured for the as-received (black-line) and the welded wires with the Curr power of 14% (blue-line).

3.2. Weld Characteristics

The applied welding parameters influenced the weld quality, shape, structure, phase, and chemical composition—thus the mechanical properties. Depending on the laser power, the weld is formed in the joined material as a one of the three types (Figure 6). The insufficient laser power melted a too-small amount of the material and the weld did not sufficiently fill the area between the connected wires (Figure 6a). On the contrary, too high power led to the perforation of the connected wires and the weld itself (Figure 6c). The optimally selected power formed the complete weld between the joined wires with the cross-sectional shape shown in Figure 6b.

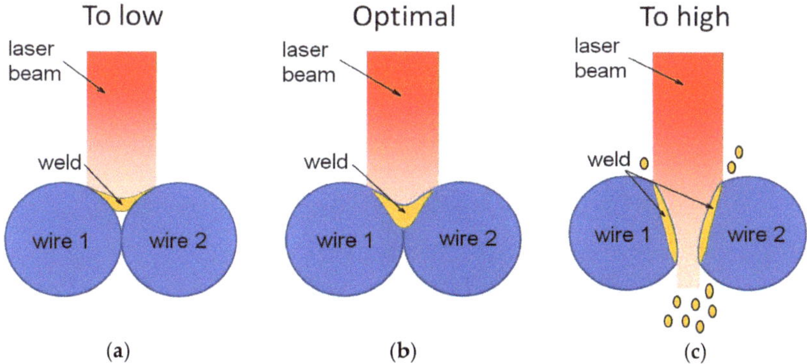

Figure 6. Scheme of the formation of the weld cross-section depending on the applied laser power.

Due to the form of the welded material—a wire of 0.4 mm in diameter,—the parameters such as pulse time and spot size were set as constants of 0.4 ms and 0.1 mm, respectively. The laser power calculated as a percentage of the device's maximum power was adopted as a variable. A relatively small spot and a short pulse time of the laser beam made it possible to remelt a small area and cool it quickly. The fast cooling process is beneficial for NiTi alloys, as it preserves the high-temperature B2 phase at room temperature without decomposing into equilibrium phases. In addition, the effect of alloy oxidation is reduced without the need for additional shielding gas. However, the disadvantage of such parameters is the poor mechanical characteristics of the weld. Hence, the formation of the appropriate welding shape (Figure 6b) depended on the laser beam power.

Figure 7 shows how the laser power formed the weld bead. Based on the microscopic images, using 3D modelling (Figure 7b), the weld's cross-section was determined (Figure 7c).

The results showed that welding with a laser power lower than 4% produced a small fusion volume (Figure 7—part relating to 4%). On the contrary, the laser power above 16% formed a welding crater. Consequently, the laser power above 50% caused the weld's gradual, power-dependent perforation. A completely different shape was obtained by applying the 8–14% power. The comparison showed that in that case the fusion profile was the closest to the optimal one (Figure 6b). That is why the welds produced with the 4–14% laser power were selected for further research.

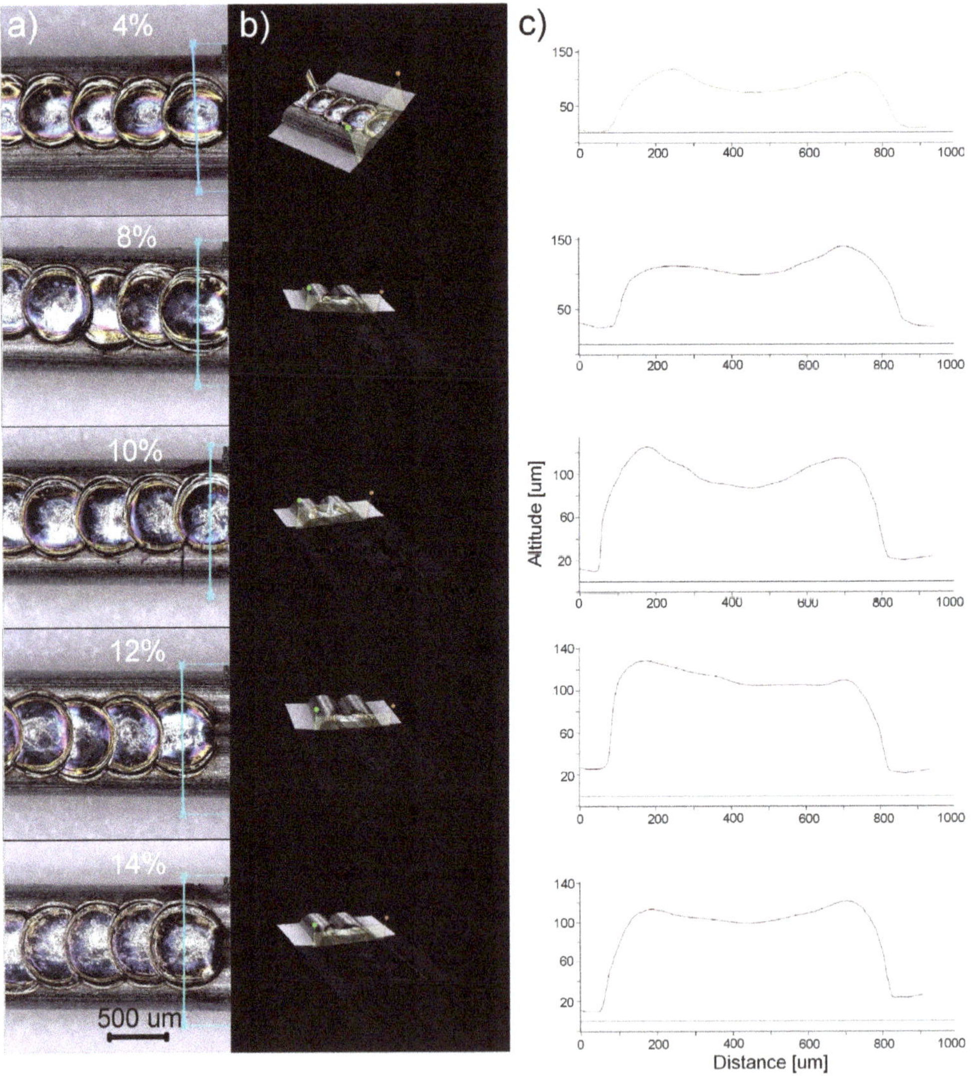

Figure 7. Microscopic image of the weld bead with an example of the weld profile modelling (marked as a blue line) (**a**), the 3D model of the welded area (**b**) and the modelled weld profile (**c**).

3.3. Martensitic Transformation in Welded Wires

The reversible martensitic transformation is a measurable effect confirming pseudoelasticity. Its course was examined based on measured thermograms (DSC). The samples containing the area of 2–3 welded points were cut out for the tests. The measured thermograms for the sample in the initial state and the welded wires were summarized in Figure 8. The determined characteristic parameters of martensitic transformation are shown in Table 2.

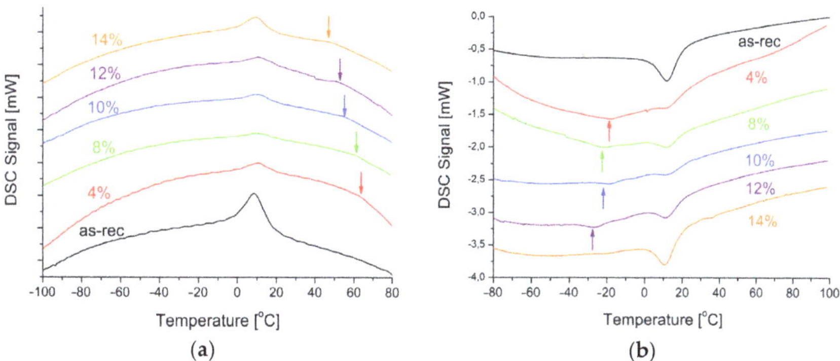

Figure 8. DSC cooling (**a**) and heating (**b**) curves measured for the as-received wire and the welded one with various laser powers.

Table 2. Transformation temperatures and enthalpies determined for the as-received wires as well as the welded one.

Power [%]	M_s [°C]	M_f [°C]	ΔH [J/g]	A_s [°C]	A_f [°C]	ΔH [J/g]
As-received	18.9	−4.9	3.39	0.3	20.6	2.85
4%	17.3	−1.8	0.9	−33.3	21.9	2.9
8%	22.3	0.2	1.84	−31.4	22.6	2.04
10%	21.6	1.6	1.58	−27.3	22.8	1.55
12%	20.3	−1.1	1.87	−36.9	21.01	2.07
14%	20.3	1.3	3.11	1.02	19.94	2.48

The occurrence of martensitic transformation and its reversibility is the most important information coming from the thermal analysis. In the DSC cooling curves, there were thermal peaks (Figure 8a) which corresponded to the peaks in the heating direction (Figure 8b). They were broadened, as compared to the ones appearing in the solid material produced via traditional casting. This phenomenon resulted from the grain refinement and/or the increase in the density of structural defects—primarily dislocations. The dislocations are characteristic for the wire production—particularly, while reducing the wire diameter. An additional effect visible on the DSC curves was the multi-stage transformation depending on the applied laser power. The DSC cooling curves showed deviations from the baseline in the 55–75 °C range (indicated by arrows in Figure 8a), and in the −40 °C to −10 °C range in the heating curves (indicated by arrows in Figure 8b). The similar thermal behavior was observed for the laser-treated wire of 100 μm in thickness [27].

Considering the nature of the martensitic transformation, those two facts are not related to each other. Let us first discuss the −40 °C to 10 °C range of heating curves. They may result from the multi-stage martensitic transformation occurring through an additional R-phase which changes the sequence B2↔B19′ to B2↔R↔B19′ [28]. During the cooling process, the B2↔R transformation shows a very narrow temperature range of about 7–10 degrees when the difference between the peak maximum temperatures is measured. This value is characteristic of the B2↔R sequence. In the B2↔B19′ transformation, the necessary difference is 20–30 degrees and it depends on the nickel content. However, in the DSC cooling curves, in the range from −10 °C to practically −100 °C, it was evident that the curves deviated from the usual straight baseline. This may be caused by local stress fields associated with the higher density of dislocations and their clustering under an additional heat treatment, e.g., laser [29]. Consequently, the areas which require cooling undergo a martensitic transformation in different temperature ranges, broadening the thermal peak by up to several dozen degrees. In the reverse transformation, the broad thermal peaks occurred in the range of −40 °C to −10 °C. Increasing the beam power to 14% provided enough energy to rebuild the material's structure. As a result, the course of the martensitic transformation was similar to the initial wire.

The deviations from the baseline occurring at higher temperatures (55–75 °C) had a completely different character and origin (marked with arrows in Figure 8a). They were not identified as peaks as they did not meet the mathematical criteria adopted to determine the thermal peak. However, it is a fact that the DSC curves change their course in this temperature range. As the delivered heat is proportional to the amount of transforming material, it can be assumed that a small volume in the welded fusion area transforms in this thermal range. Moreover, with the laser power (delivered energy) increase, the main thermal peak enthalpy decreases. This phenomenon is confirmed by a different volume of the material transforming in the discussed temperature range. In order to clarify this aspect, microscopic observations were carried out, and the chemical composition was measured in the weld areas.

The observations were performed on the transverse cross-sections of the welded wires. The exemplary images for 4% and 14% power are shown in Figure 9. First, the SEM observations confirmed the 3D modelling from the light microscope images. The SEM images did not reveal the Ti_2Ni particles in the fusion area, regardless of the applied welding power. At the wire's initial state, the Ti_2Ni particles were evenly distributed throughout the entire volume (Figures 3 and 4). In the fusion area, the precipitations completely disappeared. This effect was visible in the enlarged image of the wire welded with the 4% power—Figure 9a—an area marked as "B". The Ti_2Ni particles in the fusion zone contained almost 7–10 at.% less titanium than the particles outside the heat-affected zone. In contrast, the particles displayed a slightly bigger amount of nickel as they dissolved in the weld center. However, the particles were depleted in titanium at the periphery of the heat-affected zone. An example is the particle shown in Figure 9a—the enlarged area marked with "A", located in the heat-affected zone with the titanium content reduced to 61.2 at.%. As titanium passed into the weld area, its amount increased.

The fusion zone contained about 0.4 at.% higher amount of titanium than the rest of the wires. It is a known fact that an increase of about 0.1 at.% Ti raises martensitic transformation temperatures by about 10 °C. In contrast, an increase in the nickel content by 0.1 at.% decreases the transformation temperatures by about 10 °C. The slightly higher titanium content in the fusion zone caused the martensitic transformation. As a result, the broadened maximum appeared in the DSC cooling curves in the 55–75 °C range (Figure 8a). The comparison of the DSC cooling curves indicated the decreased titanium content in the fusion zone, depending on the laser power. The chemical composition of the weld matrix revealed the titanium content lowered to almost 49 at.% (Figure 9b—an area marked as "C"). In addition, the higher the laser power (starting from 8%), the more carbon-containing areas can be found. An example is shown in Figure 9b—the area marked "D". The inclusion particles contained almost three times more titanium than nickel and up to 55 at.% of carbon. Such inclusions were also observed by the authors [30], identifying them as titanium carbides. Such a distribution of elements proved that the nickel content increase up to 51 at.% caused the martensitic transformation at lower temperatures. It was visible in the DSC heating curves in the −40 °C to −10 °C range. The A_f temperature is the most important factor regarding pseudoelasticity. The temperature analysis (Table 2) showed that for the welded wires it was comparable to the temperature of the initial state changing slightly in the 1.5 °C range. Therefore, it can be concluded that pseudoelasticity occurred in the welded wires subjected to an external stress at room temperature.

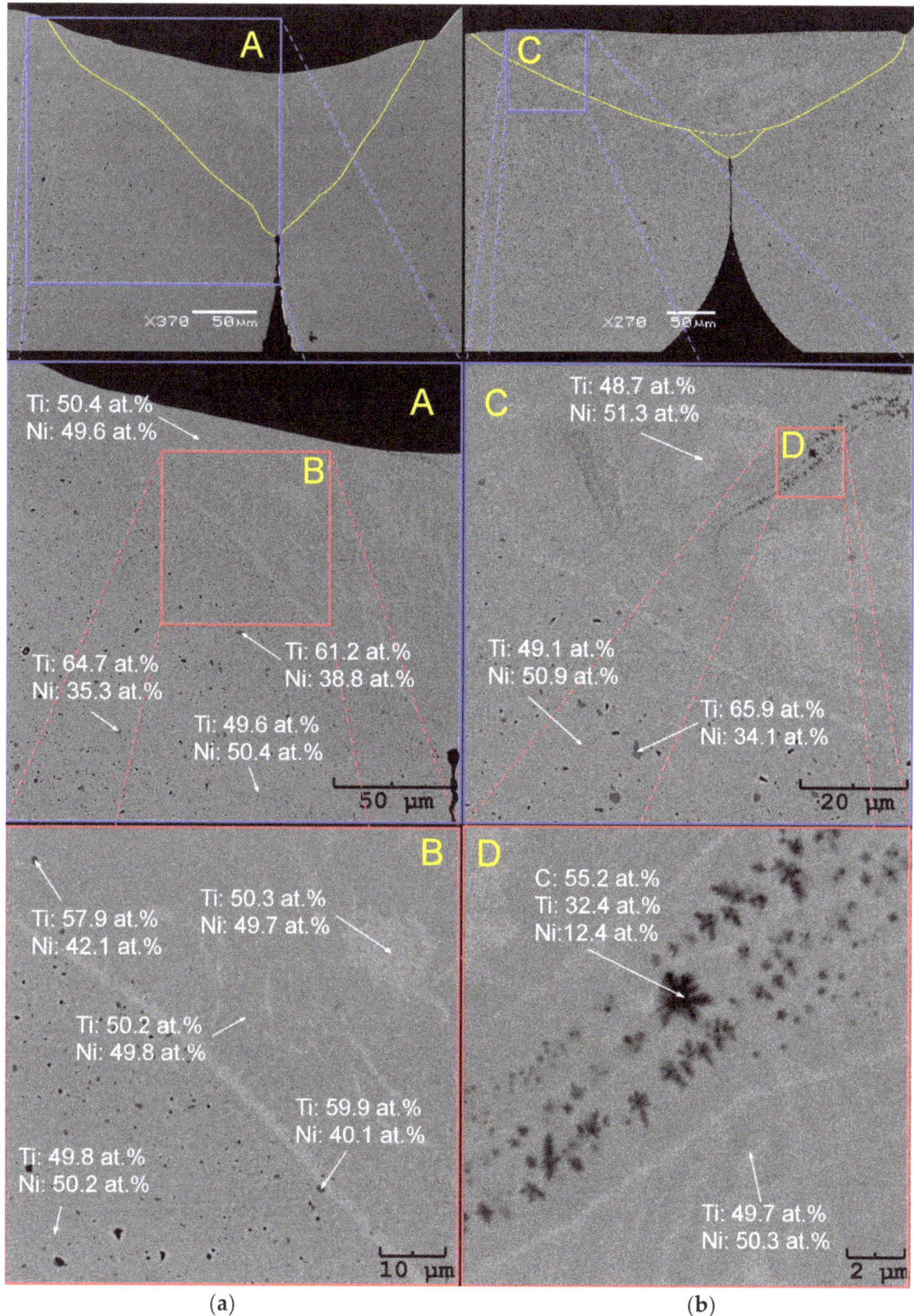

(a) (b)

Figure 9. SEM-BS images observed on the transverse cross-section of the wires welded with the power of 4% (**a**) and 14% (**b**).

3.4. The Hardness of the Welded Area

Microhardness is the material's resistance to local deformations caused by a mechanical impact on a small area. In the case of the welded wires subjected to an external stress, the microhardness test showed the possible stress concentration areas. The indentation measurements were taken on the wire's transverse cross-sections and the welded area. The measurements were based on a relationship between the test force applied to the sample—F [mN] and the h [μm] displacement. The results are presented in Figure 10 (the as-received wire—Figure 10a, the 10% laser power sample—Figure 10b). To measure the mechanical properties of the welds, the Vickers hardness HV_{IT} was determined. The values are summarized in Table 3.

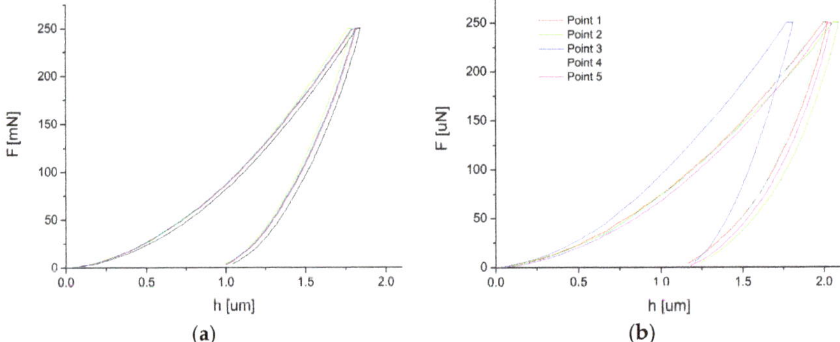

Figure 10. Results of the indentation test done on the wire's cross-section (**a**) and the weld obtained with the laser power of 10% (**b**).

Table 3. Average Vickers hardness in dependency on the applied laser power.

Laser Power [%]	Vickers Hardness HV_{IT}	Remarks
0	381.9 ± 10.2	wire
4	282.4 ± 13.8	weld
8	302.1 ± 21.1	weld
10	304.6 ± 28.2	weld
12	310.8 ± 20.6	weld
14	321.4 ± 26.4	weld

The average microhardness determined for the wire before its welding indicated a relatively high value of around 382. These values were characteristic of the B2 parent phase identified in the welded wires. Generally, the B2 phase has a higher hardness than the B19′ martensite. For example, in the laser-treated NiTi alloy which comprised of about 80% B2 phase and 20% B19″ martensite, the micro-hardness ranged from 325 to 340 [31]. Depending on the welding parameters, in the NiTi wire welded to NiTiCu, the microhardness varied from 210 to 260 [32]. Additionally, the hardness value depended on the nickel content in the NiTi alloy. In work [33], the nickel content was 50.7 to 51.27 at.%. The alloy was eviscerated via the laser bed fusion. As a result, the hardness varied from 372 to 724. The chemical composition of the reported wires indicated that the hardness measurements were done for the B2 parent phase. Increasing the nickel content above 50.5 at.% resulted in the Ni_4Ti_3 precipitates occurrence, which increased the alloy's hardness.

The determined average microhardness for the wire in the initial state equalled 381, which was slightly higher than the cited literature data. The reason behind this result might be the Ti_2Ni particles' presence. Figure 11 shows the SEM image with an exemplary indentation in the wire cross-section. The densely distributed Ti_2Ni particles can be clearly distinguished from the NiTi matrix. Their influence on the measurement cannot be avoided, as it is known that the increased density of dislocations and the decreasing grain size affect mechanical properties. Such phenomena led to the reduction of the wire diameter and the mechanical properties increase, e.g., hardness. In the fusion area, the Ti_2Ni precipitates which were characteristic of the wire microstructure in the initial state were not noted. They appeared only sporadically on the edges of the weld. In the heat-affected zone, the chemical composition differed in comparison to the internal part of the wires. The melting process eliminated structural defects such as dislocations. Both facts contributed to the lower microhardness. For the lowest welding power of 4%—the determined hardness decreased to almost 282. However, increasing the welding power caused the microhardness boost, and for 14%, it was 321. This trend was consistent with [31,32]. In our studies, the microhardness increase took place due to inclusions—the (Ti,Ni)C phase containing over 50 at.% of carbon. The particles formed clusters around the remelted weld bead as the laser power increased. For the lower laser powers (e.g., 8%), the clusters were observed only at the surface of the weld face. While increasing the welding power, the clusters appeared throughout the entire welded volume, affecting the microhardness value (Figure 10b). An exemplary microstructure of the indentation area free of inclusions is shown in Figure 12a—marked as Point 1. The representative curve measured for this indent refers to the red line marked in Figure 10b. Its shape and course are characteristic of the remaining measured points, except for Point 3—a blue line in Figure 10b. The microstructure observed for Point 3—Figure 12b—revealed a band of hard inclusions containing the (Ti,Ni)C phase, particularly in the indented area. Consequently, the microhardness determined for Point 3 was 360, which was comparable to the wire in the initial state. Increasing the volume fraction of the carbide phase in the weld boosted its hardness.

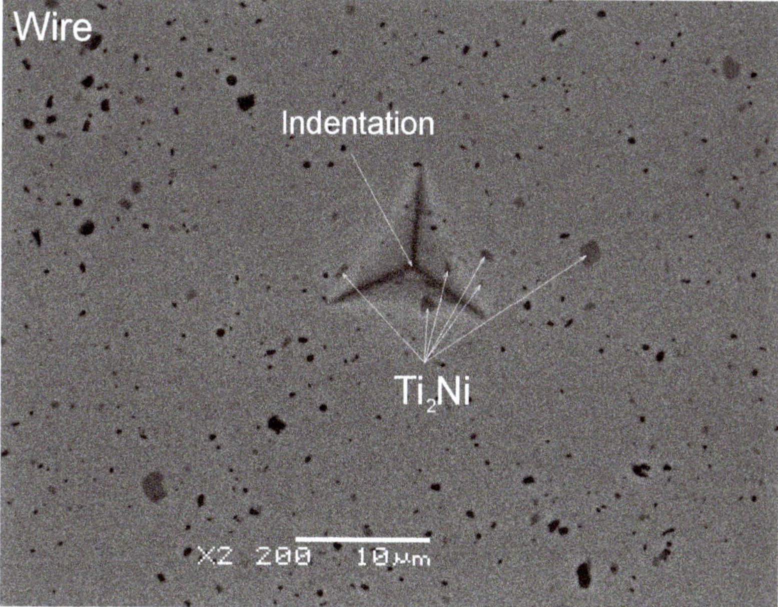

Figure 11. SEM-BS image of the exemplary indentations in the wire.

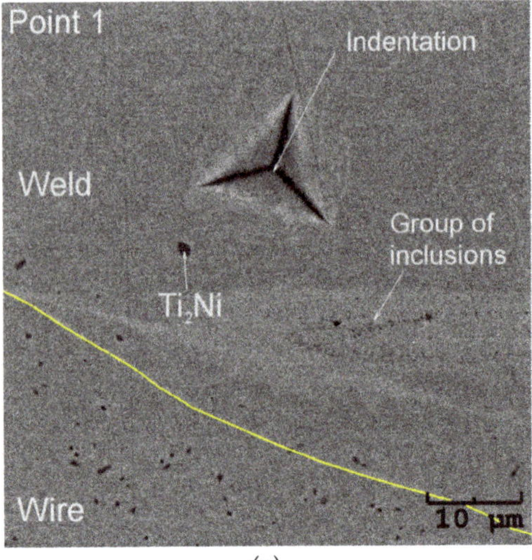

(a)

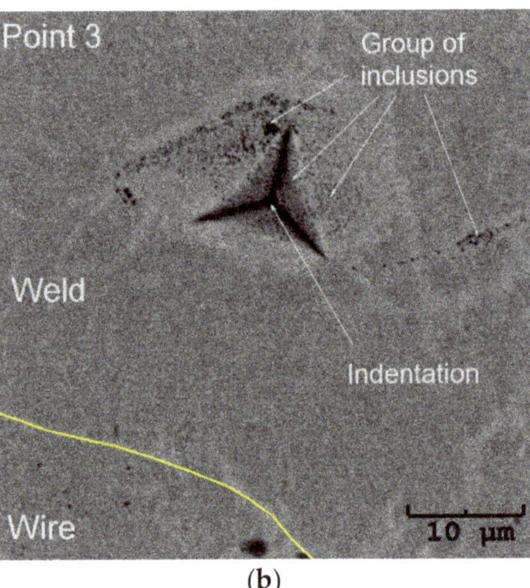

(b)

Figure 12. SEM-BS images of the indentations done in the fusion zone for Point 1 (**a**) and Point 3 in the wires welded with the laser power of 10% (**b**).

3.5. Pseudoelastic Behaviour of the Welded Wires

In order to check pseudoelasticity of the wires, a tensile test was carried out. Preliminary studies showed that the martensitic transformation was completed at a strain lower than 7.2% when the elastic deformation of martensite began. Hence, the experiments were carried out to obtain a maximum elongation of 7% for the as-received wire and 6.8% for the welded wires. The results are summarized in Figure 13. For each sample, five load/unload cycles were measured, except for the wire welded at 4%. After starting the second cycle and reaching the critical stress to trigger the martensitic transformation, the sample crashed, which proved the fusion zone's amount, size, and/or shape to be inadequate. The stresses concentrated at the boundary of the heat-affected zone and the parent wires were high enough to destroy the joint. After the first cycle, all the samples behaved similarly and the residual plastic deformation appeared, as reported in the literature [23]. The loops were closed in the remaining cycles, regardless of the applied welding power.

The critical stress to induce the martensitic transformation was calculated from the load/unload curves. The results, depending on the applied laser power, are shown in Figure 14. In the case of the forward martensitic transformation, the critical stress reached the highest value of 587 MPa for the as-received wire. Similar results were obtained by the authors in [20]. The cyclic load/unload repetition decreased the critical stress to 549 MPa for the fourth cycle, and for the fifth cycle it remained the same. This means that after four cycles, the martensitic transformation began to stabilize. The samples welded with the 4% and 8% powers revealed a similar strain of 524 MPa. The 10% and 12% welded samples required 518 MPa to induce martensite. Welding with the 14% power significantly reduced the critical stress value to 507 MPa, which was notably different from the wire's initial state. The fourth and fifth cycle load/unload gave similar results to the initial state results, regardless of the applied welding power.

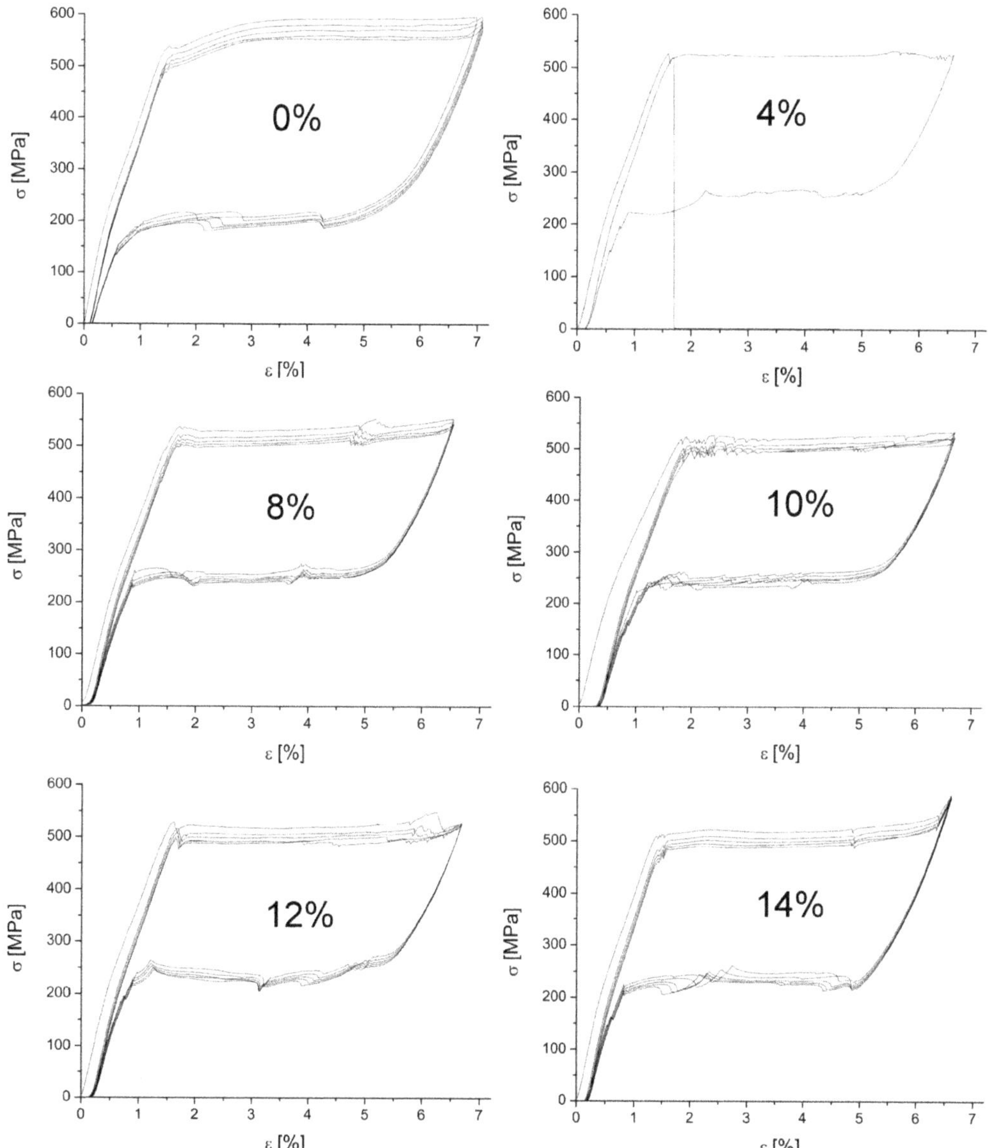

Figure 13. Stress-strain curves measured for the as-received as well as for the welded ones at the various percentages of the maximum (100%) laser beam power.

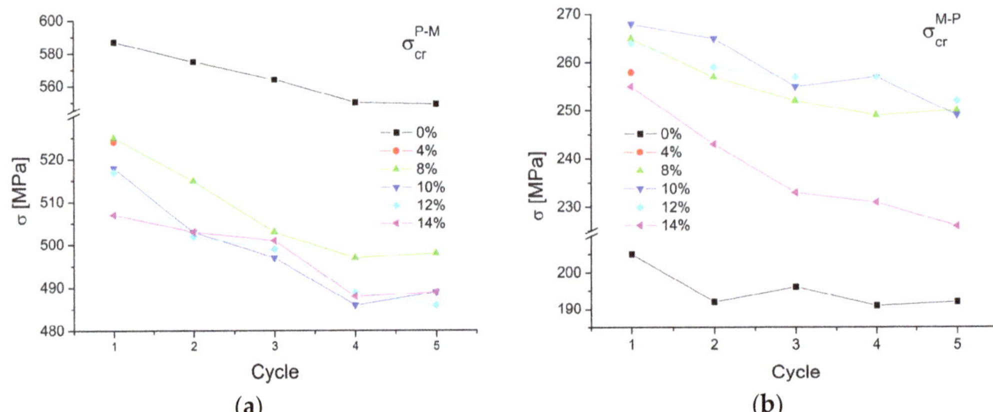

Figure 14. The critical stress triggering the martensitic transformation for the forward (**a**) and reverse course (**b**).

In the case of the reverse transformation, for the as-received wire, the stress inducing the martensitic transformation was the lowest and amounted to about 190 MPa. For the welded wires (the laser power from 4% to 12%), the first cycle required about 265 MPa to start the reverse transformation. For 14%, it was the lowest one—255 MPa. This means that for the welded specimens, the wire began to return to its original shape at a higher stress.

Summing up the critical stresses required to initiate the martensitic transformation, the value decreased by about 6% after five cycles, as compared to the first cycle. This may result from structural changes in the welded area and the heat-affected zone. The spreading heat acted locally as an annealing factor, decreasing the material's density and/or clustering structural defects. It is known that heat affects the wire structure reconstruction and the grain growth after the production stage deformation.

4. Conclusions

The studies carried out on the welded wires, obtained with a simple laser apparatus and a variable power parameter, revealed the valid dependency. The following conclusions can be drawn from the research results:

- The applied laser beam power influenced the shape of the weld. The power ranged from 12% to 14%, assuming the complete filling of the space between the welded wires was achieved. However, the laser power above 14% caused the crater formation and the perforation of the joint.
- The fusion zone received with the low laser powers (below 8%) was free of Ti_2Ni precipitations, present in the entire volume of the wire, except for the weld. However, increasing the laser power caused turbulences of the liquid weld material, accommodating inclusions of nickel titanium carbide into the weld.
- The weld microhardness was characteristic of the B2 parent phase and increased with the higher laser power from 282 (4%) to 321 (14%). This tendency resulted from the inclusions as the laser power increased.
- Welding affected the martensitic transformation course by lowering its enthalpy and extending the temperature range it occurred in.
- The higher laser power lowered the critical stress needed to induce pseudoelasticity—from 570 MPa for the initial state of the wire to 507 MPa for the wire welded with the 14% power.
- For the wire in the initial state and the welded samples—regardless of the applied welding power—the reversible martensitic transformation course stabilized after four load/unload cycles. Moreover, as pseudoelasticity was cyclic, the critical stress was reduced by 6–7%.

Author Contributions: Conceptualization, T.G.; methodology, T.G., K.G. and A.B.; software, T.G., B.S., K.G. and A.B.; validation, T.G., K.G., B.S. and A.B.; formal analysis, T.G.; investigation, T.G., K.G., B.S. and A.B.; resources, T.G., A.B. and K.G.; data curation, T.G., A.B., K.G. and B.S.; writing—original draft preparation, T.G.; writing—review and editing, T.G. and K.G.; visualization, T.G.; supervision, T.G. All authors have read and agreed to the published version of the manuscript.

Funding: The studies were financially supported by the European Regional Development Fund for 2014–2020 of the Regional Operational Program of the Malopolska Region; grant number RPMP.01.02.01-120059/19-00.

Institutional Review Board Statement: Not applicable.

Informed Consent Statement: Not applicable.

Data Availability Statement: The data presented in this study are available on request from the corresponding authors.

Conflicts of Interest: The authors declare no conflict of interest.

References

1. Jani, M.J.; Leary, M.; Subic, A.; Gibson, M.A. A review of shape memory alloy research, applications and opportunities. *Mater. Des.* **2014**, *56*, 1078–1113. [CrossRef]
2. Duerig, T.; Pelton, A.; Stöckel, D.J.M.S. An overview of nitinol medical applications. *Mater. Sci. Eng.* **1999**, *A273–275*, 149–160. [CrossRef]
3. Yahia, L. *Shape Memory Implants*; Springer: Berlin/Heidelberg, Germany, 2000.
4. Biscarini, A.; Mazzolai, G.; Tuissi, A. Enhanced Nitinol properties for biomedical applications. *Recent Pat. Biomed. Eng.* **2008**, *1*, 180–196. [CrossRef]
5. Petrini, L.; Migliavacca, F. Biomedical Applications of Shape Memory Alloys. *J. Metall.* **2011**, *2011*, 501483.
6. Morgan, N.B. Medical shape memory alloy applications-the market and its products. *Mater. Sci. Eng. A* **2004**, *378*, 16–23. [CrossRef]
7. Song, C. History and current situation of shape memory alloys devices for minimally invasive surgery. *Open Med. Devices J.* **2010**, *2*, 24–31. [CrossRef]
8. Garg, S.; Magro, M.; Serruys, P.W. Drug-Eluting Stents. *Compr. Biomater.* **2011**, *6*, 427–448.
9. Stoeckel, D.; Pelton, A.; Duerig, T. Self-expanding nitinol stents: Material and design considerations. *Eur. Radiol.* **2004**, *14*, 292–301. [CrossRef]
10. Stoeckel, D. Nitinol—A material with unusual properties. *Endovasc. Udate* **1998**, *1*, 1–5.
11. Eijk, C.V.; Fostervoll, H.; Sallom, Z.K.; Akselsen, O.M. Plasma welding of NiTi to NiTi, stainless steel and hastelloy C 276. In Proceedings of the ASM Materials Solutions 2003 Conference, Pittsburgh, PA, USA, 13–15 October 2003.
12. Tillmann, W.; Eilers, A.; Henning, T. Vacuum brazing and heat treatment of NiTi shape memory alloys. *IOP Conf. Ser. Mater. Sci. Eng.* **2021**, *1147*, 012025. [CrossRef]
13. Delobelle, V.; Delobelle, P.; Liu, Y.; Favier, D.; Louche, H. Resistance welding of NiTi shape memory alloy tubes. *J. Mater. Process. Technol.* **2013**, *213*, 1139–1145. [CrossRef]
14. Oliveira, J.P.; Barbosa, D.; Braz Fernandes, F.M.; Miranda, R.M. Tungsten inert gas (TIG) welding of Ni-rich NiTi plates: Functional behavior. *Smart Mater. Struct.* **2016**, *25*, 03LT01. [CrossRef]
15. Kramár, T.; Tauer, J.; Vondrouš, P. Welding of nitinol by selected technologies. *Acta Polytech.* **2019**, *59*, 42–50. [CrossRef]
16. Oliveira, J.P.; Miranda, R.M.; Braz Fernandes, F.M. Welding and joining of NiTi shape memory alloys: A review. *Prog. Mater. Sci.* **2017**, *88*, 412–466. [CrossRef]
17. Gugel, H.; Schuermann, A.; Theisen, W. Laser welding of NiTi wires. *Mater. Sci. Eng. A* **2008**, *481–482*, 668–671. [CrossRef]
18. Khan, M.I.; Panda, S.K.; Zhous, Y. Effects of welding parameters on the mechanical performance of laser welded nitinol. *Mater. Transit.* **2008**, *49*, 2702–2708. [CrossRef]
19. Falvo, A.; Furgiuele, F.M.; Maletta, C. Laser welding of a NiTi alloy: Mechanical and shape memory behaviour. *Mater. Sci. Eng. A* **2005**, *412*, 235–240. [CrossRef]
20. Raju, R.M.; Salamat-Zadeh, F.; Brriesci, G. Laser micro welding of nitinol for cardiovascular applications. In Proceedings of the 2nd World Congress on Electrical Engineering and Computer Systems and Science (EECSS'16), Budapest, Hungary, 16–17 August 2016.
21. Khan, M.I.; Zhou, Y. Micro-welding of shape-memory alloys. In *Joining and Assembly of Medical Materials and Devices*; A volume in Woodhead Publishing Series in Biomaterials; Zhou, Y., Breyen, M.D., Eds.; Woodhead Publishing Limited: Oxford, UK; Cambridge, MA, USA; Philadelphia, PA, USA; New Delhi, India, 2013; pp. 133–153.
22. Levi, D.S.; Kusnezov, N.; Carman, P. Smart materials applications for pediatric cardiovascular devices. *Pediatr. Res.* **2008**, *63*, 552–558. [CrossRef]

23. Chan, C.W.; Man, H.C.; Yue, T.M. Effect of postweld heat treatment on the microstructure and cyclic deformation behavior of laser-welded NiTi-shape memory wires. *Metall. Mater. Trans. A* **2012**, *43*, 1956–1965. [CrossRef]
24. Oliver, W.C.; Pharr, G.M. An improved technique for determining hardness and elastic modulus using load and displacement sensing indentation experiments. *J. Mater. Res.* **1992**, *7*, 1564–1583. [CrossRef]
25. Instrumented Indentation Testing (IIT). Available online: https://wiki.anton-paar.com/us-en/instrumented-indentation-testing-iit/ (accessed on 2 February 2023).
26. Tsuchiya, K. Mechanisms and properties of shape memory effect and pseudoelasticity in alloys and other materials: A practical guide. In *Shape Memory and Pseudoelastic Alloys-Technologies and Application*; Yamauchi, K., Ohkata, I., Tsuchiya, K., Miyazaki, S., Eds.; Woodhead Publishing Limited: Oxford, UK; Cambridge, MA, USA; Philadelphia, PA, USA; New Delhi, India, 2011; pp. 3–12.
27. Biffi, C.A.; Casati, R.; Tuissi, A. Laser shape setting of thin NiTi wires. *Smart Mater. Struct.* **2016**, *25*, 01LT02. [CrossRef]
28. Goryczka, T.; Morawiec, H. Structure studies of the R-phase using X-ray diffraction methods. *J. Alloy. Compd.* **2004**, *367*, 137–141.
29. Morawiec, H.; Ilczuk, J.; Stróż, D.; Goryczka, T.; Chrobak, D. Two-stage martensitic transformation in NiTi alloys caused by stress fields. *J. Phys IV Fr.* **1997**, *7*, C5-155–C5-159.
30. Zhang, Z.; Frenzel, J.; Neuking, K.; Eggeler, G. On the reaction between NiTi melts and crucible graphite during vacuum induction melting of NiTi shape memory alloys. *Acta Mater.* **2005**, *53*, 3971–3985. [CrossRef]
31. Marattukalam, J.J.; Singh, A.K.; Datta, S.; Das, M.; Balla, V.K.; Bontha, S.; Kalpathy, S.K. Microstructure and corrosion behavior of laser processed NiTi alloy. *Mater. Sci. Eng. C* **2015**, *57*, 309–313. [CrossRef] [PubMed]
32. Mehrpouya, M.; Gisario, A.; Barletta, M.; Broggiato, G.B. Welding strength of dissimilar laser-welded NiTi and NiTiCu shape memory wires. *Manuf. Lett.* **2019**, *22*, 25–27. [CrossRef]
33. Wen, S.; Liu, Y.; Zhou, Y.; Zhao, A.; Yan, C.; Shi, Y. Effect of Ni content on the transformation behavior and mechanical property of NiTi shape memory alloys fabricated by laser powder bed fusion. *Opt. Laser Technol.* **2021**, *134*, 106653. [CrossRef]

 materials

Article

An SMA Transducer for Sensing Tactile Sensation Focusing on Stroking Motion

Ryusei Oya [1] and Hideyuki Sawada [2,*]

1 Graduate School of Advanced Science and Engineering, Waseda University, Tokyo 169-8555, Japan
2 Faculty of Science and Engineering, Waseda University, Tokyo 169-8555, Japan
* Correspondence: sawada@waseda.jp

Abstract: The authors have developed a micro-vibration actuator using filiform SMA wire electrically driven by periodic electric current. While applying the SMA actuators to tactile displays, we discovered a phenomenon that the deformation caused by a given stress to an SMA wire generated a change in the electrical resistance. With this characteristic, the SMA wire works as a micro-force sensor with high sensitivity, while generating micro-vibration. In this paper, the micro-force sensing ability of an SMA transducer is described and discussed. Experiments are conducted by sliding the SMA sensor on the surface of different objects with different speeds, and the sensing ability is evaluated to be related with human tactile sensation.

Keywords: shape-memory alloy wire; force sensing; superelasticity; tactile sensor; human tactile perception; texture sensing

Citation: Oya, R.; Sawada, H. An SMA Transducer for Sensing Tactile Sensation Focusing on Stroking Motion. *Materials* **2023**, *16*, 1016. https://doi.org/10.3390/ma16031016

Academic Editor: Antonio Mattia Grande

Received: 18 December 2022
Revised: 17 January 2023
Accepted: 18 January 2023
Published: 22 January 2023

1. Introduction

Virtual reality (VR) and augmented reality (AR) technologies have developed rapidly, and are expected to be widely employed as the 'second world' in a variety of scenarios such as the metaverse and new forms of entertainment and business communication. These technologies provide realistic experiences mainly by controlling visual and auditory information to be perceived by users' various senses. The presentation of tactile sensation, which is recognized by our skin, is an important technology to enhance perceptions of reality in VR and AR worlds; however, such technologies are still in the research stage, and no commercially-available devices and systems have been introduced so far. If tactile senses could be controlled in a virtual world, users would be able to further enjoy highly immersive games in entertainment fields and to extend the means of flexible communication in various business opportunities. Tactile communication will also be able to open new medical care in tele-medicine and tele-surgery. In order to create a system that controls tactile senses, it will be necessary to quantitatively measure the tactile information. For dealing with visual and auditory information, commonly-used devices such as cameras, microphones and speakers are commercially available. However, no standard equipment to record and display tactile information has been made available thus far.

Tactile sensing is expected to be utilized in various industries, and has been actively studied so far. Typical sensors applicable in tactile sensing include capacitive sensors, piezoresistive sensors, optical methods, and piezoelectric devices [1–3]. Capacitive tactile sensors are designed to settle capacitors between two electrodes, so that the electrostatic capacity changes when force is applied and the distance between the electrodes changes [4,5]. The capacitive sensors have the advantages of high frequency response, precise load measurement, and wide dynamic range, but they tend to be susceptible to external noise. Piezoresistive sensors, on the other hand, are robust against electric noise, since they utilize the piezoresistive effect [6,7]. However, they require a high-voltage power source for driving, and the frequency response is comparatively low. Optical tactile sensors are operated by converting a load into light intensity or refractive index [8,9]. They have the advantages of high frequency response, precise load measurement, and wide dynamic range. However, they are constructed with larger peripheral systems with optical sensors. Piezoelectric tactile sensors are operated by converting a load into an electrical signal via a piezoelectric element [10,11]. The sensors have the advantage of high frequency response, but their spatial resolution is comparatively low. Other tactile sensing techniques using ultrasonic [12] and acceleration [13] elements have also been proposed recently, but technical difficulties such as high energy consumption and complex system structures have been reported when considering practical applications.

A shape-memory alloy (SMA) is well-known as an alloy that remembers its original shape. An SMA is able to memorize its shape as an original one in austenite phase at higher temperature. The SMA changes the state from austenite to martensite upon cooling. It is deformable by the application of external force in the martensite phase at lower temperature, and returns to its preliminary-remembered shape when heated to transit into austenite phase [14,15]. Shape-memory alloys are widely applied as thermo-mechanical actuators in devices such as shape-memory springs, thermo-responsive valves and catheters, and highly efficient energy converters in the biomedical [16,17], aerospace [18,19], automotive [20,21], robotics [22,23] and construction fields [24,25], as well as human-machine interfaces [26,27]. Owing to the characteristics of their light weight, compact size and generatable force, the SMA actuators are expected to be alternatives to conventional electronic actuators, electric motors, and pneumatic and hydraulic actuators. Shape-memory alloys can also be utilized as sensors and have been applied to temperature sensors [28,29], magnetic sensors [30,31], and strain sensors [32,33]. Their transformation speed is, however, comparatively slow, since the phase transformation between martensite phase and austenite phase is led by the temperature, which is conducted by giving heat to the material body or radiating heat to the surrounding environment.

A filiformed SMA wire with a diameter of 50–100 μm presents unique characteristics, swiftly responding to temperature changes related to the martensite and austenite phases. By applying weak current to an SMA wire, heat is generated by the internal electrical resistance, and the wire instantly shrinks up to 5% lengthwise. When the current is stopped and the temperature drops, it returns to its original length by heat radiating to the air. The SMA wire is thin and flexible enough to be cooled down right after the current stops, and it returns to the initial length as the temperature shifts from the austenite to the martensite phase. This means that the contraction and return of the SMA wire can be precisely controlled by a properly-prepared pulse current, and the phenomena is physically recognized as micro-vibration, having different frequencies and amplitudes.

The authors have developed a micro-vibration actuator electrically driven by periodic current, generated by a current control circuit for tactile displays. A vibration actuator is composed of a 5 mm-long SMA wire with a diameter of 50 μm, and micro-vibration with an amplitude of 1–2 μm is generated, which is perceived by the human body as various tactile sensations in accordance with the different frequencies. We employ pulse-width modulated (PWM) current to control the vibration mode of the SMA wire using a specially-designed current controller. The pulse has an amplitude of H volts, a width of W msec and a period of L msec. The duty ratio W/L determines the heating and cooling time of the SMA. The value H × W, which is equivalent to the calories exchanged, determines the amplitude of a vibration, and the vibration frequency is completely controlled by regulating L. The deformation of an SMA wire in motion was recorded by using a high-speed camera, and we verified that the SMA wire perfectly synchronized with the ON/OFF pulse current.

The authors have also discovered that the deformation caused by stress applied to an SMA wire generates a change in the electrical resistance. With this characteristic, the SMA wire works as a force sensor with high sensitivity, while generating micro-vibration. We have developed a sensor structure to effectively conduct a given stress to an SMA wire. In this paper, the micro-force-sensing ability of the SMA transducer is studied and discussed. Experiments are conducted by sliding the SMA transducer on the surface of different objects with different speeds, and the sensing ability will be evaluated so as to be related to human tactile sensations. The new transducer using SMA wires will be applied to tactile displays and sensors to measure and record various tactile sensations, and also to be effectively employed as physical feedback in VR and AR environments.

2. SMA Sensor and Tactile Sensing System

2.1. Physical Properties of SMAs

SMA has two typical physical properties related to body temperature and force: shape memory effect and superelasticity [34], respectively, as shown in Figure 1.

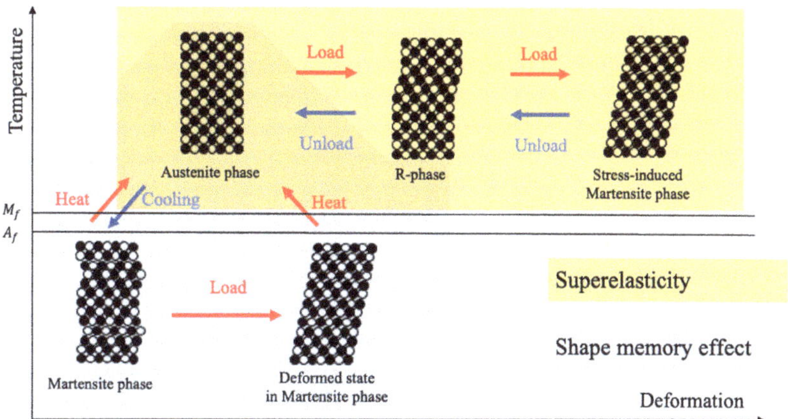

Figure 1. Physical properties of the deformation of an SMA.

The shape memory effect presented by the background colored in blue is observed by exchanging heat in the body. An SMA in martensite is deformable by the application of load, and the shape returns to its original form by receiving heat to transit to austenite. The transition between martensite and austenite is reversible by the heat application to the body or the heat radiation from the body.

The superelasticity, on the other hand, is the phenomenon observed in the austenite phase as shown in the orange color. A loaded SMA in austenite deforms to transit to martensite, which is called the stress-induced martensite phase, and the strain is released by removing the load. This transformation exhibits the change of electrical resistance, and especially the transformation between austenite and R-phase shows the quick response in time.

2.2. Filiform SMAs for Micro-Vibration Actuators

The authors have developed a micro-vibration actuator using a filiformed SMA wire electrically driven by pulsed current. Figure 2 shows a vibration actuator composed of a 5 mm-long SMA wire with a diameter of 0.05 mm. By applying weak current to the alloy, the temperature rises to T_2 due to the generated heat inside the wire body, and the alloy shrinks up to 5% lengthwise relative to the original length. When the current stops and the temperature drops to T_1 due to the heat radiation, the alloy returns to its original length. Figure 3 shows the temperature characteristics of the SMA wire employed in this study having the specific temperatures T_1 = 68 and T_2 = 72 degrees Celsius [35].

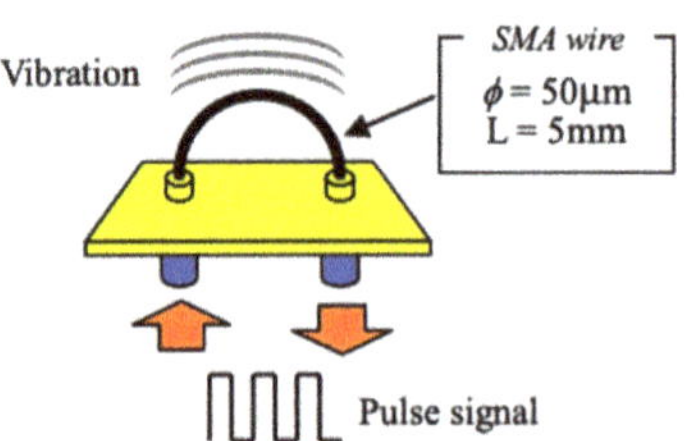

Figure 2. SMA vibration actuator.

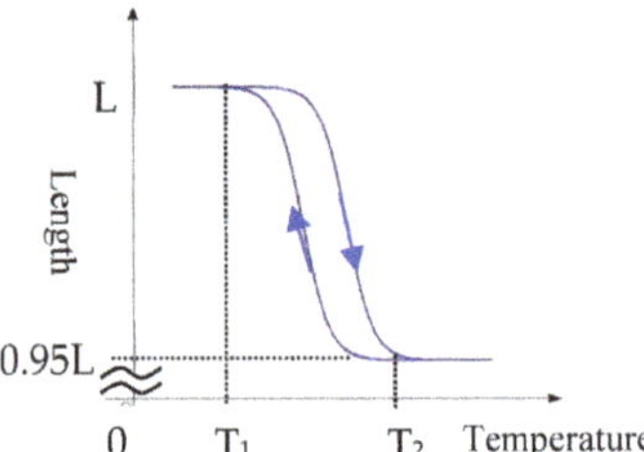

Figure 3. Temperature characteristics of SMA wire.

The SMA wire is so thin that it rapidly cools after the current stops, and returns to its original length when the temperature shifts from T_2 to T_1. This means that the shrinkage and the return to initial length of the SMA wire can be controlled by the pulse current. By driving the SMA wire with properly-prepared pulse current, micro-vibration with an amplitude of several micrometers is generated, which is perceived by the human body as tactile sensation, although the vibration is invisible. We employ pulse-width modulated (PWM) current having different frequencies and duty ratios generated from a specially-designed current controller to control the vibration mode of the SMA wire. Through our studies so far, we discovered that the vibration frequencies were controlled up to 1 kHz under duty ratios of approximately 2–20%.

2.3. SMA Wires for Force Sensing

Owing to the superelasticity, the deformation caused by a given tensile stress to an SMA wire generates the change of the electrical resistance. By measuring the resistance change, the amount of stress applied to a wire can be estimated.

Figure 4 illustrates the sensor structure using an SMA wire, which consists of a 75 μm (diameter) × 3 mm (length) SMA wire and a 1.5 mm (diameter) × 5 mm (length) round-head pin made of stainless steel. The SMA wire used for the tactile sensor is BMF75 (Toki Corporation), and its physical properties are shown in Table 1. BMF75 is a one-way shape-memory alloy, and is composed of Ti-Ni-Cu, which is characterized by stable self-expanding properties. The tip of the pin is soldered at the middle of the SMA wire, so that force applied to the other end of the pin is efficiently conducted to the SMA wire, as shown in Figure 5. Two lead wires are connected to both ends of the SMA wire to measure the electrical resistance by employing a specially-designed control circuit. The circuit provides weak electrical current to the SMA wire to keep the SMA in the austenite phase. When force is applied to the metal pin, tensile force is conducted to the wire to induce the stress-induced martensite phase, which is measured as the change in the electrical resistance.

We also tested the fatigue of the SMA wire against tensile force, which was anticipated due to the sensor structure. A 10 cm wire was prepared, and one end was fixed to a metal frame. At the other end, weights were hung until the wire was cut off. We confirmed the SMA wire was cut off at the weight of 1.3 kg, and the wire also properly worked after applying a tensile force of 1 kg.

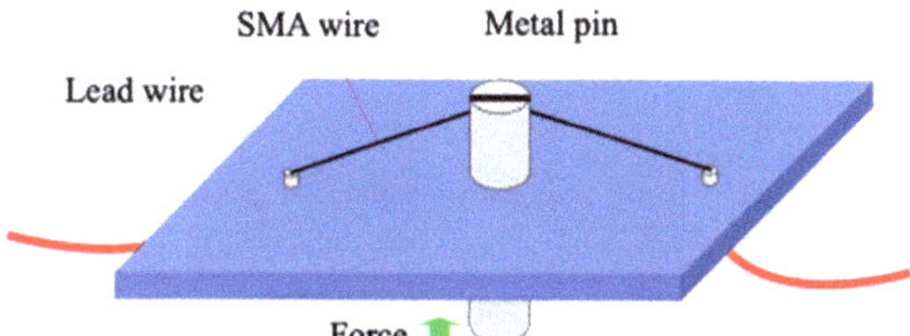

Figure 4. SMA sensor for sensing micro-force.

Table 1. Physical properties of BMF75.

Physical Property	Value
Standard diameter (μm)	75
Practical force produced (load) (gf)	35
Practical kinetic strain (%)	4.0
Standard drive current (mA)	140
Standard drive voltage (V/m)	35.4
Standard power (W/m)	4.63
Standard resistance (Ω/m)	236
Tensile strength (Kgf)	0.45
Weight (mg/m)	28

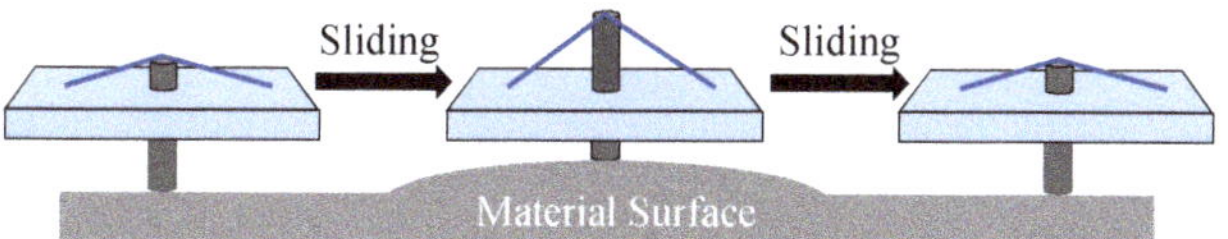

Figure 5. Sliding the SMA sensor on material surface.

2.4. *Structure of SMA Tactile Sensing System*

We have constructed a tactile sensing system, as shown in Figure 6. The system consists of an SMA transducer mounted on a liner plotter (AxiDraw SE/A3), a transducer control circuit, and a deep learning-based classifier. In this experiment, we consider a situation where a human feels a tactile sensation of an object by sliding their fingertip on the object surface. When we try to feel an object by hand, we may place our fingers on the surface and stroke our hand with different pressures and speeds to properly understand the textures. Different objects have different textures with various softness, roughness and elasticity. To suitably perceive the textures, the touching pressure and the sliding speed may be suitably adjusted to understand the different physical structures on the surface.

To realize the human-like recognition of tactile sensation, the SMA sensor is mounted on a linear plotter to control the sliding speed, as shown in Figure 7. A weight of 100 g is placed on the actuator so that the pressure can be applied while sliding the sensor on an object surface. Since the SMA wire keeps its performance with the strain up to 1 kg, the sensor works properly with a 100 g weight.

The resistance value of the SMA wire R_{sma} is measured by the following calculation:

$$V_{out} = 3.0 \cdot \frac{10}{10 + R_{sma}} \tag{1}$$

using the circuit shown in Figure 8. Time-series data obtained by the sensing system is given to a classifier using the Transformer [36,37].

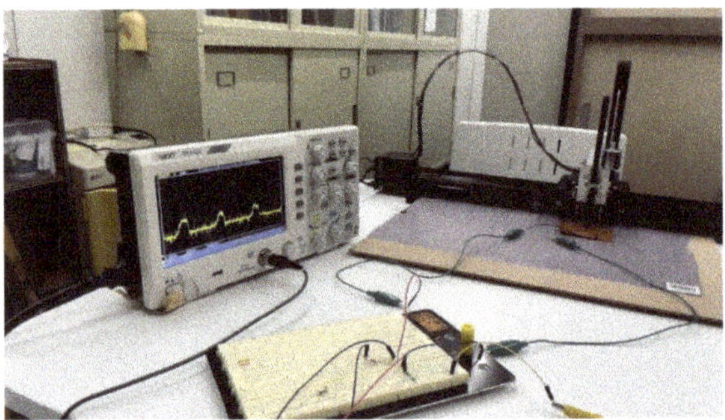

Figure 6. Tactile sensing system using a SMA sensor.

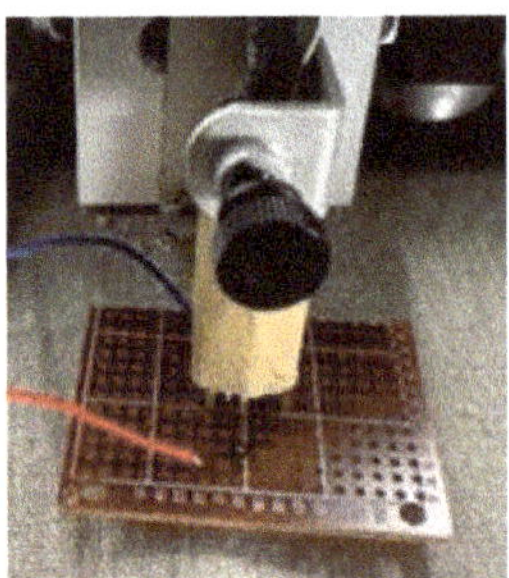

Figure 7. SMA sensor mounted on a linear plotter.

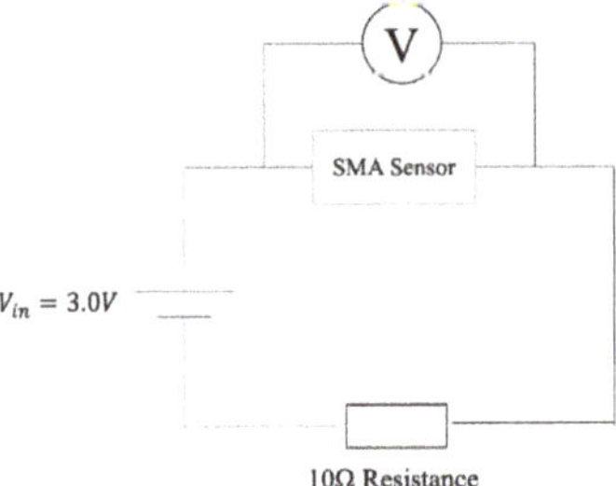

Figure 8. SMA driving circuit.

3. Preliminary Experiment

To find the sensing characteristics of the tactile transducer system, we firstly conducted a preliminary experiment. Five materials having different textures, namely a rubber mat, a plastic mesh, cotton fabric, a Velcro tape and a plastic sheet, were selected to test the sensing ability of the sensor. The relationship of the texture features is shown in Figure 9. For example, a rubber mat has a less bumpy surface with a periodic pattern, while by contrast, a Velcro tape is made up of random and rough surfaces. All the materials were prepared in a size of 30 cm × 40 cm, and were spread on a table. The SMA sensor unit mounted on the linear plotter stroked each material from one end to the other at eight different speeds, which were set at 1, 3, 5, 7.5, 10, 12.5, 15 and 20 cm/s. A weight of 100 g was placed on the sensor for applying initial stress to the SMA wire.

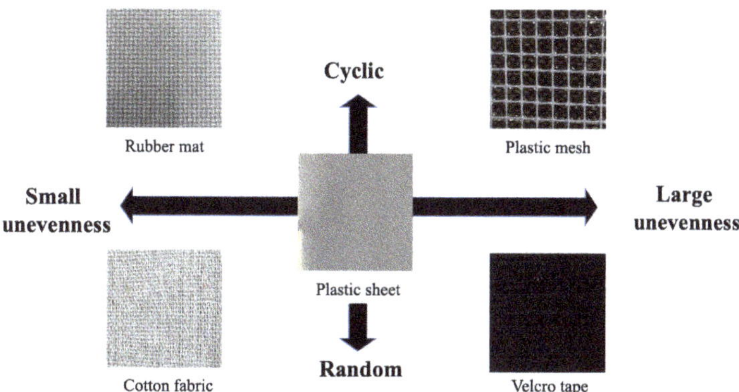

Figure 9. Selected five materials with different textures.

Sampling rate was set at 5000 Hz, and 50 stroking trials were conducted at each speed for each material. Figure 10 shows the examples of sensing data in one second obtained by changing the stroking speed. At each speed, different surfaces present different signal features, having particular periods and amplitudes. By changing the sliding speed, frequencies found in sensing data increase in accordance with the texture patterns and sliding speed. A plastic sheet, for example, has a smooth surface with low friction, and the obtained data have small amplitude without particular patterns. A plastic mesh, on the other hand, has a clear periodic pattern, and in the obtained data, clear repeating patterns are observed for all the different speeds.

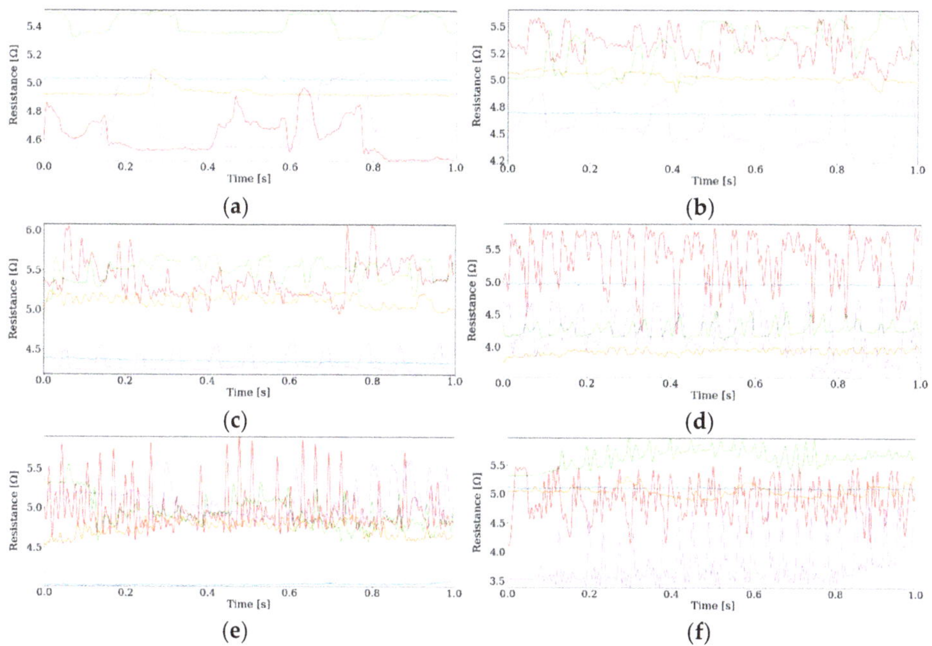

Figure 10. *Cont.*

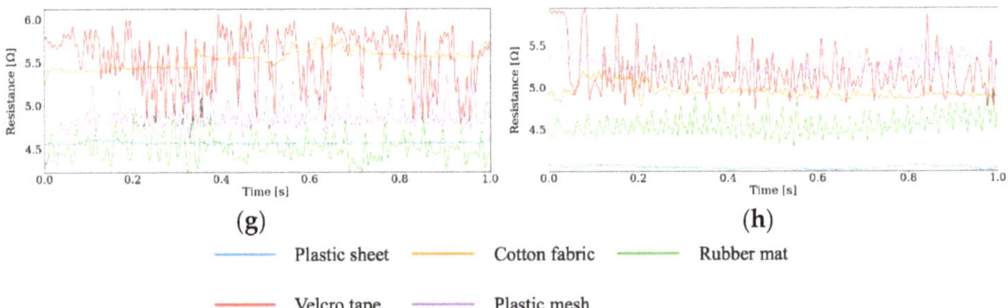

Figure 10. Examples of sensing data with different stroking speeds: (**a**) sliding speed 1 cm/s, (**b**) sliding speed 3 cm/s, (**c**) sliding speed 5 cm/s, (**d**) sliding speed 7.5 cm/s, (**e**) sliding speed 10 cm/s, (**f**) sliding speed 12.5 cm/s, (**g**) sliding speed 15 cm/s, (**h**) sliding speed 20 cm/s.

4. Tactile Classification Using SMA Sensing System

In the preliminary experiment, we discovered that the SMA sensor was able to output reasonable signals by the stroking actions on different material surfaces. Here, we conducted tactile classification by using ten materials with different textures, including the five materials used in the preliminary experiment. Figure 11 shows the pictures of the material surfaces, which were a rubber mat, a plastic mesh, cotton fabric, a Velcro tape, a plastic sheet, a sheet made of bamboo, a plastic-made tile, a plastic mat with small patterns, a plastic mat with large patterns, and a leather sheet with studs.

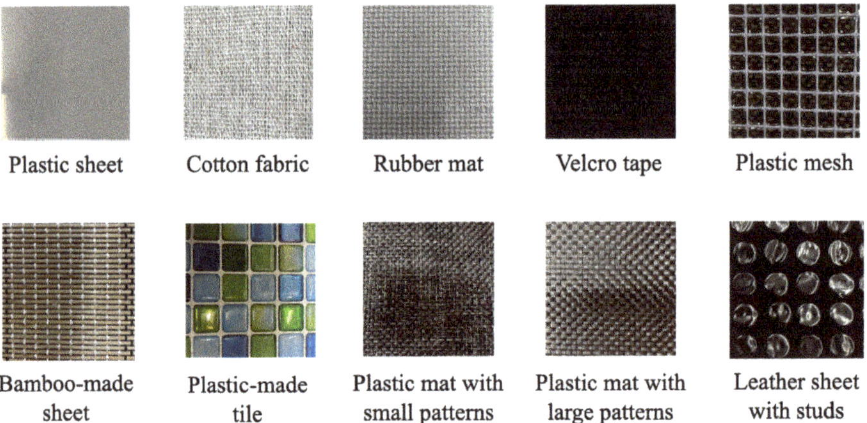

Figure 11. Ten materials for tactile classification experiment.

For each material, the sliding action was conducted for 2 s, which was repeated 50 times. Sensing data were obtained with a sampling rate of 5 kHz, and the data were then down sampled into 100 Hz after the low-pass filtering. Data with a stroking-time duration of 500 ms were fed to the classifier. A deep learning-based classifier using Transformer was employed for the recognition of the materials. Eighty percent of the samples obtained from each material at each speed were used for the network training, and the other 20% of the samples were used for the validation.

Classification results at the sliding speed of 10 cm/s are summarized as a confusion matrix in Figure 12. The horizontal axis of the confusion matrix is the predicted result for each material obtained by the discriminator, and the vertical axis represents the correct label for each material. The average recognition rate for the ten materials was 86.2%. The materials with distinctive textures, such as the Velcro tape, bamboo sheet and plastic tile, presented high accuracy of 100%. On the other hand, the materials having similar smoothness, such as the rubber mat and the two plastic mats, were confused with one another. Since the diameter of the tactile pin is 1.5 mm, the size of the pin tip is smaller than the texture patterns of the mats, as shown in Figure 13. When the pin tip contacted the edge of the texture patterns, similar signals were output for those three materials, which might have given greater effect to the classification.

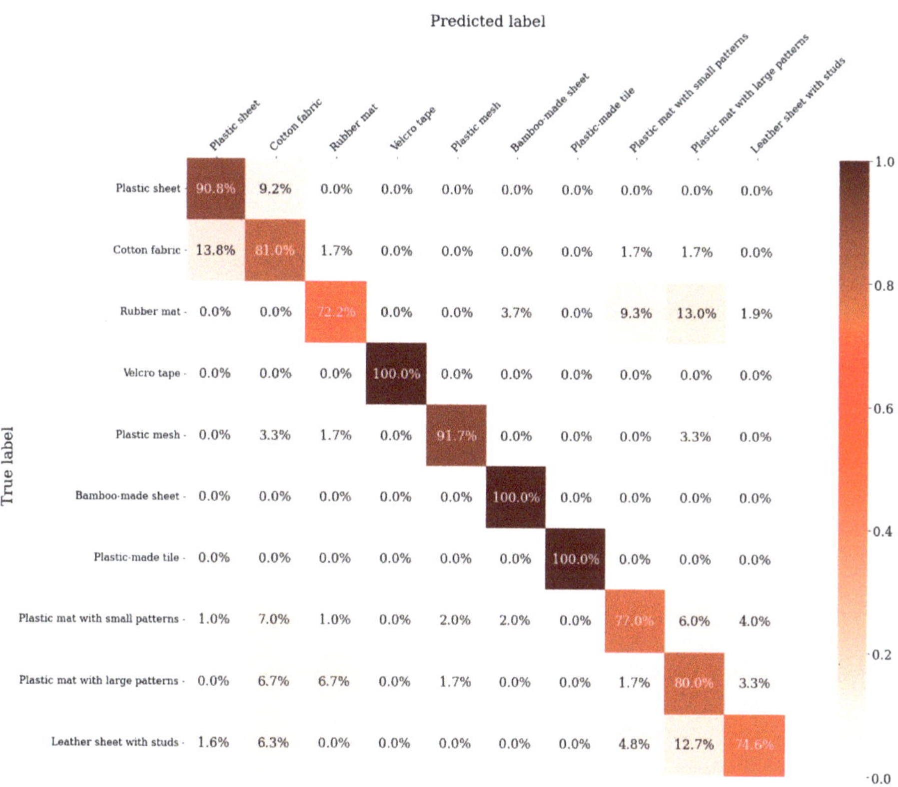

Figure 12. Confusion matrix of classification for ten different materials at sliding speed 10 cm/s.

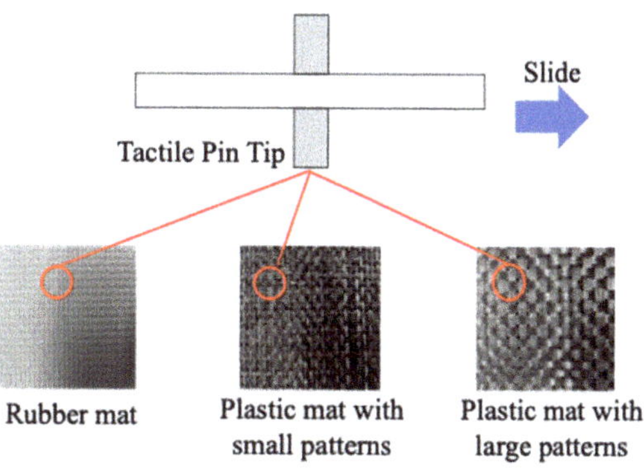

Figure 13. Contact situation between a tactile pin tip and material surface.

To avoid such misrecognition, we will consider the shape of the pin tip and the pin material to improve the classification ability. The softness of human skin, together with the structure of fingerprints, are considered to be important in how the perception of tactile sensation is processed. By referring to the skin structure, our tactile sensor will be improved in terms of how it conducts the physical stimuli to the SMA wire.

5. Conclusions

In this paper a tactile transducer using an SMA wire was introduced. We discovered that the resistance change against the force application to an SMA wire was caused by the superelasticity, and was given as the physical property to the tactile sensing. By introducing the special structure employing a tactile pin, the physical stimuli obtained by the sliding motion on a material surface was efficiently conducted to the SMA wire to cause the resistance change. By examining the output signals, the characteristic patterns were obtained from different material surfaces, in accordance with the stroking speed. We also employed the deep learning-based classifier, and verified the classification performance against ten different textures. An average of 86.2% recognition rate was achieved. However, classification performance was slightly decreased among the materials that had similar textures. We will further improve the shape of the tactile pin, together with the compositing material of the pin.

Author Contributions: Conceptualization, R.O. and H.S.; Methodology, R.O. and H.S.; Writing—draft, R.O. and manuscript writing, H.S. All authors have read and agreed to the published version of the manuscript.

Funding: This work was supported by JSPS KAKENHI Grants-in-Aid for Scientific Research on Innovative Areas (Research in a proposed research area) 18H05473 and 18H05895, and by JSPS KAKENHI Grant-in-Aid for Scientific Research (B) 20H04214.

Institutional Review Board Statement: Not applicable.

Informed Consent Statement: Not applicable.

Data Availability Statement: Not applicable.

Conflicts of Interest: The authors declare no conflict of interest.

References

1. Tiwana, M.I.; Redmond, S.J.; Lovell, N.H. A review of tactile sensing technologies with applications in biomedical engineering. *Sens. Actuators A Phys.* **2012**, *179*, 17–31. [CrossRef]
2. Claver, U.P.; Zhao, G. Recent progress in flexible pressure sensors based electronic skin. *Adv. Eng. Mater.* **2021**, *23*, 2001187. [CrossRef]
3. Zou, L.; Ge, C.; Wang, Z.; Cretu, E.; Li, X. Novel tactile sensor technology and smart tactile sensing systems: A review. *Sensors* **2017**, *17*, 2653. [CrossRef] [PubMed]
4. Lee, H.-K.; Chang, S.-L.; Yoon, E. A Flexible Polymer Tactile Sensor: Fabrication and Modular Expandability for Large Area Deployment. *J. Microelectromechanical Syst.* **2006**, *15*, 1681–1686. [CrossRef]
5. Mannsfeld, S.; Tee, B.; Stoltenberg, R.; Chen, C.; Barman, S.; Muir, B.; Sokolov, A.; Reese, C.; Bao, Z. Highly sensitive flexible pressure sensors with microstructured rubber dielectric layers. *Nat. Mater.* **2010**, *9*, 859–864. [CrossRef]
6. Zhu, S.-E.; Krishna Ghatkesar, M.; Zhang, C.; Janssen, G.C.A.M. Graphene based piezoresistive pressure sensor. *Appl. Phys. Lett.* **2013**, *102*, 161904. [CrossRef]
7. Stassi, S.; Cauda, V.; Canavese, G.; Pirri, C.F. Flexible tactile sensing based on piezoresistive composites: A review. *Sensors* **2014**, *14*, 5296–5332. [CrossRef]
8. Ward-Cherrier, B.; Pestell, N.; Cramphorn, L.; Winstone, B.; Giannaccini, M.E.; Rossiter, J.; Lepora, N.F. The TacTip Family: Soft Optical Tactile Sensors with 3D-Printed Biomimetic Morphologies. *Soft Robot.* **2018**, *5*, 216–227. [CrossRef]
9. Sferrazza, C.; D'Andrea, R. Design, motivation and evaluation of a full-resolution optical tactile sensor. *Sensors* **2019**, *19*, 928. [CrossRef]
10. Dargahi, J. Piezoelectric tactile sensor with three sensing elements for robotic, endoscopic and prosthetic applications. *Sens. Actuators* **2000**, *80*, 23–30. [CrossRef]
11. Spanu, A.; Pinna, L.; Viola, F.; Seminara, L.; Valle, M.; Bonfiglio, A.; Cosseddu, P. A high-sensitivity tactile sensor based on piezoelectric polymer PVDF coupled to an ultra-low voltage organic transistor. *Org. Electron. Phys. Mater. Appl.* **2016**, *36*, 57–60. [CrossRef]

12. Chuang, C.H.; Weng, H.K.; Cheng, J.W.; Shaikh, M.O. Ultrasonic tactile sensor integrated with TFT array for contact force measurements. In Proceedings of the 2017-19th International Conference on Solid-State Sensors, Actuators and Microsystems (TRANSDUCERS), Kaohsiung, Taiwan, 18–22 June 2017; pp. 512–515.
13. Orii, H.; Tsuji, S.; Kouda, T.; Kohama, T. Tactile texture recognition using convolutional neural networks for time-series data of pressure and 6-axis acceleration sensor. In Proceedings of the IEEE International Conference on Industrial Technology, Toronto, ON, Canada, 22–25 March 2017; pp. 1076–1080.
14. Gollerthan, S.; Young, M.L.; Baruj, A.; Frenzel, J.; Schmahl, W.W.; Eggeler, G. Fracture mechanics and microstructure in NiTi shape memory alloys. *Acta Mater.* **2009**, *57*, 1015–1025. [CrossRef]
15. Gu, H.; Bumke, L.; Chluba, C.; Quandt, E.; James, R.D. Phase engineering and supercompatibility of shape memory alloys. *Mater. Todays* **2018**, *21*, 265–277. [CrossRef]
16. Sun, F.; Hao, Y.L.; Nowak, S.; Gloriant, T.; Laheurte, P.; Prima, F. A thermo-mechanical treatment to improve the superelastic performances of biomedical Ti-26Nb and Ti-20Nb-6Zr (at.%) alloys. *J. Mech. Behav. Biomed. Mater.* **2011**, *4*, 1864–1872. [CrossRef]
17. Niinomi, M.; Nakai, M.; Hieda, J. Development of new metallic alloys for biomedical applications. *Acta Biomater.* **2012**, *8*, 3888–3903. [CrossRef]
18. Singh, K.; Sirohi, J.; Chopra, I. An Improved Shape Memory Alloy Actuator for Rotor Blade Tracking. *J. Intell. Mater. Syst. Struct.* **2003**, *14*, 767–786. [CrossRef]
19. Costanza, G.; Tata, M.E. Shape Memory Alloys for Aerospace, Recent Developments, and New Applications: A Short Review. *Materials* **2020**, *13*, 1856. [CrossRef]
20. Bellini, A.; Colli, M.; Dragoni, E. Mechatronic design of a shape memory alloy actuator for automotive tumble flaps: A case study. *IEEE Trans. Ind. Electron.* **2009**, *56*, 2644–2656. [CrossRef]
21. Williams, E.A.; Shaw, G.; Elahinia, M. Control of an automotive shape memory alloy mirror actuator. *Mechatronics* **2010**, *20*, 527–534. [CrossRef]
22. Sreekumar, M.; Nagarajan, T.; Singaperumal, M.; Zoppi, M.; Molfino, R. Critical review of current trends in shape memory alloy actuators for intelligent robots. *Ind. Robot.* **2007**, *34*, 285–294. [CrossRef]
23. Noh, M.; Kim, S.-W.; An, S.; Koh, J.-S.; Cho, K.-J. Flea-inspired catapult mechanism for miniature jumping robots. *IEEE Trans. Robot.* **2012**, *28*, 1007–1018.
24. Cladera, A.; Weber, B.; Leinenbach, C.; Czaderski, C.; Shahverdi, M.; Motavalli, M. Iron-based shape memory alloys for civil engineering structures: An overview. *Constr. Build. Mater.* **2014**, *63*, 281–293. [CrossRef]
25. Shahverdi, M.; Czaderski, C.; Motavalli, M. Iron-based shape memory alloys for prestressed near-surface mounted strengthening of reinforced concrete beams. *Constr. Build. Mater.* **2016**, *112*, 28–38. [CrossRef]
26. Alhuda Hamdan, N.; Wagner, A.; Voelker, S.; Steimle, J.; Borchers, J. Springlets: Expressive, flexible and silent on-skin tactile interfaces. In Proceedings of the Conference on Human Factors in Computing Systems, Glasgow, UK, 4–9 May 2019; pp. 1–14.
27. Muthukumarana, S.; Elvitigala, D.S.; Forero Cortes, J.P.; Matthies, D.J.C.; Nanayakkara, S. Touch me Gently: Recreating the Perception of Touch using a Shape-Memory Alloy Matrix. In Proceedings of the Conference on Human Factors in Computing Systems, Honolulu, HI, USA, 25–30 April 2020; pp. 1–12.
28. Formentini, M.; Lenci, S. An innovative building envelope (kinetic façade) with Shape Memory Alloys used as actuators and sensors. *Autom. Constr.* **2018**, *85*, 220–231. [CrossRef]
29. Wang, W.; Zeng, W.; Weitong, R.; Eric, M.; Sonkusale, S. Thermo-Mechanically Trained Shape Memory Alloy for Temperature Recording With Visual Readout. *IEEE Sens. Lett.* **2021**, *5*, 9292656. [CrossRef]
30. Ambrosino, C.; Capoluongo, P.; Campopiano, S.; Cutolo, A.; Giordano, M.; Davino, D.; Visone, C.; Cusano, A. Fiber Bragg Grating and Magnetic Shape Memory Alloy: Novel High-Sensitivity Magnetic Sensor. *IEEE Sens. J.* **2007**, *7*, 228–229. [CrossRef]
31. Kumar, A.; Pawar, S.; Pandey, A.; Dutta, S.; Kaur, D. Anisotropic magnetoelectric functionality of ferromagnetic shape memory alloy heterostructures for MEMS magnetic sensors. *J. Phys. D Appl. Phys.* **2020**, *53*, 395302. [CrossRef]
32. Nahm, S.H.; Kim, Y.I.; Kim, J.M.; Yoon, D.J. A study on the application of Ni-Ti shape memory alloy as a sensor. *Mater. Sci. Forum* **2005**, *475–479*, 2043–2046. [CrossRef]
33. Nagai, H.; Oishi, R. Shape memory alloys as strain sensors in composites. *Smart Mater. Struct.* **2006**, *15*, 493–498. [CrossRef]
34. Bhattacharya, K. *Microstructure of Martensite: Why it Forms and How it Gives Rise to the Shape-Memory Effect*; Oxford University Press: Oxford, UK, 2003.
35. Mizukami, Y.; Sawada, H. Tactile information transmission by apparent movement phenomenon using shape-memory alloy device. *Int. J. Disabil. Hum. Dev.* **2006**, *5*, 277–284. [CrossRef]
36. Vaswani, A.; Shazeer, N.; Parmar, N.; Uszkoreit, J.; Jones, L.; Gomez, A.N.; Kaiser, L.; Polosukhin, I. Attention is all you need. *Adv. Neural Inf. Process. Syst.* **2017**, *30*, 5999–6009.
37. Wu, N.; Green, B.; Ben, X.; O'Banion, S. Deep Transformer Models for Time Series Forecasting: The Influenza Prevalence Case. *arXiv* **2020**, arXiv:2001.08317.

 *materials*

Article

Production, Mechanical and Functional Properties of Long-Length TiNiHf Rods with High-Temperature Shape Memory Effect

Roman Karelin [1,2,*], Victor Komarov [1,2], Vladimir Cherkasov [2], Vladimir Yusupov [1], Sergey Prokoshkin [2] and Vladimir Andreev [1]

[1] Baikov Institute of Metallurgy and Materials Science RAS, Moscow 119334, Russia
[2] National University of Science and Technology MISIS, Moscow 119049, Russia
* Correspondence: rdkarelin@gmail.com

Abstract: In the present work, the possibility of manufacturing long-length TiNiHf rods with a lowered Hf content and a high-temperature shape memory effect in the range of 120–160 °C was studied. Initial ingots with 1.5, 3.0 and 5.0 at.% Hf were obtained by electron beam melting in a copper water-cooled stream-type mold. The obtained ingots were rotary forged at the temperature of 950 °C, with the relative strain from 5 to 10% per one pass. The obtained results revealed that the ingots with 3.0 and 5.0 at.% Hf demonstrated insufficient technological plasticity, presumably because of the excess precipitation of $(Ti,Hf)_2Ni$-type particles. The premature destruction of ingots during the deformation process does not allow obtaining high-quality long-length rods. A long-length rod with a diameter of 3.5 mm and a length of 870 mm was produced by rotary forging from the ingot with 1.5 at.% Hf. The obtained TiNiHf rod had relatively high values of mechanical properties (a dislocation yield stress σ_y of 800 MPa, ultimate tensile strength σ_B of 1000 MPa, and elongation to fracture δ of 24%), functional properties (a completely recoverable strain of 5%), and a required finishing temperature of shape recovery of 125 °C in the as-forged state and of 155 °C after post-deformation annealing at 550 °C for 2 h.

Keywords: shape memory alloys; rotary forging; structure; mechanical properties; NiTiHf

Citation: Karelin, R.; Komarov, V.; Cherkasov, V.; Yusupov, V.; Prokoshkin, S.; Andreev, V. Production, Mechanical and Functional Properties of Long-Length TiNiHf Rods with High-Temperature Shape Memory Effect. *Materials* **2023**, *16*, 615. https://doi.org/10.3390/ma16020615

Academic Editor: Dezső Beke

Received: 18 November 2022
Revised: 23 December 2022
Accepted: 4 January 2023
Published: 9 January 2023

1. Introduction

Ti-Ni-based shape memory alloys (SMAs) are functional materials, actively used for the manufacturing of various devices and construction elements for engineering or medical applications [1–5]. Today, permanent technological development leads to the formation of new special requirements for the operational characteristics of applied alloys [6–11]. For example, great attention and high demand are observed for Ti-Ni-based alloys with a high-temperature shape memory effect.

Ternary high-temperature Ti-Ni-based SMAs (TiNiX), where X is a precious metal, such as Pd, Pt, etc., or where X is Zr or Hf, are actively developed and studied [12–15]. However, the first group of alloys are too expensive due to the use of precious metals as alloying elements, and the second group exhibits insufficient deformability. TiNiHf alloys seem to be more attractive for wide practical application because they perform required mechanical and functional properties in combination with relatively low cost and greater stability of the operational characteristics as compared to TiNiZr alloys [15]. Therefore, to date, numerous studies of TiNiHf alloys revealed the main features of the formation of their structural state and properties [16–34]. Ternary TiNiHf alloys are usually divided into two groups, one with increased Ni content and one with reduced Ni content, similarly to binary TiNi SMAs. According to studies [16–20], alloys with low Ni content (less than 49.5 at.%) have insufficient technological plasticity due to the precipitation of the (Ti, $Hf)_2Ni$-type phase. An increase in the Ni content leads to a decrease in the characteristic temperatures of martensitic transformations; therefore, in order to obtain a high-temperature state in the alloy, the content of Hf also should be increased [16,21]. That leads to the increase in the production cost and makes it difficult to obtain a temperature range of shape recovery between 100 and 200 °C. For example, the most commonly applied $Ti_{29.7}Ni_{50.3}Hf_{20}$ alloy has the finishing temperature of shape recovery of about 300 °C or higher, in dependence with the applied melting and deformation modes [16,17,19]. Binary TiNi near-equiatomic or Ti-enriched alloys also do not provide such a temperature range of shape recovery. The finishing temperature of the reverse MT in binary alloys hardly exceeds 110 °C. Additionally, in Ni-rich TiNiHf alloys, the aging process is developed during thermal treatment in the temperature range of 400–650 °C [23–28]. That allows precisely changing the combination of properties but may affect the functional characteristics of applied devices during the operation at higher temperatures. This factor must be considered during the development of technological schemes for the production of various devices from TiNiHf SMAs. Another problem, as was already mentioned, is the sufficiently low deformability of TiNiHf alloys. Therefore, the problem of obtaining high-quality long-length rods from TiNiHf alloys with low Ni and Hf contents and with a finishing temperature of shape recovery of about 150 °C is unsolved.

The melting processes of TiNi-based SMAs from pure charge components are often associated with some difficulties, such as an increased concentration of gas impurities due to the high reactivity of pure titanium. For TiNiHf alloys, an increase in the concentration of gas impurities, in particular oxygen, is especially critical because of an increase in the concentration of embrittling $(Ti, Hf)_2Ni$-type excess phases with a high concentration of Ti and Hf, as well as HfO_2 in alloys with a high concentration of Hf and Ni [16–20]. Therefore, the application of finished Ti-Ni billets as a charge component for melting TiNiHf alloys may decrease the concentration of gas impurities due to the absence of pure titanium during the melting process. The application of this concept also expands the possibilities of recycling TiNi binary alloys.

Based on the foregoing, the aim of the present work consisted in the production of high-quality long-length rods from TiNiHf alloys with a reduced content of nickel and hafnium, providing high mechanical and functional properties, in combination with a finishing temperature of the reverse martensitic transformation of about 120–160 °C through the application of finished Ti-Ni billets and high-purity hafnium wire during the melting process and various modes of thermomechanical treatment.

2. Materials and Methods

A polished TiNi SMA rod with a diameter of 12 mm, manufactured by industrial center MATEK SMA Ltd., Moscow, Russia and cold-worked hafnium wire with a diameter of 2 mm, manufactured by E.A. Yudin Chemical and Metallurgical Plant, Novosibirsk, Russia were chosen as the charge materials for the melting of TiNiHf initial ingots. The chemical composition of the applied TiNi rod and Hf wire is shown in Tables 1 and 2, respectively.

Table 1. Chemical composition of the TiNi SMA rod with a diameter of 12 mm, used as a charge material for melting of TiNiHf initial ingots.

Ni, wt.%	Ti, wt.%	Impurities, wt.%				
		C	O	N	H	Other
55.15	Balance	0.035	0.030	0.002	0.001	<0.1

Table 2. Chemical composition of the hafnium wire with a diameter of 2 mm, used as a charge material for melting of TiNiHf initial ingots.

Hf, wt.%	Impurities, wt.%						
	C	O	N	H	Si	Fe	Zr
Balance	0.008	0.020	<0.005	0.001	<0.005	0.030	0.58

Initial TiNiHf ingots were obtained by electron beam melting (EBM) in a furnace with a power of 60 kW at a vacuum of 1×10^{-5} in a copper water-cooled stream-type crystallizer. The dimensions, weight, and chemical composition of the obtained ingots are shown in Table 3.

Table 3. Chemical composition of the melted TiNiHf ingots.

No.	Dimensions, mm	Weight, g	Composition (without Impurities)					
			wt.%			at.%		
			Ti	Ni	Hf	Ti	Ni	Hf
1	11 × 16 × 162	102	42.9	52.7	4.4	49.0	49.5	1.5
2	10 × 15 × 230	160	40.4	49.6	10.0	48.5	48.5	3.0
3	11 × 17 × 200	168	38.2	47.0	14.8	47.5	47.5	5.0

After melting, the ingots were homogenized at 1000 °C for 1 h in a vacuum and rotary forged at the temperature of 950 °C, with a relative strain from 5 to 10% per one pass. After RF, post-deformation annealing (PDA) at 550 °C for 2 h was applied. The microstructure was studied using a *UNION* optical microscope and the scanning electron microscope (SEM) Scios (Thermo Fisher Scientific, Waltham, MA, USA) at an accelerating voltage of 10 kV. The investigation of the chemical composition by energy dispersive X-ray spectroscopy (EDS) was carried out using the EDX system from *EDAX*. The study of the temperatures of forward (M_s, M_f) and reverse (A_s, A_f) martensitic transformations (MTs) was carried out using a Mettler Toledo differential scanning calorimeter in the cooling-heating cycle at a rate of 10 K/min in the temperature range from minus 40 to 200 °C. The phase composition was studied using a Dron-3 X-ray diffractometer in $Cu_{K\alpha}$ radiation in the 2Θ angle range from 20 to 80° at room temperature. The Vickers hardness measurements were carried out at room temperature using a LECOM 400-A tester under a load of 1 N. The mechanical properties were determined at room temperature by uniaxial tensile tests using the universal tensile machine INSTRON 3382, with a deformation rate of 2 mm/min. The following mechanical parameters were determined: critical stress for martensite reorientation (transformation yield stress) σ_{cr}, dislocation yield stress σ_y, ultimate tensile strength σ_B, and elongation to failure δ. The measurement of functional properties, such as total completely recoverable strain (ε_{rt}) and the temperature range of shape recovery (TRSR), was carried out using the thermomechanical method by deformation in bending. The bending of preliminary cut samples with the size of 0.4 × 0.6 × 20 mm was conducted at room temperature around mandrels with various diameters. The deformation value was determined based on the following relationship: $\varepsilon = d/(D + d)$, where d is the diameter of the sample, and D is the diameter of the mandrel. The shape recovery rate (SRR) was determined as a ratio of recovered strain to induced strain, $\varepsilon_{rt}/\varepsilon_i*100\%$ [35]. After the strain induction, the sample was heated in oil to implement the shape memory effect and determine the ε_{rt} and TRSR.

3. Results and Discussion

3.1. The Features of the Manufacturing Process of Initial Ingots and Long-Length Rods from TiNiHf Alloys

Initial ingots were produced by EBM. EBM has a number of advantages in comparison to induction and electric arc melting, including the effective purification of metals from gas and metallic impurities; the absence of defects of shrinkage origin in ingots because of the possibility of a smooth change in the power of the electron beam and complete filling of the shrinkage cavity; and the possibility of using charge materials in any form. The shape of the obtained ingots in a stream-type mold allows applying subsequent thermomechanical treatment immediately after melting, without additional mechanical operations. Images of the obtained ingots are shown in Figure 1.

Figure 1. Images of TiNiHf initial ingots obtained by electron beam melting in a copper water-cooled stream-type mold: 1.5 at.% of Hf (**a**), 3.0 at.% of Hf (**b**), and 5.0 at.% of Hf (**c**).

Albeit hot rotary forging, the most favorable forming technique for the processing of hard-to-deform metallic materials [36], was applied to produce long-length rods from initial ingots, only the first ingot with 1.5 at.% HF was successfully forged to a diameter of 3.5 mm and a length of 870 mm. Ingots with 3.0 and 5.0 at.% Hf were destructed (Figure 2) during the first few passes because of insufficient technological plasticity, presumably because of the increase in the amount of $(Ti,Hf)_2Ni$-type brittle phase with the decrease in Ni content. The premature destruction of ingots during the process of rotary forging does not allow obtaining high-quality long-length rods. Therefore, further investigation of mechanical and functional properties, except the temperature range of martensitic transformations, was carried out only for the TiNiHf rod with 1.5. at.% of Hf.

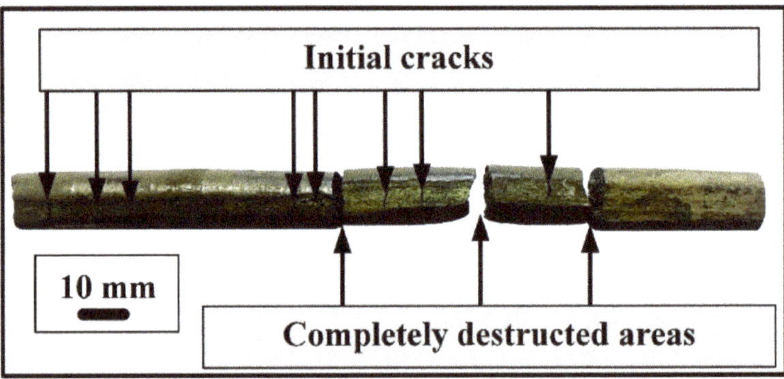

Figure 2. Image of TiNiHf sample with 3.0 at.% Hf after several passes of rotary forging.

3.2. Microstructural and Phase Analysis

The X-ray diffractograms of TiNiHf rods with a diameter of 3.5 mm in the as-forged state and after additional PDA at 550 °C for 2 h are shown in Figure 3.

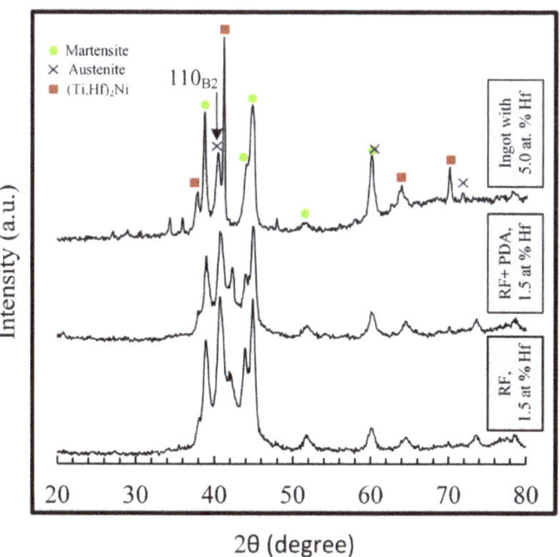

Figure 3. X-ray diffractograms of TiNiHf SMA rods with a diameter of 3.5 mm after RF and RF + PDA 550 °C for 2 h.

The obtained results revealed that after rotary forging the studied TiNiHf alloys at room temperature, the main phase was B19′-martensite, and some amount of B2-austenite phase is also presented. X-ray lines corresponding to the $(Ti,Hf)_2Ni$-type particles of excess phase were clearly observed. PDA at 550 °C for 2 h led to the narrowing of X-ray line peaks, indicating the decrease in the density of structural defects and accompanied by a slight increase in the proportion of the B2-austenite phase, and some increase in the intensity of the 110_{B2} line can be observed. The phase composition, however, was practically the same as after RF.

Analyses of the microstructure revealed that after rotary forging, a dynamically recrystallized structure with an average grain size in the range of 20–30 microns and a huge number of precipitated $(Ti,Hf)_2Ni$-type particles was formed in the sample (Figure 4a). Post-deformation annealing did not lead to changes in the structure, noticeable by light microscopy (Figure 4b).

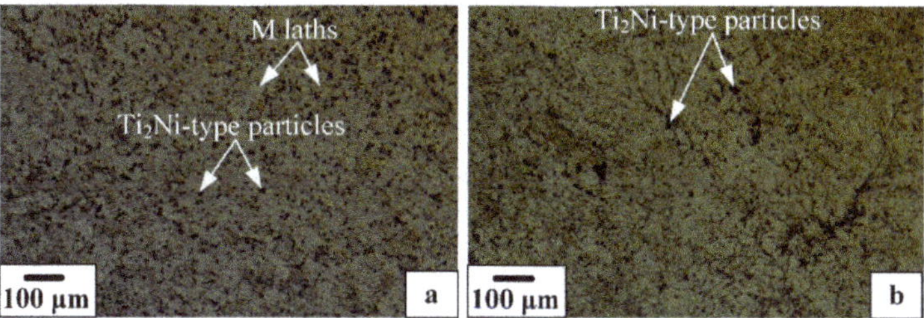

Figure 4. Microstructure of the TiNiHf SMA rods with a diameter of 3.5 mm after RF (**a**) and RF + PDA 550 °C for 2 h (**b**). Light optical microscopy.

The SEM images, obtained after RF and RF + PDA, are shown in Figure 5. The black rounded and elongated inclusions are associated with the $(Ti,Hf)_2Ni$-type phase based on the results of the EDS analysis (Figure 5). The definition of structural elements and grain/subgrain boundaries is complicated due to the martensitic state of the studied rods at room temperature. The difference in the structural state with the application of PDA was not determined by the SEM, just like after light microscopy. The elemental maps, shown in Figure 6a, indicate the elemental distribution of Ni, Ti and Hf in the obtained rods. Elemental EDS mapping revealed the practically homogeneous distribution of elements, albeit indicating small Ti-rich and Ni-poor areas, presumably corresponding to the $(Ti,Hf)_2Ni$-type particles. The elemental maps, taken with higher resolutions in order to study the elemental distribution along the precipitates and their surrounding matrix, confirm this presumption (Figure 6b).

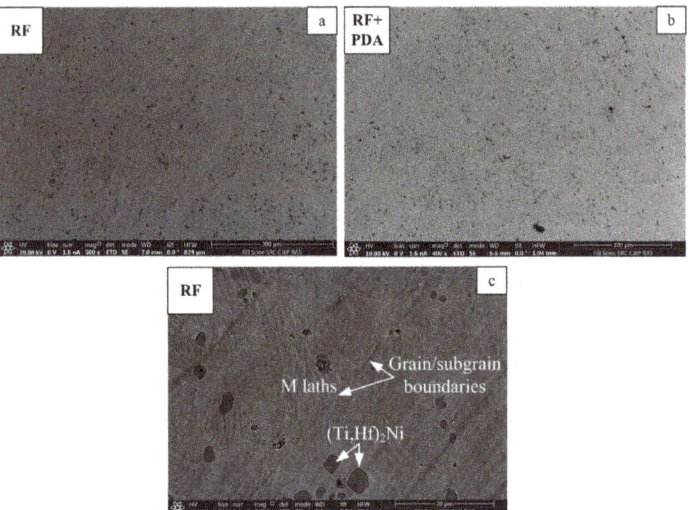

Figure 5. SEM images of TiNiHf rods after RF (**a**,**c**) and RF + PDA (**b**).

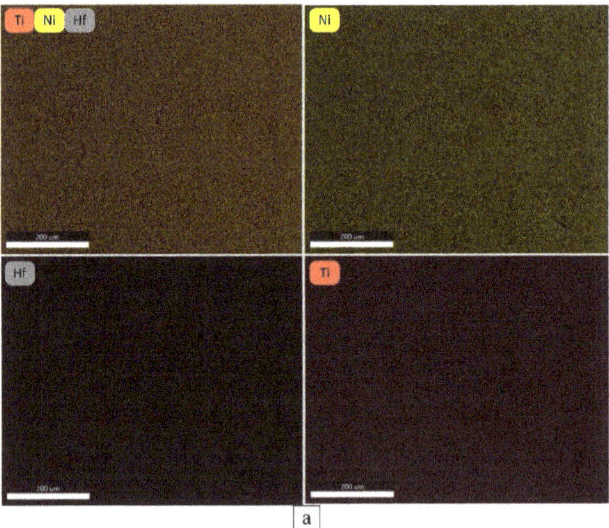

Figure 6. *Cont.*

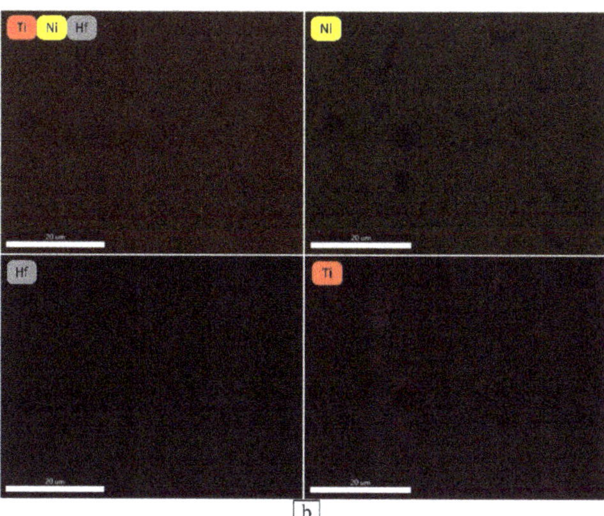

Figure 6. Elemental EDS mapping of Ti, Ni and Hf in TiNiHf rods after RF (**a**) and with higher resolution (**b**).

3.3. Temperature Ranges of Martensitic Transformations

The results of differential scanning calorimetry are presented in Figure 7.

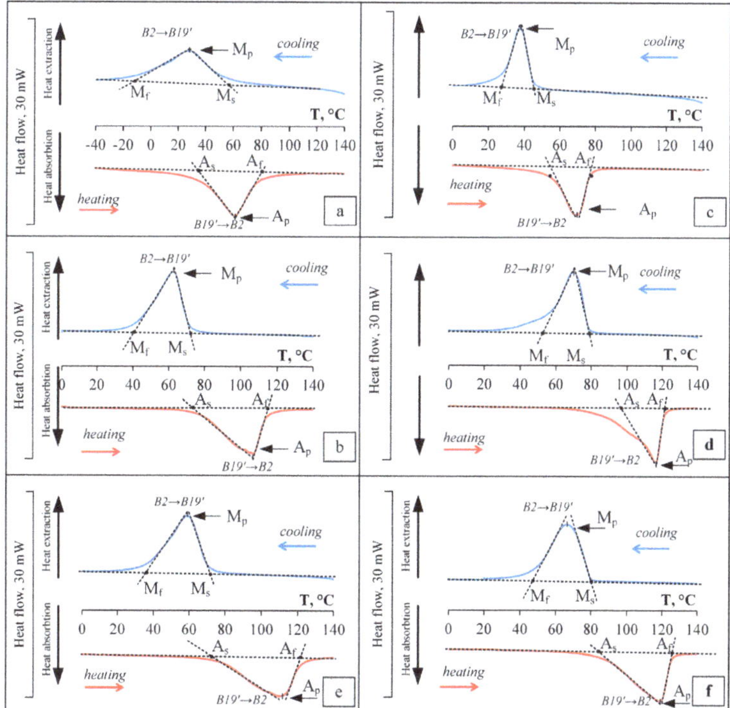

Figure 7. Calorimetric curves of the TiNiHf initial ingots in as-cast state and after annealing at 1000 °C for 1 h: 1.5 at.% Hf—(**a**,**c**), 3.0 at.% Hf—(**b**,**d**), and 5.0 at.% Hf—(**e**,**f**).

Based on the analysis of the obtained calorimetric curves, summary Table 4 was compiled with the characteristic temperatures of the forward and reverse MTs for the samples of the TiNiHf system.

Table 4. Characteristic temperatures of forward and reverse MTs of TiNiHf initial ingots in as-cast state and after annealing at 1000 °C for 1 h.

Hf Content	Annealing	M_s, °C	M^P, °C	M_f, °C	A_s, °C	A^P, °C	A_f, °C	$A_s - A_f$, °C	$M_s - M_f$, °C
1.5 at.% Hf	-	54	28	−12	32	62	79	46	66
1.5 at.% Hf	+1000 °C, 1 h	45	38	27	52	70	79	26	18
3.0 at.% Hf	-	71	60	40	71	107	114	43	31
3.0 at.% Hf	+1000 °C, 1 h	78	70	53	99	117	121	22	25
5.0 at.% Hf	-	71	55	35	76	110	123	47	36
5.0 at.% Hf	+1000 °C, 1 h	80	62	47	86	120	126	40	33

The results of the MT study showed that an increase in the Hf content in the initial cast state led to an increase in the finishing temperature of reverse MT A_f, from 79 °C with an Hf content of 4.4 wt.%, up to 126 °C with 14.8 wt.%. The temperature range of the forward MT also shifted towards higher temperatures with the increase in Hf content. Homogenization annealing at 1000 °C for 1 h led to a noticeable narrowing of the hysteresis of MT due to an increase in the homogeneity of the structural state, and did not lead to the noticeable shift of the temperature range of the reverse MT. The DSC curves of TiNiHf alloy rods after RF and PDA are shown in Figure 8.

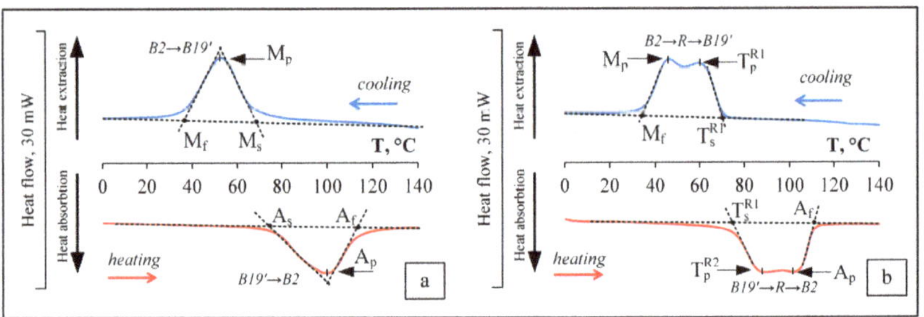

Figure 8. Calorimetric curves of the TiNiHf rods after RF (**a**) and RF + PDA at 550 °C for 2 h (**b**).

Based on the analysis of the obtained calorimetric curves, summary Table 5 was compiled with the characteristic temperatures of the forward and reverse MTs for the samples of the TiNiHf system.

Table 5. Characteristic temperatures of forward and reverse MTs of TiNiHf rods after RF and RF + PDA at 550 °C for 2 h.

TMT	T_s^{R1}, °C	T_p^{R1}, °C	T_f^{R1}, °C	M_s, °C	M^P, °C	M_f, °C	T_s^{R2}, °C	T_p^{R2}, °C	T_f^{R2}, °C	A_s, °C	A^P, °C	A_f, °C	$M_s/T_s^{R1} - A_f$, °C	$M_s/T_s^{R2} - M_f$, °C
RF	-	-	-	67	54	36	-	-	-	76	99	113	37	31
RF + 550 °C, 2 h	70	57	-	-	45	35	75	89	-	-	102	116	25	44

The obtained results showed that RF led to a noticeable increase in the finishing temperature of reverse MT A_f as compared to the cast state, from 79 to 113 °C. The application of PDA led to the change in the sequence of the forward and reverse MTs: the transformations proceeded through the intermediate R-phase, while after RF, the transformations proceeded in one stage. The change in the stages of MT can be associated with the influence of static aging processes that develop during annealing and, as a consequence, an increase in the density of structure defects due to the precipitation of the Ti_3Ni_4 excess phase (H-phase). This result also showed that the Ni content in the sample was higher than was suggested (more than 50.2 at.%). In this case, there was no noticeable change in the characteristic temperatures of the forward and reverse MTs after PDA.

3.4. Temperature Ranges of Martensitic Transformations

The mechanical properties of TiNiHf SMA rods are shown in Table 6. Representative stress–strain diagrams are shown in Figure 9.

Table 6. Mechanical properties of TiNiHf rods after RF and RF + PDA at 550 °C for 2 h.

TMT	σ_{cr}, MPa	σ_y, MPa	$\Delta\sigma$, MPa	σ_B, MPa	δ, %	HV
RF	214	800	586	1000	24	210
RF + 550 °C, 2 h	202	840	638	990	29	215

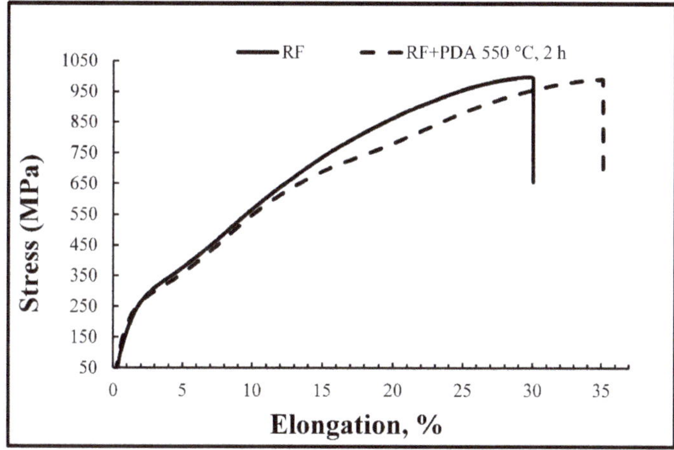

Figure 9. Representative stress–strain diagrams of TiNiHf rods after RF and RF + PDA at 550 °C for 2 h.

The obtained rods showed high strength characteristics: σ_y = 800 MPa, σ_B = 1000 MPa after RF and σ_y = 840 MPa, σ_B = 990 MPa after PDA at 550 °C for 2 h, and, simultaneously, sufficiently high ductility: δ = 24% after RF and 29% after PDA. Results of the Vickers hardness test revealed that after RF, the value of hardness was 210 HV. The application of PDA did not noticeably change the hardness value.

The results of determining the temperature range of shape recovery (TRSR) after strain inducing are shown in Table 7. A comparison of the temperature range of the reverse martensitic transformation obtained by DSC (Table 5) and TRSR (Table 7) reveals that strain inducing led to an increase in the temperature of starting and finishing the reverse martensitic transformation.

Table 7. Total completely recoverable strain and TRSR of TiNiHf rods after RF, RF + PDA at 550 °C for 2 h, and RF + PDA at 1000 °C for 1 h.

TMT	Induced Strain, %	Total Completely Recoverable Strain, %	SRR, %	TRSR, °C
RF	2.0	2.0	100	65–125
RF	6.0	5.0	83	100–155
RF + 550 °C, 2 h	2.0	2.0	100	110–155
RF + 550 °C, 2 h	6.0	4.0	67	70–180
RF + 1000 °C, 1 h	6.0	4.0	67	120–155

Based on the presented results, it can be concluded that after RF, the finishing temperature of shape recovery after 2% of induced strain was 125 °C. A induced strain of 6% led to an increase in the A_f temperature to 155 °C. The SRR, after 6% of induced strain, was 83%, as the value of total completely recoverable strain was 5%. PDA at a temperature of 550 °C for 2 h allowed increasing A_f to 180 °C, while the SRR decreased to 67%. The difference between RF and RF + PDA states consisted in the development of the softening processes during PDA and corresponding increase in TRSR. PDA at a temperature of 1000 ° C for 1 h did not lead to a noticeable change in the A_f temperature in comparison to the hot-forged state, while the SRR also decreased to 67%.

4. Conclusions

The study of the possibility of the production of long-length rods from TiNiHf alloys, with a lowered Hf content and a high-temperature shape memory effect in the range of 120–160 °C, was performed in the present work. The following conclusions can be drawn:

1. Initial ingots with 1.5, 3.0 and 5.0 at.% Hf were successfully obtained by electron beam melting from Ti-Ni billets and pure Hf wire, used as raw materials.
2. Ingots with 3.0 and 5.0 at.% Hf demonstrated insufficient technological plasticity, presumably because of the excess precipitation of $(Ti,Hf)_2Ni$-type particles with the decrease in Ni content. The premature destruction of ingots during the process of rotary forging does not allow obtaining high-quality long-length rods.
3. A good-quality rod with a diameter of 3.5 mm and a length of 870 mm was obtained from the ingot with 1.5 at.% Hf. The obtained TiNiHf rod had relatively high values of mechanical properties: a dislocation yield stress of 800 MPa, ultimate tensile strength of 1000 MPa, and elongation to fracture of 24%.
4. The obtained rods provided the required values of functional properties: a completely recoverable strain of 5%, and a finishing temperature of shape recovery after 2% of induced strain of 125 °C in the as-forged state and of 155 °C after post-deformation annealing at 550 °C for 2 h.

Author Contributions: Conceptualization, R.K. and V.A.; methodology, V.K.; validation, V.Y. and S.P.; formal analysis, V.Y.; investigation, R.K., V.K. and V.C.; writing—original draft preparation, R.K.; writing—review and editing, V.A. and V.K.; supervision, V.A. and S.P. All authors have read and agreed to the published version of the manuscript.

Funding: This study was carried out within the framework of the state task of IMET RAS No. 075-00715-22-00, and the part of the SEM characterization was carried out with the financial support of the strategic project, "Biomedical materials and bioengineering", within the framework of the Strategic Academic Leadership Program "Priority 2030" at NUST «MISiS».

Institutional Review Board Statement: Not applicable.

Informed Consent Statement: Not applicable.

Data Availability Statement: Not applicable.

Conflicts of Interest: The authors declare no conflict of interest.

References

1. Jani, J.M.; Leary, M.; Subic, A.; Gibson, M.A. A review of shape memory alloy research, applications and opportunities. *Mater. Des.* **2014**, *56*, 1078–1113. [CrossRef]
2. Resnina, N.; Palani, I.A.; Belyaev, S.; Prabu, S.M.; Liulchak, P.; Karaseva, U.; Manikandan, M.; Jayachandran, S.; Bryukhanova, V.; Sahu, A.; et al. Structure, martensitic transformations and mechanical behaviour of NiTi shape memory alloy produced by wire arc additive manufacturing. *J. Alloys Compd.* **2021**, *851*, 156851. [CrossRef]
3. Resnina, N.; Rubanik, V. *Shape Memory Alloys: Properties, Technologies, Opportunities*; Trans Tech Publishing: Wollerau, Switzerland, 2015; p. 640.
4. Ryklina, E.; Korotitskiy, A.; Khmelevskaya, I.; Prokoshkin, S.; Polyakova, K.; Kolobova, A.; Soutorine, M.; Chernov, A. Control of phase transformations and microstructure for optimum realization of one-way and two-way shape memory effects in removable surgical clips. *Mater. Des.* **2017**, *136*, 174–184. [CrossRef]
5. Sun, Q.; Matsui, R.; Takeda, K.; Pieczyska, E. *Advances in Shape Memory Materials: In Commemoration of the Retirement of Prof. Hisaaki Tobushi*; Springer: Berlin/Heidelberg, Germany, 2017; p. 241. [CrossRef]
6. Karelin, R.D.; Khmelevskaya, I.Y.; Komarov, V.S.; Andreev, V.A.; Perkas, M.M.; Yusupov, V.S.; Prokoshkin, S.D. Effect of quasi-continuous equal-channel angular pressing on structure and properties of Ti-Ni shape memory alloys. *J. Mater. Eng. Perform.* **2021**, *30*, 3096–3106. [CrossRef]
7. Komarov, V.; Khmelevskaya, I.; Karelin, R.; Prokoshkin, S.; Zaripova, M.; Isaenkova, M.; Korpala, G.; Kawalla, R. Effect of biaxial cyclic severe deformation on structure and properties of Ti-Ni alloys. *J. Alloys Compd.* **2019**, *797*, 842–848. [CrossRef]
8. Komarov, V.; Khmelevskaya, I.; Karelin, R.; Kawalla, R.; Korpala, G.; Prahl, U.; Yusupov, V.; Prokoshkin, S. Deformation Behavior, Structure, and Properties of an Aging Ti-Ni Shape Memory Alloy after Compression Deformation in a Wide Temperature Range. *JOM* **2021**, *73*, 620–629. [CrossRef]
9. Khmelevskaya, I.Y.; Karelin, R.D.; Prokoshkin, S.D.; Isaenkova, M.G.; Perlovich, Y.A.; Fesenko, V.A.; Komarov, V.S.; Zaripova, M.M. Features of nanostructure and functional properties formation in Ti-Ni shape memory alloys subjected to quasi-continuous equal channel angular pressing. *IOP Conf. Ser. Mater. Sci. Eng.* **2019**, *503*, 012024. [CrossRef]
10. Valiev, R.; Estrin, Y.; Horita, Z.; Langdon, T.; Zehetbauer, M.; Zhu, Y. Producing bulk ultrafine-grained materials by severe plastic deformation. *JOM* **2006**, *58*, 33–39. [CrossRef]
11. Sabirov, I.; Enikeev, N.; Murashkin, M.; Valiev, R. *Bulk Nanostructured Materials with Multifunctional Properties*; Springer: Berlin/Heidelberg, Germany, 2015; p. 118. [CrossRef]
12. Khan, M.I.; Kim, H.Y.; Miyazaki, S. A review of TiNiPdCu alloy system for high temperature shape memory applications. *Shape Mem. Superelasticity* **2015**, *1*, 85–106. [CrossRef]
13. Tan, C.L.; Tian, X.H.; Ji, G.J.; Gui, T.L.; Cai, W. Elastic property and electronic structure of TiNiPt high-temperature shape memory alloys. *Solid State Commun.* **2008**, *147*, 8–10. [CrossRef]
14. Noebe, R.; Biles, T.; Padula, S.; Soboyeio, W.; Srivastan, T. *Advanced Structural Materials: Properties, Design Optimization, and Applications*; CRC Press: Boca Raton, FL, USA, 2007; p. 145.
15. Karakoc, O.; Atli, K.C.; Evirgen, A.; Pons, J.; Santamarta, R.; Benafan, O.; Noebe, R.D.; Karaman, I. Effects of training on the thermomechanical behavior of NiTiHf and NiTiZr high temperature shape memory alloys. *Mater. Sci. Eng. A* **2020**, *794*, 139857. [CrossRef]
16. Tong, Y.; Shuitcev, A.; Zheng, Y. Recent development of TiNi-based shape memory alloys with high cycle stability and high transformation temperature. *Adv. Eng. Mater.* **2020**, *22*, 1900496. [CrossRef]
17. Young, A.W.; Wheeler, R.W.; Ley, N.A.; Benafan, O.; Young, M.L. Microstructural and thermomechanical comparison of Ni-rich and Ni-lean NiTi-20 at.% Hf high temperature shape memory alloy wires. *Shape Mem. Superelasticity* **2019**, *5*, 397–406. [CrossRef]
18. Belbasi, M.; Salehi, M.T. Influence of chemical composition and melting process on hot rolling of NiTiHf shape memory alloy. *J. Mater. Eng. Perform.* **2014**, *23*, 2368–2372. [CrossRef]
19. Babacan, N.; Bilal, M.; Hayrettin, C.; Liu, J.; Benafan, O.; Karaman, I. Effects of cold and warm rolling on the shape memory response of Ni50Ti30Hf20 high-temperature shape memory alloy. *Acta Mater.* **2018**, *157*, 228–244. [CrossRef]
20. Kim, J.H.; Park, C.H.; Kim, S.W.; Hong, J.K.; Oh, C.S.; Jeon, Y.M.; Kim, K.M.; Yeom, J.T. Effects of microstructure and deformation conditions on the hot formability of Ni-Ti-Hf shape memory alloys. *J. Nanosci. Nanotechnol.* **2014**, *14*, 9548–9553. [CrossRef] [PubMed]
21. Chang-Long, T.; Wei, C.; Xiao-Hua, T. First-principles study on the effect of Hf content on martensitic transformation temperature of TiNiHf alloy. *Chin. Phys.* **2006**, *15*, 2718. [CrossRef]
22. Pushin, V.G.; Kuranova, N.N.; Pushin, A.V.; Uksusnikov, A.N.; Kourov, N.I.; Kuntsevich, T.E. Structural and phase transformations, mechanical properties, and shape-memory effects in quasibinary Ni50Ti38Hf12 alloy obtained by quenching from the melt. *Phys. Met. Metallogr.* **2016**, *117*, 1251–1260. [CrossRef]
23. Javadi, M.M.; Belbasi, M.; Salehi, M.T.; Afshar, M.R. Effect of aging on the microstructure and shape memory effect of a hot-rolled NiTiHf alloy. *J. Mater. Eng. Perform.* **2011**, *20*, 618–622. [CrossRef]
24. Karaca, H.E.; Saghaian, S.M.; Ded, G.; Tobe, H.; Basaran, B.; Maier, H.J.; Noebe, R.D.; Chumlyakov, Y.I. Effects of nanoprecipitation on the shape memory and material properties of an Ni-rich NiTiHf high temperature shape memory alloy. *Acta Mater.* **2013**, *61*, 7422–7431. [CrossRef]

25. Amin-Ahmadi, B.; Pauza, J.G.; Shamimi, A.; Duerig, T.W.; Noebe, R.D.; Stebner, A.P. Coherency strains of H-phase precipitates and their influence on functional properties of nickel-titanium-hafnium shape memory alloys. *Scr. Mater.* **2018**, *147*, 83–87. [CrossRef]
26. Amin-Ahmadi, B.; Gallmeyer, T.; Pauza, J.G.; Duerig, T.W.; Noebe, R.D.; Stebner, A.P. Effect of a pre-aging treatment on the mechanical behaviors of Ni50. 3Ti49. 7− xHfx (x≤ 9 at.%) Shape memory alloys. *Scr. Mater.* **2018**, *147*, 11–15. [CrossRef]
27. Tagiltsev, A.I.; Panchenko, E.Y.; Timofeeva, E.E.; Chumlyakov, Y.I.; Fatkullin, I.D.; Marchenko, E.S.; Karaman, I. The effect of stress-induced martensite aging in tension and compression on B2–B19′ martensitic transformation in Ni50. 3Ti32. 2Hf17. 5 high-temperature shape memory alloy. *Smart Mater. Struct.* **2021**, *30*, 025039. [CrossRef]
28. Meng, X.L.; Cai, W.; Chen, F.; Zhao, L.C. Effect of aging on martensitic transformation and microstructure in Ni-rich TiNiHf shape memory alloy. *Scr. Mater.* **2006**, *54*, 1599–1604. [CrossRef]
29. Kasimtsev, A.; Volodko, S.; Yudin, S.; Sviridova, T.; Cheverikin, V. Synthesis of powder alloys based on the Ti-Ni-Hf system via a calcium hydride reduction process. *Inorg. Mater.* **2019**, *55*, 449–457. [CrossRef]
30. Volodko, S.; Yudin, S.; Cheverikin, V.; Kasimtsev, A.; Markova, G.; Sviridova, T.; Karpov, B.; Goncharov, S.; Alimov, I. Structure and Properties of Ti28Ni50Hf22 Powder Alloy. *Inorg. Mater. Appl. Res.* **2020**, *11*, 1165–1172. [CrossRef]
31. Catal, A.A.; Bedir, E.; Yilmaz, R.; Canadinc, D. Design of a NiTiHf shape memory alloy with an austenite finish temperature beyond 400 °C utilizing artificial intelligence. *J. Alloys Compd.* **2022**, *904*, 164135. [CrossRef]
32. Shuitcev, A.; Gunderov, D.V.; Sun, B.; Li, L.; Valiev, R.Z.; Tong, Y.X. Nanostructured Ti29.7Ni50.3Hf20 high temperature shape memory alloy processed by high-pressure torsion. *J. Mater. Sci. Technol.* **2020**, *52*, 218–225. [CrossRef]
33. Shuitcev, A.; Vasin, R.N.; Balagurov, A.M.; Li, L.; Bobrikov, I.A.; Tong, Y.X. Thermal expansion of martensite in Ti29.7Ni50.3Hf20 shape memory alloy. *Intermetallics* **2020**, *125*, 106889. [CrossRef]
34. Akgul, O.; Tugrul, H.O.; Kockar, B. Effect of the cooling rate on the thermal and thermomechanical behavior of NiTiHf high-temperature shape memory alloy. *J. Mater. Res.* **2020**, *35*, 1572–1581. [CrossRef]
35. Komarov, V.; Khmelevskaya, I.; Karelin, R.; Postnikov, I.; Korpala, G.; Kawalla, R.; Prahl, U.; Yusupov, V.; Prokoshkin, S. Deformation Behavior, Structure and Properties of an Equiatomic Ti–Ni Shape Memory Alloy Compressed in a Wide Temperature Range. *Trans. Ind. Inst. Met.* **2021**, *74*, 2419–2426. [CrossRef]
36. Sheremetyev, V.; Lukashevich, K.; Kreitcberg, A.; Kudryashova, A.; Tsaturyants, M.; Galkin, S.; Andreev, V.; Prokoshkin, S.; Brailovski, V. Optimization of a thermomechanical treatment of superelastic Ti-Zr-Nb alloys for the production of bar stock for orthopedic implants. *J. Alloys Compd.* **2022**, *928*, 167143. [CrossRef]

 *materials*

Article

Evolution of Structure and Properties of Nickel-Enriched NiTi Shape Memory Alloy Subjected to Bi-Axial Deformation

Victor Komarov [1,2,3,*], Roman Karelin [1,2], Irina Khmelevskaya [2], Vladimir Cherkasov [2], Vladimir Yusupov [1], Grzegorz Korpala [3], Rudolf Kawalla [3], Ulrich Prahl [3] and Sergey Prokoshkin [2]

1 Baikov Institute of Metallurgy and Materials Science RAS, 119334 Moscow, Russia
2 National University of Science and Technology MISIS, 119049 Moscow, Russia
3 Institute of Metal Forming, TU Bergakademie Freiberg, 09599 Freiberg, Germany
* Correspondence: vickomarov@gmail.com

Abstract: The effect of a promising method of performing a thermomechanical treatment which provides the nanocrystalline structure formation in bulk NiTi shape memory alloy samples and a corresponding improvement to their properties was studied in the present work. The bi-axial severe plastic deformation of Ti-50.7at.%Ni alloy was carried out on the MaxStrain module of the Gleeble system at 350 and 330 °C with accumulated true strains of e = 6.6–9.5. The obtained structure and its mechanical and functional properties and martensitic transformations were studied using DSC, X-ray diffractometry, and TEM. A nanocrystalline structure with a grain/subgrain size of below 80 nm was formed in bulk nickel-enriched NiTi alloy after the MaxStrain deformation at 330 °C with e = 9.5. The application of MaxStrain leads to the formation of a nanocrystalline structure that is characterized by the appearance of a nano-sized grains and subgrains with equiaxed and elongated shapes and a high free dislocation density. After the MaxStrain deformation at 330 °C with e = 9.5 was performed, the completely nanocrystalline structure with the grain/subgrain size of below 80 nm was formed in bulk nickel-enriched NiTi alloy for the first time. The resulting structure provides a total recoverable strain of 12%, which exceeds the highest values that have been reported for bulk nickel-enriched NiTi samples.

Keywords: shape memory alloys; NiTi; severe plastic deformation; nanostructured materials

Citation: Komarov, V.; Karelin, R.; Khmelevskaya, I.; Cherkasov, V.; Yusupov, V.; Korpala, G.; Kawalla, R.; Prahl, U.; Prokoshkin, S. Evolution of Structure and Properties of Nickel-Enriched NiTi Shape Memory Alloy Subjected to Bi-Axial Deformation. *Materials* **2023**, *16*, 511. https://doi.org/10.3390/ma16020511

Academic Editor: Francesco Iacoviello

Received: 20 November 2022
Revised: 22 December 2022
Accepted: 3 January 2023
Published: 5 January 2023

1. Introduction

One of the most rapidly developing fields of the modern materials science is associated with the smart functional materials with a shape memory effect (SME). NiTi alloys are the most frequently used smart functional materials with the best functional properties compared to all of the shape memory alloys (SMA). Consequently, NiTi SMA are widely applied in the aircraft industry, mechanical engineering, and particularly in medicine [1–3]. The most significant medical products manufactured from NiTi SMA are vascular stents, clips of various designs, and devices for orthodontics, etc. [4,5] They are commonly produced from NiTi SMA containing more than 50.5 at.% Ni, which provides shape recovery at body temperature [5–7]. The improvement of the functional characteristics of NiTi SMA will improve the reliability and durability of applied devices, and it will also contribute to the design of new ones. Thus, it is a relevant technological and scientific task that requires additional research.

The mechanical and functional properties of NiTi SMA are structure sensitive and depend on the modes of the thermomechanical treatment (TMT) applied for the production of the alloy. Conventional manufacturing technologies of NiTi SMA usually include hot deformation by rolling or forging at recrystallization temperatures and above. Elevated deformation temperatures lead to the formation of the recrystallized structure with a size of about 25–30 microns and a relatively low values of the completely recoverable strain of 4% and lower [8,9].

Severe plastic deformation (SPD) is one of the most commonly applied methods of performing a thermomechanical treatment, allowing one to obtain a true nanocrystalline structure (NCS) in NiTi SMA and considerably improve their operational characteristics [10–13]. However, an NCS can only be obtained in thin NiTi samples after multi-pass cold rolling or high-pressure torsion and subsequent post-deformation annealing (PDA) [10,14]. Therefore, the further development of SPD methods is focused on the search for new deformation modes for manufacturing bulk NiTi samples with NCS [6–8].

In our previous studies, the true nanocrystalline structure was formed for the first time in severely deformed bulk NiTi SMA. The application of bi-axial SPD on the MaxStrain (MS) module for the processing of equiatomic Ti-50.0at.% Ni resulted in the development of a mixed nanocrystalline structure, with the average size of the grain/subgrains being 55 nm, and a very high completely recoverable strain of 9% [15–17]. The present work continues the previous studies, and it is focused on the investigation of the effect of bi-axial deformation on the structural phase state and properties of bulk nickel-enriched NiTi SMA. In these alloys, deformation in the temperature range of dynamic polygonization is accompanied by dynamic strain aging with the precipitation of Ni_4Ti_3 particles, which cannot occur in non-aging equiatomic NiTi SMA, and it has a great impact on the structure's formation and properties [18–21]. The precipitated particles restrict the grain growth, however, they can lead to earlier sample destruction compared to that of equiatomic alloy and the accumulation of a lowered value of true strain. Thus, the aim of this work is to continue studying the effect of bi-axial MS deformation on the structure and properties of NiTi SMA, focusing on the nickel-enriched alloy.

2. Materials and Methods

Hot-rolled nickel-enriched NiTi (50.7 at.% Ni) SMA plates, with a height of 15 mm, were chosen as the study material. The hot-rolled NiTi rod was subjected to a reference treatment (RT), consisting of annealing at 900 °C for 30 min and subsequent cooling in water.

Then, from the reference-treated NiTi plate, samples with a grip zone of 15 × 15 × 150 mm^3 and a central bulk deformable zone of 10 × 10 × 11 mm^3 were cut using an electric discharge machine (Figure 1). For the precise control of the temperature change in the deformable zone during the deformation process, a technological channel for the thermocouple was drilled, as is shown in Figure 1.

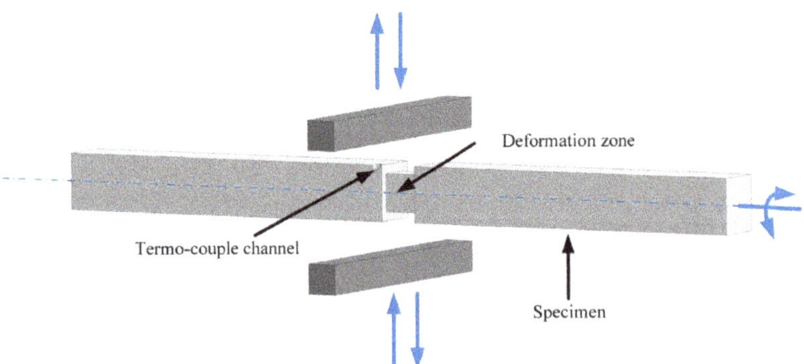

Figure 1. Deformation scheme on the MaxStrain module.

Bi-axial MS deformation was performed using the *Gleeble HDS-V40* physical simulation system. During the MS deformation, the central zone of the specimen was deformed by a compression between two strikers with the rotation of the sample by 90° around its longitudinal axis after each strike. The heating of the deformation zone was supplied by the electric current.

In the present study, the following regimes of MS deformation were applied: the deformation cycle was repeated for 10 or 14 times at 350 and 330 °C at an anvil speed of 0.5 mm/s. Accumulated true strains were calculated as the sum of the logarithmic strains for each compression, and these equaled e = 6.6 and 9.5.

The samples used for the analyses of the obtained structural phase state and its mechanical and functional properties were cut using the electrical discharge machine. Characteristic temperatures of martensitic transformations (MT) were measured using the *Mettler-Toledo* Differential scanning calorimeter, *DSC-3* with a cooling–heating rate of 10 K/min in the temperature range from −100 to +100 °C. The structure was studied using a *D8 Advance* X-ray diffractometer, a *VEGA 3 TESCAN* Scanning electron microscope (SEM), and a *JEM-2100* Transmission electron microscope (TEM). The samples for the structure analysis were cut from the sample in the direction that was perpendicular to the last compression.

The total recoverable strain ε_{rt} was determined by employing a thermomechanical method using a bending mode for strain involving cold water (the R-phase state). The size of the sample was 0.5 × 0.5 × 10 mm^3. The method consists of the following sequence of operations: consecutively increasing the induced strain; determining the value of the residual strain after unloading and the difference between the induced strain and residual strain (elasticity and superelasticity effect); heating it to above the A_s temperature for the strain recovery; indicating the values of residual strain and recovered strain (shape memory effect) [9]. The total recoverable strain ε_{rt} is defined when the residual strain is 0%, and consequently, the shape recovery rate is 100%. The mechanical properties were estimated by the Vickers hardness measurements using the *LECOM 400-A* hardness tester at room temperature under 1N load for 10 s.

3. Results and Discussion

3.1. MS Deformation

The images of the sample and the scheme of MaxStrain deformation are shown in Figure 1. While MS deformation is biaxial, a certain part of a metal flows out from the deformation zone to the specimen head after each compression due to a large plastic strain component. Reducing the volume of the deformable material makes it difficult to build accurate stress–strain curves, and therefore, we focused on the strain force. During the first compression, the sample was reduced to a height of 7 mm (e = 0.36). After the next passes, the reduced height was greater than the previous one was by 0.1–0.2 mm (e = 0.6–0.7) in order to ensure uniform strain distribution. The final height of the sample with e = 9.5 was 5.5 mm.

The maximum deformation force increased from 160 to 180 kN as the temperature decreased from 350 to 330 °C. During the deformation of the equiatomic NiTi SMA, these values were 15–20 kN lower, which could have been caused both by a change in the phase composition of the alloy and by dispersion hardening associated with the precipitation of the Ni_4Ti_3 phase particles.

At a temperature of 350 °C, the NiTi SMA had good deformability, which allowed it to accumulate the required values of a true strain. The decrease in the deformation temperature to 330 °C was accompanied by an appearance of cracks at the point of transition from the deformed zone to the grip one (Figure 2). Cracks appeared during the last compression at e = 9.5, thus, a further increase in the value of accumulated strain at a given temperature was essentially achieved.

The analysis of a fracture reveals a lot of pits of different sizes and depths on the fracture surface (Figure 2). This surface structure is explained by the fact that when the maximum hardened states are reached in local volumes, microvoids appear in areas that are obstacles to continuous deformation. As the stress increases, the microvoids grow and merge, leading to complete destruction, with the formation pits at the fracture, which are interconnected by bridges. Such a pitted microstructure is characteristic of a ductile fracture.

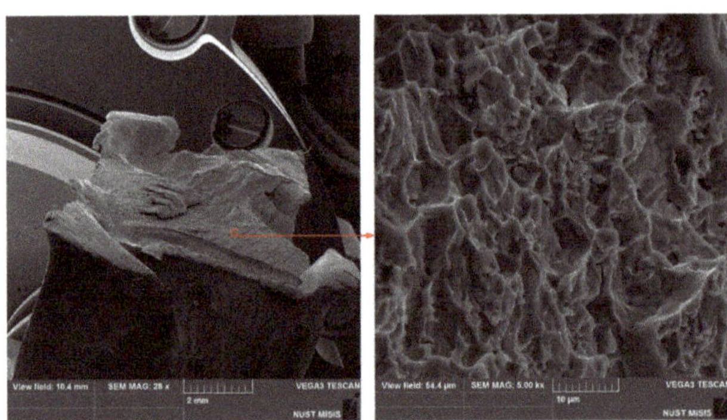

Figure 2. Scanning electron microscopy image of the NiTi SMA sample in the place of destruction after deformation at 330 °C with e = 9.5.

Thus, the critical mode of MS deformation can be defined as deformation at 330 °C with e = 9.5. In an equiatomic alloy, the fracture occurred at 250 °C with e = 11, which can be explained by the absence of precipitates of the hardening Ni_4Ti_3 phase.

3.2. XRD Study

The representative X-ray diffraction patterns of NiTi SMA after the RT and MS deformation, which were obtained at room temperature, are shown in Figure 3a.

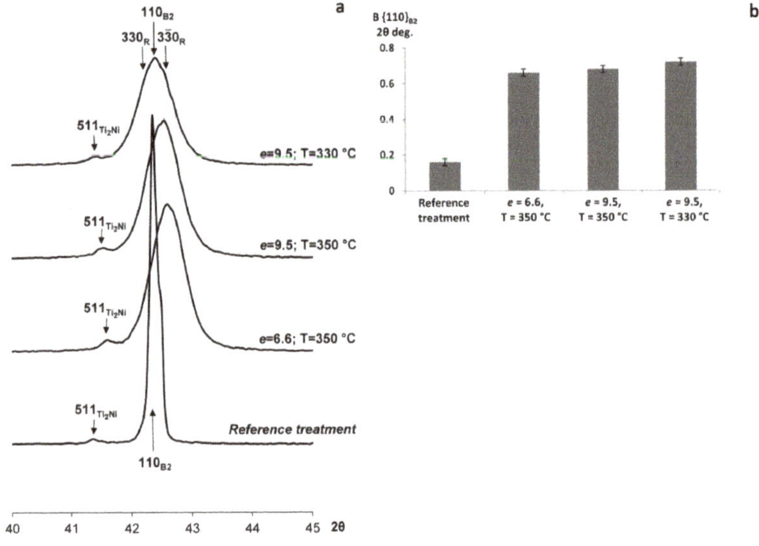

Figure 3. X-ray line$\{110\}_{B2}$ profiles (**a**) and FWHM (**b**) of NiTi at room temperature after various treatments.

The X-ray analysis reveals the B2-austenite line and a certain amount of Ti_2Ni phase (see $\{511\}_{Ti_2Ni}$ line) at room temperature after the RT. Deformation leads to an increase in the characteristic temperatures of martensitic transformation and the corresponding appearance of the intermediate R-phase at room temperature. This increase is proven by the results of DSC.

The measurements of the X-ray line width of B2-austenite B_{hkl} (full width at half maximum (FWHM)) allow us to estimate changes in the crystal lattice's defectiveness. MS deformation is accompanied by a dramatic increase in the B_{110} line FWHM, which indicates a significant increase in the degree of the lattice's defectiveness and the appearance of the R-phase (Figure 3b). The R-phase may contribute to the line broadening, but considering the observed changes, this contribution can be omitted.

3.3. TEM Study

In the studied alloy, after the RT, the average size of the austenite grains is about 25–30 microns. To study the structure and phase composition of the NiTi alloy, the TEM images were obtained after MS deformation with accumulated strains of $e = 6.6$ and 9.5 at 350 and 330 °C, respectively. In the temperature range of 300–500 °C, the dynamic polygonization processes must develop in the nickel-enriched NiTi SMA, as was shown in [18].

The bright and dark field TEM images reveal that MS deformation at 350 °C leads to the formation of a nanocrystalline structure that is characterized by the appearance of a nano-sized grains and subgrains (50–100 nm in transverse direction) with equiaxed and elongated shapes of about 10^{11} cm^{-2} and a high free dislocation density (Figure 4). The cellular substructure and submicron-sized areas with a uniform distribution of dislocations can be observed as well. It can be consequently concluded that the dynamic recovery and polygonization processes in nickel-enriched NiTi SMA in the same thermomechanical conditions are less developed compared to those of an equiatomic NiTi alloy. This can be explained by the retarding effect of the finely dispersed Ni_4Ti_3 phase particles that were precipitated during dynamic strain aging [19,20].

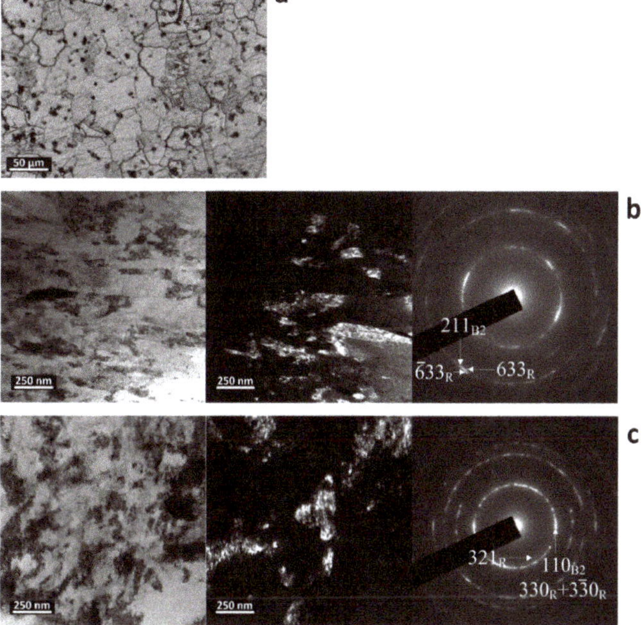

Figure 4. Structure of NiTi alloy after reference treatment, light optical microscopy (**a**), after MS deformation at 350 °C, $e = 6.6$ (**b**), and 330 °C, $e = 9.5$ (**c**); both of them involved transmission electron microscopy. Left, bright-field images; center, dark-field images; right, selected area electron diffraction patterns.

After the analysis of the SAED pattern, only the reflexes of B2-austenite and weak reflexes of the R-phase were determined. The observed parallel plates are crystals of the R-phase, which are formed upon cooling after MS deformation. The TEM study does not reliably reveal the reflexes or images of the Ni_4Ti_3 phase due to there being a very high degree of lattice defectiveness and extremely small Ni_4Ti_3 particle sizes.

MS deformation at 330 °C, e = 9.5, also leads to the formation of the nanograined/nanosubgrained structure in combination with a highly dislocated substructure (Figure 4b). The dislocation density in this case is visually higher than it is after the MS deformation at 350 °C; the grains/subgrains are more equiaxed and smaller. Thus, under MS deformation at 330 and 350 °C, the processes of dislocations redistribution are limited due to the retarding effect of the Ni_4Ti_3 phase particles, and the nanograined/nanosubgrained structure is less developed than it is in the non-aging NiTi alloy [16,17].

3.4. DSC Study

The results of the DSC show that after the RT, one-stage B2 $\leftrightarrow$ B19′ forward and reverse MTs proceed, which are indicating by a single calorimetric peak on the calorimetric curves (Figure 1). The obtained characteristic temperatures of the MT are presented in Table 1: Ms, M_f and A_s, A_f are start and finish temperatures of the forward and reverse MTs; T_R—starting temperature of B2→R MT; M_p; A_p—peak temperatures of forward and reverse MTs. The increase in the strain leads to the appearance of a two-stage B2→R→B19′ transformation, which is indicated by the broadening and multiplication of the peaks in both of the MTs (Figure 5; Table 1).

Table 1. Characteristic temperatures of martensitic transformation of NiTi SMA.

Treatment	M_f, °C	M_p,°C	M_s, °C	T_R,°C	A_s, °C	A_p,°C	A_f, °C
Reference	−32	−17	−4	-	−20	−13	1
T = 350 °C, e = 6.6	<−100	−70	−40	45	−26	−10	42
T = 350 °C, e = 9.5	<−100	−68	−32	45	−22	−8	35
T = 330 °C, e = 9.5	<−100	−70	−35	49	−23	−8	41

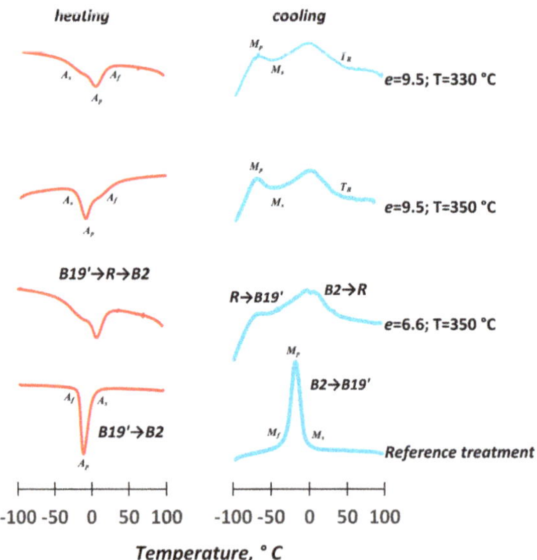

Figure 5. DSC curves of NiTi at room temperature after various treatments.

The strain hardening in combination with the strain aging has the highest influence on the MT sequence and temperatures during the MS deformation. The formation of B19′-martensite is inhibited by the strain hardening which also facilitates the formation of the R-phase from B2-austenite. The dynamic precipitation of Ni_4Ti_3 phase particles indicates, indirectly, significant shifts in the temperature range of the forward martensitic transformation which are below minus 100 °C and cannot be explained only by the deformation hardening. The formation of the Ni_4Ti_3 phase particles, because of the strain aging, also inhibits the R-phase formation. The increase in the strain rate from 6.6 to 9.5 at 350 °C does not significantly change the calorimetric peak positions.

3.5. Functional Properties and Hardness

The strain hardening of the alloys induced by MS deformation increases the Vickers hardness and the total completely recoverable strain ε_{rt}, which is a main functional property of SMA (Table 2). The higher the resistance is to plastic deformation, the higher the strain is that can be induced without involving a dislocation slip and then recovered by shape memory mechanisms, such as thermoelastic MT and martensite reorientation [22].

Table 2. Hardness and total maximum completely recoverable strain of NiTi SMA.

Treatment	$\overline{HV_1}$	ε_{rt}
Reference	242 ± 4	4.0 ± 0.3
T = 350 °C, e = 6.6	325 ± 7	10.9 ± 0.3
T = 350 °C, e = 9.5	341 ± 7	11.7 ± 0.3
T = 330 °C, e = 9.5	362 ± 6	12.0 ± 0.3

The MS deformation is accompanied by significant increase in hardness compared to that which occurs during the RT (Table 2). This trend continues with an increase in the accumulated strain and a decrease in the deformation temperature. We note that the hardness of the previously studied equiatomic alloy is lower compared to that of the Ti-50.7at.%Ni alloy by 15–30 HV under the same MS deformation conditions [18]. This result may be caused by different deformation mechanisms involved in the indentation process. In the nickel-enriched NiTi SMA, indentation is performed at room temperature, i.e., which is far above the A_S temperature, and recovery upon unloading includes the elastic and superelastic strain components, whereas the equiatomic alloy indented below M_f undergoes only elastic recovery upon unloading [22].

The calorimetric study of the Ti-50.7at.%Ni alloy shows that MS deformation leads to the change in the MT sequence, which proceeds through the intermediate R-phase, and the significant extension of the R-phase(Table 1). Therefore, during ε_{rt} measurement, the strain induction starts in the R-phase state. Such a strain-inducing condition allows it to reaching maximum values of the recoverable strain compared to that of the strain-inducing condition in the B19′-martensite state because of the difference between the critical stress for B19′-martensite formation in the R-phase (transformation yield stress) and the reorientation stress of thermal B19′-martensite.

The ε_{rt} value consists of three components: elastic strain and superelastic strain, which recover upon unloading, and the recovery strain of the shape memory effect, which recovers while heating it above A_f. The results of the ε_{rt} measurement for Ti-50.7at.%Ni alloy are shown in Table 2. After all of the studied regimes of the MS deformation were conducted, the ε_{rt} was 11–12%, which is comparable with the highest value obtained in previous studies [13,17,22], and it is much higher compared to that of the RT.

The total recoverable strain's tendency to increase with an increase in the accumulated strains and decrease in the MS deformation temperatures correlates with the increase in the hardness, which indicates an increase in the MS deformation hardening (Table 2). This tendency has also been observed earlier for the equiatomic NiTi alloy [13,16–18]. The increase in hardness is accompanied by the increase in the dislocation yield stress. Therefore, the difference between the dislocation and transformation yield stresses increases as the accumulated strain increases and the deformation temperature decreases. This leads to the later involvement of the irrecoverable dislocation slip in the deformation process, and consequentially, an improvement of the shape recovery characteristics [17,22].

4. Conclusions

The severe plastic deformation of the nickel-enriched Ti-50.7at.%Ni shape memory alloy was carried out for the first time using the MaxStrain (MS) mode at temperatures of 350 and 330 °C with accumulated true strains of e = 6.6 and 9.5, respectively. The critical regime of MS deformation accompanied by the destruction of a sample was defined as T = 330 °C, with e = 9.5. As a result of MS deformation, a nanocrystalline structure with a high dislocation density and an average grain/subgrain size that was below 100 nm was formed, providing a significant increase in the total completely recoverable strain of the nickel-enriched NiTi SMA of up to 12% compared to 4% which occurred after the reference treatment. Traces of the Ni_4Ti_3 phase particles were not observed directly by the TEM and X-ray studies because of their extremely fine size, which were presumably less than 5 nm. However, a dynamic deformation-induced precipitation of Ni_4Ti_3 phase particles was indicated indirectly by the DSC results. After the MS deformation at 330 and 350 °C of nickel-enriched NiTi alloys, the processes of dynamic recovery and polygonization, including dislocation redistribution and subgrains formation, were limited due to the retarding effect of the Ni_4Ti_3 phase particles, and the nanograined/nanosubgrained structure was less developed compared to that of the near-equiatomic NiTi alloys.

Author Contributions: Conceptualization, V.K. and I.K.; methodology, I.K. and V.Y.; validation, S.P. and V.Y.; formal analysis, I.K. and R.K. (Rudolf Kawalla); investigation, V.K., V.C., R.K. (Roman Karelin) and G.K.; resources, U.P. and G.K.; data curation, G.K. and R.K. (Rudolf Kawalla); writing—original draft preparation, V.K. and R.K. (Roman Karelin); writing—review and editing, S.P., I.K. and U.P.; visualization, V.C.; supervision, S.P. and U.P. All authors have read and agreed to the published version of the manuscript.

Funding: The reported study was funded by RFBR, project number 19-33-60090. The structure characterization was carried out during the implementation of the strategic project, "Biomedical materials and bioengineering", within the framework of the Strategic Academic Leadership Program "Priority 2030" at NUST «MISiS».

Institutional Review Board Statement: Not applicable.

Informed Consent Statement: Not applicable.

Data Availability Statement: Not applicable.

Conflicts of Interest: The authors declare no conflict of interest.

References

1. Jani, J.M.; Leary, M.; Subic, A.; Gibson, M.A. A review of shape memory alloy research, applications and opportunities. *Mater. Des.* **2014**, *56*, 1078–1113. [CrossRef]
2. Zareie, S.; Issa, A.S.; Seethaler, R.J.; Zabihollah, A. (2020, October). Recent advances in the applications of shape memory alloys in civil infrastructures: A review. *Elsevier Struct.* **2020**, *27*, 1535–1550. [CrossRef]
3. Jani, J.M.; Leary, M.; Subic, A. Shape memory alloys in automotive applications. Applied Mechanics and Materials. *Trans. Tech. Publ. Ltd.* **2014**, *663*, 248–253. [CrossRef]
4. Morgan, N.B. Medical shape memory alloy applications—The market and its products. *Mater. Sci. Eng. A* **2004**, *378*, 16–23. [CrossRef]
5. Yoneyama, T.; Miyazaki, S. *Shape Memory Alloys for Biomedical Applications*; Elsevier: Amsterdam, The Netherlands, 2008; p. 352.
6. Otsuka, K.; Wayman, C.M. *Shape Memory Materials*; Cambridge Univ. Press: Cambridge, UK, 1999; p. 300.
7. Ryklina, E.; Polyakova, K.; Murygin, S.; Komarov, V.; Andreev, V. On stress- and strain-temperature behavior of titanium nickelide with various grain/subgrain size. *Mater. Lett.* **2022**, *328*, 133–135. [CrossRef]
8. Yang, R.; Ma, W.; Wang, C.; Wang, T.M.; Wang, Q.H. Effect of hot rolling on microstructure and tribology behaviors of Ti-50.8 Ni alloy. *Trans. Nonferrous Met. Soc. China* **2021**, *31*, 967–979. [CrossRef]
9. Komarov, V.; Khmelevskaya, I.; Karelin, R.; Postnikov, I.; Korpala, G.; Kawalla, R.; Prokoshkin, S. Deformation behavior, structure and properties of an equiatomic Ti–Ni shape memory alloy compressed in a wide temperature range. *Trans. Indian Inst. Met.* **2021**, *74*, 2419–2426. [CrossRef]
10. Brailovski, V.; Prokoshkin, S.D.; Khmelevskaya, I.Y.; Inaekyan, K.E.; Demers, V.; Dobatkin, S.V.; Tatyanin, E.V. Structure and properties of the Ti–50.0 at% Ni alloy after strain hardening and nanocrystallizing thermomechanical processing. *Mater. Trans.* **2006**, *47*, 795–804. [CrossRef]

11. Valiev, R.Z.; Estrin, Y.; Horita, Z.; Langdon, T.G.; Zehetbauer, M.J.; Zhu, Y. Producing bulk ultrafine-grained materials by severe plastic deformation: Ten years later. *JOM* **2016**, *68*, 1216–1226. [CrossRef]
12. Gunderov, D.V.; Polyakov, A.V.; Semenova, I.P.; Raab, G.I.; Churakova, A.A.; Gimaltdinova, E.I.; Valiev, R.Z. Evolution of microstructure, macrotexture and mechanical properties of commercially pure Ti during ECAP-conform processing and drawing. *Mater. Sci. Eng. A* **2013**, *562*, 128–136. [CrossRef]
13. Karelin, R.D.; Khmelevskaya, I.Y.; Komarov, V.S.; Andreev, V.A.; Perkas, M.M.; Yusupov, V.S.; Prokoshkin, S.D. Effect of quasi-continuous equal-channel angular pressing on structure and properties of Ti-Ni shape memory alloys. *J. Mater. Eng. Perform.* **2021**, *30*, 3096–3106. [CrossRef]
14. Shuitcev, A.; Gunderov, D.V.; Sun, B.; Li, L.; Valiev, R.Z.; Tong, Y.X. Nanostructured Ti29. 7Ni50. 3Hf20 high temperature shape memory alloy processed by high-pressure torsion. *J. Mater. Sci. Technol.* **2020**, *52*, 218–225. [CrossRef]
15. Khmelevskaya, I.Y.; Kawalla, R.; Prokoshkin, S.D.; Komarov, V.S. Effect of multiaxial deformation Max-strain on the structure and properties of Ti-Ni alloy. In Proceedings of the IOP Conference Series: Materials Science and Engineering, Jakarta, Indonesia, 29–31 March 2014; Volume 63, p. 012108. [CrossRef]
16. Khmelevskaya, I.; Komarov, V.; Kawalla, R.; Prokoshkin, S.; Korpala, G. Effect of biaxial isothermal quasi-continuous deformation on structure and shape memory properties of Ti-Ni alloys. *J. Mater. Eng. Perform.* **2017**, *26*, 4011–4019. [CrossRef]
17. Komarov, V.; Khmelevskaya, I.; Karelin, R.; Prokoshkin, S.; Zaripova, M.; Isaenkova, M.; Kawalla, R. Effect of biaxial cyclic severe deformation on structure and properties of Ti-Ni alloys. *J. Alloy. Compd.* **2019**, *797*, 842–848. [CrossRef]
18. Komarov, V.; Khmelevskaya, I.; Karelin, R.; Kawalla, R.; Korpala, G.; Prahl, U.; Yusupov, V.; Prokoshkin, S. Deformation Behavior, Structure, and Properties of an Aging Ti-Ni Shape Memory Alloy after Compression Deformation in a Wide Temperature Range. *JOM* **2021**, *73*, 620–629. [CrossRef]
19. Cao, S.; Nishida, M.; Schryvers, D. Quantitative three-dimensional analysis of Ni4Ti3 precipitate morphology and distribution in polycrystalline Ni–Ti. *Acta Mater.* **2011**, *59*, 1780–1789. [CrossRef]
20. Komarov, V.; Karelin, R.; Khmelevskaya, I.; Yusupov, V.; Gunderov, D. Effect of Post-Deformation Annealing on Structure and Properties of Nickel-Enriched Ti-Ni Shape Memory Alloy Deformed in Various Initially Deformation-Induced Structure States. *Crystals* **2022**, *12*, 506. [CrossRef]
21. Poletika, T.M.; Girsova, S.L.; Lotkov, A.I.; Kudryachov, A.N.; Girsova, N.V. Structure and Multistage Martensite Transformation in Nanocrystalline Ti-50.9 Ni Alloy. *Metals* **2021**, *11*, 1262. [CrossRef]
22. Prokoshkin, S.D.; Brailovski, V.; Inaekyan, K.E.; Demers, V.; Khmelevskaya, I.Y.; Dobatkin, S.V.; Tatyanin, E.V. Structure and properties of severely cold-rolled and annealed Ti–Ni shape memory alloys. *Mater. Sci. Eng. A* **2008**, *481*, 114–118. [CrossRef]

Article

Effect of Severe Plastic Deformation and Post-Deformation Heat Treatment on the Microstructure and Superelastic Properties of Ti-50.8 at.% Ni Alloy

Tae-Jin Lee and Woo-Jin Kim *

Department of Materials Science and Engineering, Hongik University, Mapo-gu, Sangsu-dong 72-1, Seoul 121-791, Korea
* Correspondence: kimwj@hongik.ac.kr

Abstract: Severe plastic deformation via high-ratio differential speed rolling (HRDSR) was applied to the Ni-rich Ti-50.8Ni alloy. Application of HRDSR and a short annealing time of 5 min at 873 K leads to the production of a partially recrystallized microstructure with a small grain size of 5.1 μm. During the aging process for the annealed HRDSR sample at 523 K for 16 h, a high density of Ni_3Ti_4 particles was uniformly precipitated over the matrix, resulting in the formation of an R phase as the major phase at room temperature. The aged HRDSR sample exhibits excellent superelasticity and superelastic cyclability. This achievement can be attributed to an increase in strength through effective grain refinement and particle strengthening by Ni_3Ti_4 and a decrease in the critical stress for stress-induced martensite (B19′) due to the presence of the R-phase instead of B2 as a major phase at room temperature. The currently proposed method for using HRDSR and post-deformation heat treatment allows for the production of Ni-rich NiTi alloys with excellent superelasticity in sheet form.

Keywords: shape memory; superelasticity; aging; severe plastic deformation; grain refinement; superelastic cyclability

Citation: Lee, T.-J.; Kim, W.-J. Effect of Severe Plastic Deformation and Post-Deformation Heat Treatment on the Microstructure and Superelastic Properties of Ti-50.8 at.% Ni Alloy. *Materials* **2022**, *15*, 7822. https://doi.org/10.3390/ma15217822

Academic Editor: Salvatore Saputo

Received: 19 September 2022
Accepted: 26 October 2022
Published: 5 November 2022

1. Introduction

Ni-rich NiTi alloys containing more than 50.6 at.% nickel exhibit superior superelasticity [1,2]. Superelasticity occurs at temperatures above the austenite finish temperature (A_f) upon loading, and a stress hysteresis forms in the tensile or compressive stress–strain curve due to phase transformation during loading and unloading [1,2]. Superelasticity of NiTi alloys has been extensively studied for applications in biomedical and engineering fields [1,3–5]. However, control of the superelastic properties of Ni-rich NiTi alloys is difficult because the characteristics of the phase transformation sensitively vary with small changes in microstructure and composition [2]. The shape memory functions of Ni-rich NiTi alloys are known to be greatly affected by aging due to the formation of Ni_4Ti_3 precipitates [6–8]. After aging, the martensite start temperature (M_s) and austenite finish temperature (A_f) tend to increase with increasing aging time due to Ni depletion of the matrix by precipitation of the Ni_4Ti_3 phase [9]. Thus, aging is an effective way of controlling the superelastic characteristics of Ni-rich NiTi alloys. Grain size also greatly affects the transformation stress, transformation strain and hysteresis loop area of Ni-rich NiTi alloys [10–13]. From a mechanical viewpoint, hardening by Ni_4Ti_3 precipitates and grain-size reduction in Ni-rich NiTi alloys can increase the critical stress for the occurrence of slip, which is beneficial for achieving good superelasticity. This is because if the critical stress for slip is low, there is a high chance of slip by dislocations during loading.

Recently, interest in superelasticity originating from multistep phase transformation has increased [14–16]. During the multistep transformation process, an intermediate phase, the R-phase with a rhombohedral crystal structure that grows mainly on the {111} plane of B2 austenite, is formed, showing a two-step change for the phase structure:

B2 → R → B19′ [15,16] or a three-step change for the phase structure: B2 → $R1$, B2 → $R2$, $R1/R2$ → B19′ [7] upon cooling. The R-phase transformation can be introduced depending on the thermomechanical treatment conditions [7,14–16]. During cooling, R is formed before the formation of B19′ because B19′ is thermodynamically preferred, but the R phase is kinetically advantageous, having low activation energy [17]. As the B2 to R phase transformation accompanies a significantly smaller transformation strain (less than 1% [16–18]) compared to the B2 to B19′ phase transformation (approximately 8–10% [17,19,20]), this characteristic is useful for small amplitude but higher frequency actuator and damping applications [21].

Heavy plastic deformation or severe plastic deformation (SPD) has been demonstrated to be effective in the grain refinement of NiTi alloys [22–30]. To date, many studies have been conducted to investigate the effect of grain size on superelastic properties [28–30]. Malard et al. [28] and Delville et al. [29] showed that a Ni-rich NiTi alloy, which has been cold worked and then annealed by a pulsed electric current, is highly resistant to dislocation slip at grain sizes < 100 nm, while that with fully recrystallized microstructures and grain sizes exceeding 200 nm is prone to dislocation slip. The authors showed that grain refinement to the grain size <100 nm renders the alloy have superior superelasticity. Tong et al. [30] applied equal channel angular pressing (ECAP) to a Ni-rich NiTi alloy with subsequent annealing at 573–873 K. The sample with a grain size of 0.3 μm exhibited the best superelasticity and cycling stability.

Superelasticity of NiTi alloys is not often determined by a single microstructural factor but by a combined effect involving the grain size, Ni_4Ti_3 precipitate, texture and dislocation density. The combined effect of grain-size reduction by SPD and aging on the superelasticity of Ni-rich NiTi alloys has been relatively rarely studied. In this work, a Ni-rich NiTi alloy was processed by high-ratio differential speed rolling (HRDSR), which is a severe plastic deformation method applicable for materials in sheet form [22,31], and aging was applied to the deformed samples. The effects of grain size, dislocation density, texture and Ni_4Ti_3 precipitates on phase transformation and superelastic behavior were examined.

2. Experimental Procedures

Ti-50.8% at. Ni alloy plates with 3 mm × 2.6 mm × 100 mm were purchased from SMA Co., Ltd (Seoul, Republic of Korea). The purchased plate was heat-treated at 1023 K for 15 min to relieve any residual stress. This sample will be referred to as the as-received (AR) sample. For applying severe plastic deformation at cryogenic temperatures, which is known to be more effective in microstructural refinement compared to SPD at room temperature [22,32], the AR sample was immersed into liquid nitrogen for 10 min, removed from the liquid nitrogen bath and then immediately subjected to differential speed rolling with a speed ratio of 2:1 between the upper and lower rolls. After two passes, the thickness was reduced to a value of 1.55 mm, corresponding to a thickness reduction of 40%. This sample will be hereafter referred to as the HRDSR sample. The AR and HRDSR samples were annealed for 5 min and 120 min at a temperature of 873 K, respectively, in an argon atmosphere. Some of the annealed AR and HRDSR samples were aged for 16 h at 523 K in an argon atmosphere.

To evaluate the superelastic properties of the AR, HRDSR, annealed AR, annealed HRDSR, aged AR and aged HRDSR samples, a cyclic test through tensile loading–unloading was performed. A 6% strain was applied at a test temperature of 298 K.

The microstructure of the samples was observed using field emission scanning electron microscopy (BSE of FE-SEM (SU-5000, Hitachi, Tokyo, Japan), field emission transmission electron microscopy (FE-TEM (JEM 2001 F, 200 keV, JEOL, Tokyo, Japan) and electron backscattering diffraction (EBSD) analysis (Velocity Super, EDAX, Mahwah, United States). For the FE-TEM observation, samples were jet polished with a solution composed of 60% methyl alcohol (CH_3OH), 30% glycerin ($C_3H_8O_3$) and 10% nitric acid (HNO_3) and then ion milled. For the EBSD observation, the sample with a transverse cross section along the RD was mechanically ground using SiC paper and polished using 3 μm and 1 μm diamond suspensions and OP-S suspensions in order. EBSD data were analyzed with a step size of 0.2 μm using TSL-OIM analysis software, excluding data with a confidence index value of 0.1 or less. The average grain size was determined with a grain tolerance angle of 5°, and the fraction of recrystallized grains was determined using the grain orientation spread (GOS) method. The GOS of a recrystallized grain was assumed to be less than 2°.

A differential scanning calorimeter, DSC (DSC 200 F3 Maia, NETZSCH, Selb, Germany), was used in the temperature range of 123 K to 373 K at a heating and cooling rate of 5 K/min to determine the phase transformation temperature.

High-resolution X-ray diffraction (HR-XRD, SmartLab, Rigaku, Tokyo, Japan) with a CuKa (λ = 1.5412 Å) target was used for phase identification of the deformed and heat-treated samples at scan angles ranging from 35 to 80 degrees.

3. Results

3.1. Initial Material

Figure 1a–f show the EBSD inverse pole figure (IPF), grain boundary (GB) and kernel average misorientation (KAM) maps for the AR and HRDSR samples. The grain size of the AR sample is 10.9 μm. The microstructure of the AR sample is composed of equiaxed grains, some of which contain dislocation substructure. The HRDSR sample shows a heavily deformed microstructure. The fraction of low-angle grain boundaries is as high as 0.86. The <110>//ND_{B2} texture in the austenite phase of the AR sample is retained after deformation by HRDSR.

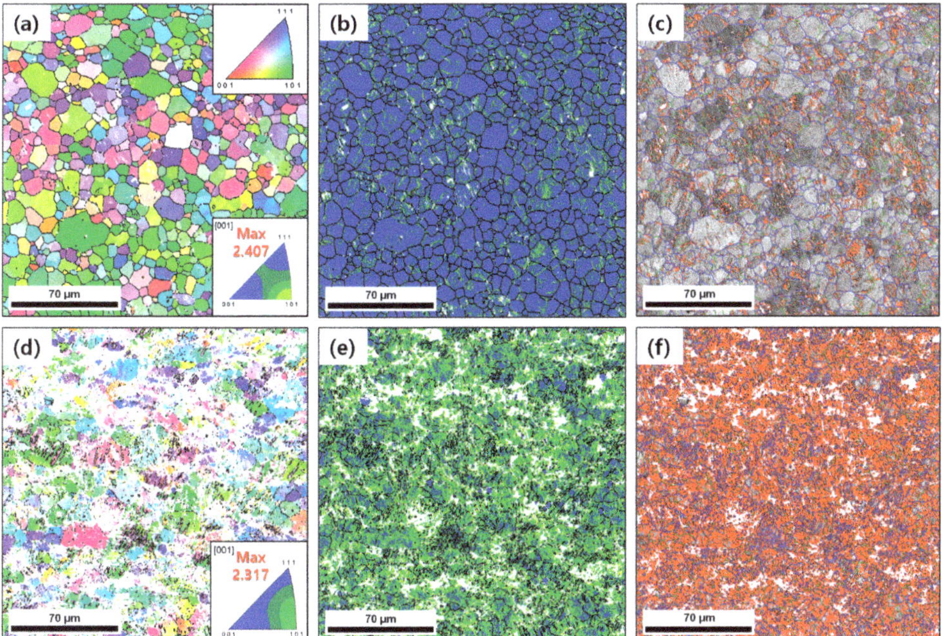

Figure 1. The inverse pole figure, KAM and GB maps for the (**a–c**) as-received (AR) and (**d–f**) HRDSR samples. In the GB map, low-angle boundaries (2–5°) are in red, intermediate angle boundaries (5–15°) are in green and high-angle boundaries (>15°) are in blue.

Figure 2a shows the DSC results for the AR and HRDSR samples. The phase transformation temperatures determined from the DSC curves are summarized in **Table 1**. The AR sample exhibits a two-stage phase transformation (B2 austenite → *R*-martensite → B19′ martensite) upon cooling. The *R*-phase peak is relatively small and broad compared to the B19′ peak. Upon heating, the peak associated with the *R*-phase does not appear, and only a single phase transformation from B19′ to B2 is observed. This type of asymmetric *R*-phase transformation is known to occur when B19′ is energetically preferred over *R* at all temperatures, but *R*-phase formation has a lower kinetic barrier upon cooling [33]. The HRDSR sample does not show any phase transformation during heating and cooling, indicating that the introduction of a high dislocation density by SPD suppresses the martensitic transformation upon cooling.

Table 1. DSC test results for the AR and HRDSR samples before and after various heat treatments.

Samples	Transformation Temperature (K)							
	Cooling				Heating			
	R_s	R_f	M_s	M_f	R_s	R_f	A_s	A_f
As-purchased			246.7	227.3			262.1	277.7
AR	268.9	229.3	207.7	187.3	-	-	245.9	257.5
annealed at 873 K for			□				□	
5 min	-	-	236.1	204.7	-	-	245.4	264.4
120 min	-	-	215.6	208.2	-	-	236.2	258.4
5 min + aged at 523 K	313.5	303.5	189.2	136.2	253.8	267.4	311.1	318.3
120 min + aged at 523 K	309.4	304.2	189.3	159.3	253.3	265.1	310.7	313.7
HRDSR	-	-	-	-	-	-	-	-
annealed at 873 K for			□				□	
5 min	-	-	274.7	181	-	-	218.5	267.6
120 min	-	-	217.7	195.8	-	-	238.8	256
5 min + aged at 523 K	310.2	303.5	189.9	137.4	249.6	260.6	308.9	313.9
120 min + aged at 523 K	322.3	308.4	210.6	187.8	264.9	273.8	314.5	319.6

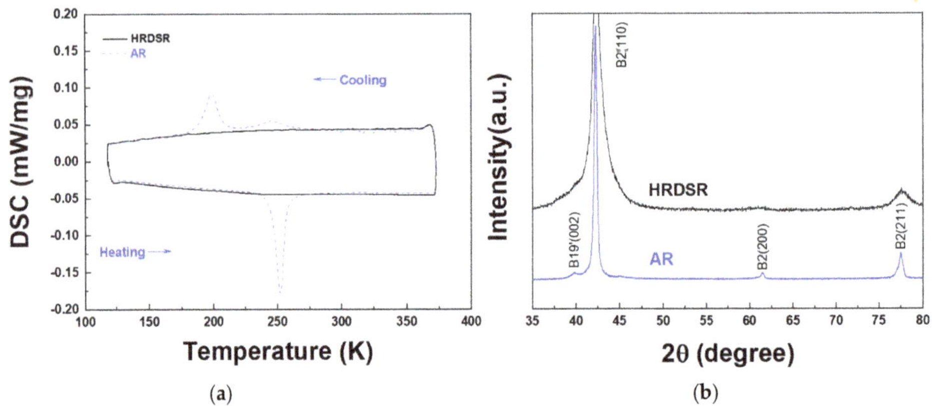

Figure 2. (**a**) The DSC curve for the AR and HRDSR samples. (**b**) The XRD curves for the AR and HRDSR samples. Identification of phases was made based on the data from JCPDF cards (01−076 −3614, 01−076−7519 and 01−076−4263).

Figure 2b shows the XRD curves for the AR and HRDSR samples, respectively. For both samples, only the B2 austenite phase is identified, which agrees with the prediction from the DSC result. Compared to the AR sample, the HRDSR sample has considerably broader peaks, indicating that the dislocation density and the ratio of amorphous and nanosized grains are greatly increased after SPD by HRDSR.

3.2. Materials Processed by HRDSR

3.2.1. Microstructures

Figure 3a–h shows the EBSD inverse pole figure maps for the annealed and aged AR and HRDSR samples. The grain size and fraction of recrystallized grains determined based on the EBSD data are plotted in **Figure 4a,b**, respectively. The grain size is increased with annealing time. For the AR sample, the grain size is increased from 10.9 to 13.9 μm after annealing for 120 min. For the HRDSR sample, the grain sizes after annealing for 5 min and 120 min are 5.1 μm and 8.7 μm, respectively. During the subsequent aging process, noticeable grain growth occurs in the AR and HRDSR samples annealed for 5 min, while limited grain growth occurs in the AR and HRDSR samples annealed for 120 min. The fraction of recrystallized grains in the AR sample increases from 0.33 to 0.43 and 0.78 after annealing for 5 and 120 min, respectively. For the HRDSR sample, the fraction of recrystallized grains dramatically increases after 5 min of annealing (from 0.03 to 0.68) and is further increased to 0.78 after annealing for 120 min. Inverse pole figures, given as insets, show that the <110>//ND_{B2} texture component in the AR sample is retained after annealing and aging. For the HRDSR sample, after annealing, a new texture component (<111>//ND)$_{B2}$ develops, supporting the occurrence of recrystallization during the heat treatment, and this texture component retains after aging. For the aged AR and HRDSR samples, the fraction of recrystallized grains decreases (rather than increases) after aging (except for the AR sample annealed for 5 min), and this decrease is most pronounced in the HRDSR sample annealed for 5 min. This unexpected result is most likely due to the precipitation of Ni_4Ti_3 particles that creates coherency strain fields surrounding them and induces the formation of *R*-phase around the particles within grain interiors, leading to an increase in GOS to a value of over 2° in the recrystallized grains obtained during the annealing process. This will be discussed later.

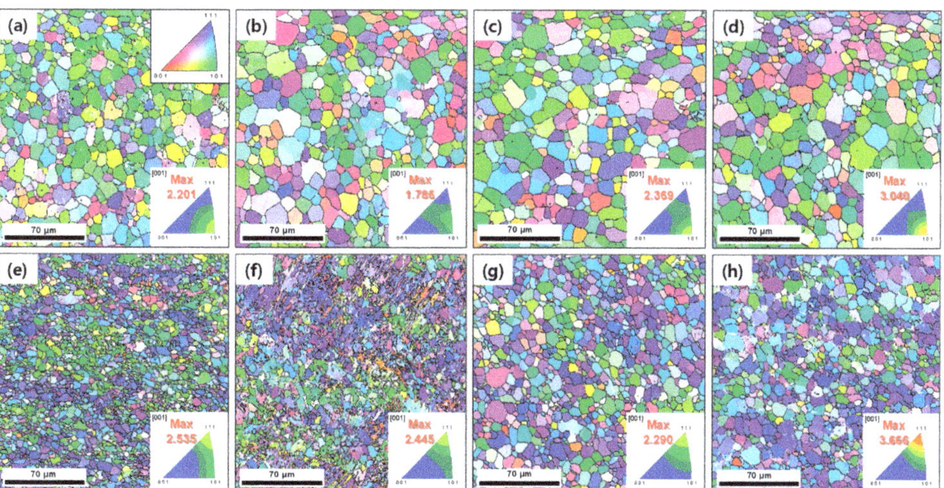

Figure 3. The IPF maps for the AR samples (**a**) annealed for 5 min at 873 K, (**b**) annealed for 5 min at 873 K and then aged at 523 K for 16 h, (**c**) annealed for 120 min at 873 K, (**d**) annealed for 120 min at 873 K and then aged at 523 K for 16 h. The IPF maps for the HRDSR samples (**e**) annealed for 5 min at 873 K, (**f**) annealed for 5 min at 873 K and then aged at 523 K for 16 h, (**g**) annealed for 120 min at 873 K, (**h**) annealed for 120 min at 873 K and then aged at 523 K for 16 h.

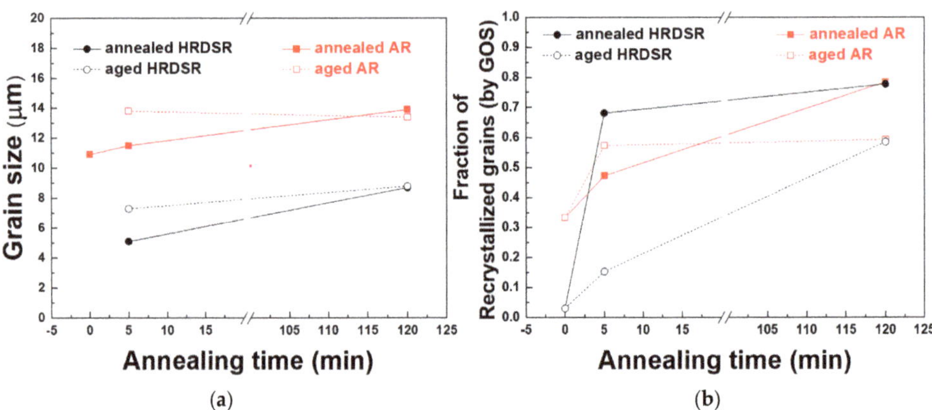

Figure 4. (**a**) The grain size of the AR and HRDSR samples before and after heat treatment. (**b**) The fraction of recrystallized grains in the AR and HRDSR samples before and after heat treatment determined using GOS.

3.2.2. Phase Transformation Temperatures

Figure 5a–c shows the DSC results for the annealed and aged AR and HRDSR samples. The phase transformation temperatures determined from the DSC curve are summarized in **Table 1**. The HRDSR sample, which does not show the austenite–martensite transformation peak upon cooling (**Figure 2a**), shows a small and broad austenite–martensite transformation peak after annealing for 5 min (**Figure 5a**). The B2→B19′ transformation becomes more obvious after prolonged annealing for 120 min. This observation indicates that the increased degree of recrystallization promotes the B2–B19′ transformation because the density of dislocations is further decreased with increasing amount of recrystallization. After aging, two-stage transformations occur in both the AR and HRDSR samples. Two-stage transformation occurs upon cooling as well as heating in both alloys. This type of symmetric *R*-phase transformation is known to occur when there is a temperature window in which *R* is thermodynamically favored over both B2 and B19′ [33]. The peaks for the B2 → *R* and *R* → B2 transformations in all the aged samples are located at approximately 300 K and 315 K upon cooling and heating, respectively. It is noted from the DSC curves that the transformation temperatures for B2 → *R* and *R* → B2 are less sensitive to the microstructure compared to the transformation temperatures for *R*→B19′ and B19′ → *R*. **Figure 5c** shows the magnified DSC curves for the aged AR and HRDSR samples during cooling in the temperature range between 220 and 300 K. Unlike the aged AR samples, small peaks appear in the aged HRDSR samples. This result suggests the possibility of the occurrence of B2 → *R*2 transformation in the aged HRDSR samples. It was claimed that the phase transformation of B2 → *R*1 occurs due to the generation of Ni_4Ti_3 near high-energy grain boundaries, and B2 → *R*2 phase transformation occurs when the dislocation networks existing inside the grains act as nucleation sites for Ni_4Ti_3 precipitation [6,7].

3.2.3. Precipitates

Figure 6a,b shows the XRD curves for the annealed and aged AR and HRDSR samples. Only the B2 austenite phase is observed in both the annealed AR and HRDSR samples. After aging treatment, *R*-phase peaks can be observed, which is clearly evidenced by a splitting of the (1 1 0) austenitic peak [7] in both the AR and HRDSRed samples, indicating that the microstructures of the aged AR and HRDSR samples are composed of a mixture of B2 and *R* phases at room temperature, in agreement with the result expected from the DSC curves at 298 K.

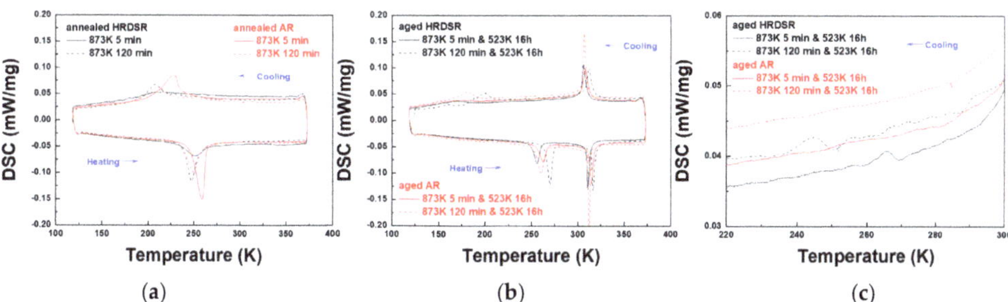

(a) (b) (c)

Figure 5. The DSC curves for (**a**) the annealed and (**b**) the aged AR and HRDSR samples. (**c**) The DSC curves for the aged AR and HRDSR samples magnified in the temperature range between 220 and 300 K upon cooling.

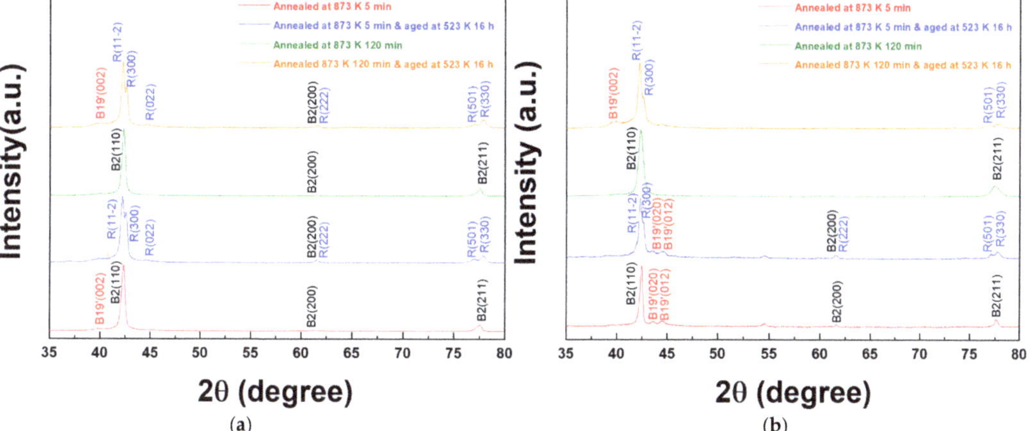

(a) (b)

Figure 6. The XRD curves for the (**a**) AR and (**b**) HRDSR samples after annealing or annealing plus aging. Identification of phases was made based on the data from JCPDF cards (01−076−3614, 01−076−7519 and 01−076−4263).

Figure 7a–h shows the KAM maps for the annealed and aged AR and HRDSR samples, and **Figure 8** shows a plot of their average KAM values. The KAM value indicates the average misorientation value between the measurement point and its surrounding points, which indicates the local strain [34]. Therefore, the density of geometrically necessary dislocations, which are necessary for preserving lattice continuity, increases with the KAM value [35]. The KAM value decreases after annealing for 5 min in both the AR and HRDSR samples, but a more significant decrease occurs in the HRDSR sample, which is due to the occurrence of recrystallization in the HRDSR sample. A large increase in KAM after aging occurs in the HRDSR sample annealed for 5 min; however, the KAM value does not change much after aging of the AR sample annealed for 5 min. The large increase in the KAM value over the matrix of the HRDSR sample after aging can be attributed to precipitation of a large amount of Ni_4Ti_3 particles during aging, which occurs because the large grain boundary areas and dislocation substructure within interiors of small partially recrystallized grains provide the preferred sites for nucleation of Ni_4Ti_3 particles and fast atomic diffusion paths. The formation of the Ni_4Ti_3 precipitates promotes the formation of the *R* phase by impeding the B2→B19′ transformation, but as Ni_4Ti_3 creates high-strain energy, the nucleation and growth of *R* preferentially occur near the grain boundary and dislocation substructure where the nucleation barrier for the *R* phase is low [36]. The fraction of low-angle grain boundaries of the HRDSR sample (annealed for 5 min) has increased from 0.23 to 0.75 after aging, and this is most likely due to the creation of many interfaces between the *R* and B2 phases. For the HRDSR sample annealed for 120 min, the increase in the average KAM value after aging is relatively small, and the large KAM values are confined to near grain boundaries. This results because the grain boundary area and the density of dislocation substructure largely decreased during the long-time annealing of 120 min.

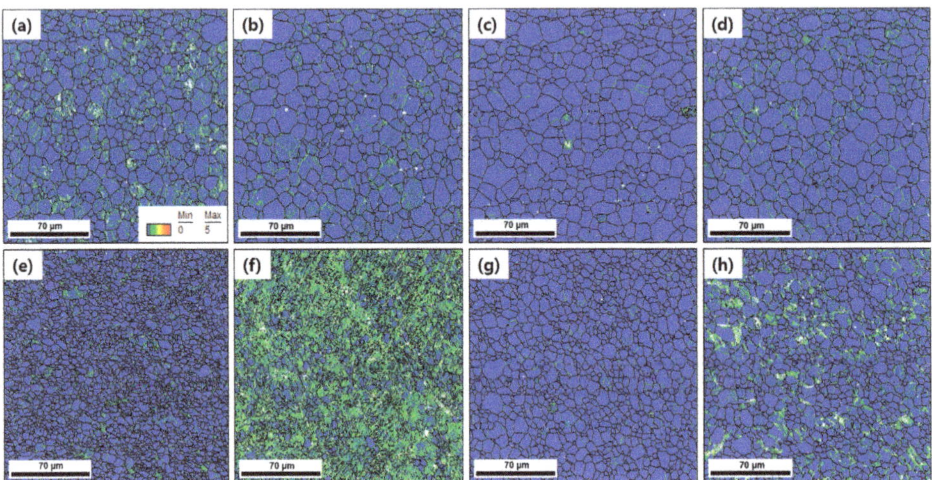

Figure 7. The KAM maps for the AR samples (**a**) annealed for 5 min at 873 K, (**b**) annealed for 5 min at 873 K and then aged at 523 K for 16 h, (**c**) annealed for 120 min at 873 K, (**d**) annealed for 120 min at 873 K and then aged at 523 K for 16 h. The IPF maps for the HRDSR samples (**e**) annealed for 5 min at 873 K, (**f**) annealed for 5 min at 873 K and then aged at 523 K for 16 h, (**g**) annealed for 120 min at 873 K, (**h**) annealed for 120 min at 873 K and then aged at 523 K for 16 h.

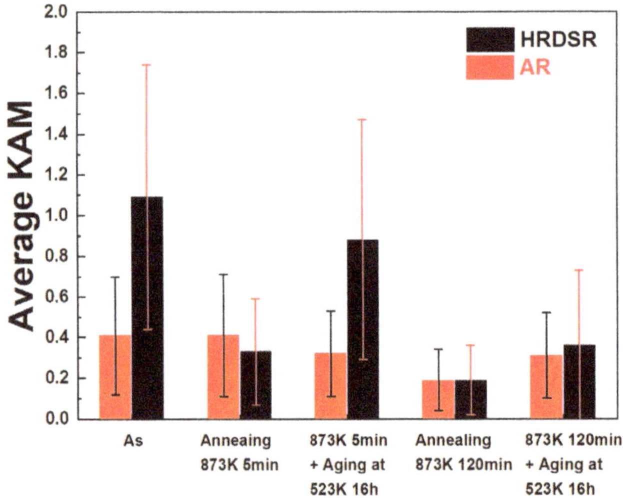

Figure 8. The average KAM values of the AR and HRDSR samples after annealing or annealing plus aging.

Figure 9a,b shows TEM micrographs for the aged HRDSR samples (after annealing for 5 min). It is evident that Ni_4Ti_3 with a typical lenticular shape is nucleated within grain interiors as well as near grain boundaries. The *R*-phase is often observed near Ni_4Ti_3 phases, supporting that Ni_4Ti_3 phase promotes the occurrence of the *R* phase. **Figure 9c** shows a TEM micrograph of the aged HRDSR sample (after annealing for 120 min). It is noted that the size of the Ni_4Ti_3 precipitate is larger than that observed for the aged HRDSR sample (after annealing for 5 min). **Figure 9d** shows an SEM micrograph of the aged HRDSR sample (after annealing for 120 min), where most of the Ni_4Ti_3 particles are observed to nucleate and grow near grain boundaries. At the same SEM magnification, it is hard to find Ni_4Ti_3 particles in the matrix of the aged HRDRS sample (after annealing for 5 min) (not shown here), indicating that Ni_4Ti_3 precipitate particles in the aged HRDRS sample (after annealing for 5 min) are much smaller than those in the aged HRDRS sample (after annealing for 120 min).

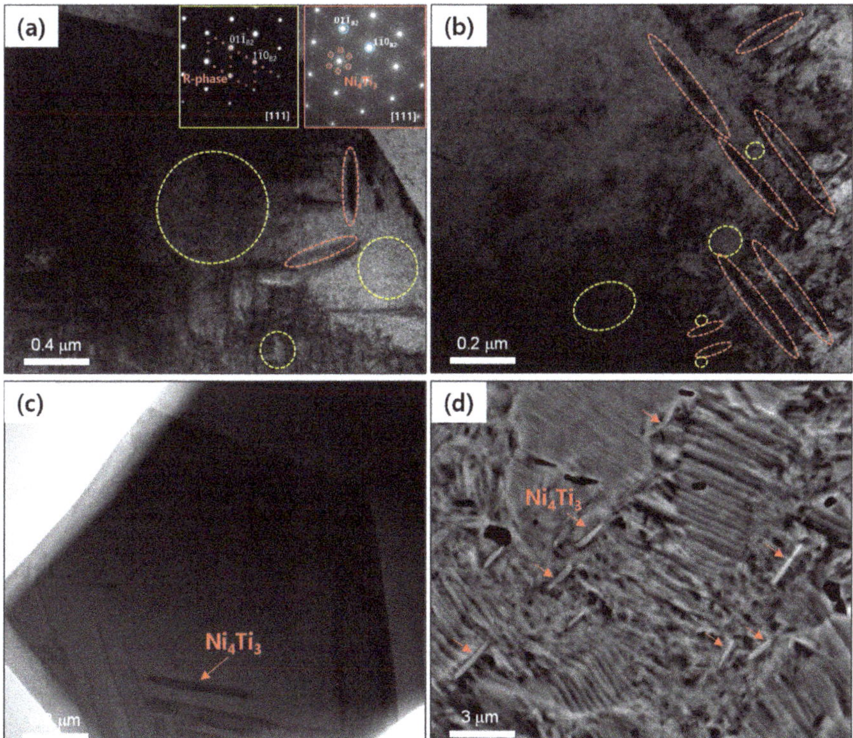

Figure 9. The TEM micrographs for the aged HRDSR sample (after annealing for 5 min): (**a**) near grain boundaries and (**b**) grain interior. The (**c**) TEM and (**d**) SEM micrographs for the aged HRDSR sample (after annealing for 120 min). The regions marked by yellow and red circles represent the regions where B2+*R* phases and B2+Ni_4Ti_3 phases are identified to exist, respectively.

3.2.4. Texture

The superelastic strain of NiTi alloys depends on the crystal orientation. Miyazaki et al. [37] showed that $[233]_{B2}$, $[111]_{B2}$ and $[011]_{B2}$ have high superelastic strains of 10.7%, 9.8% and 8.4% in a single crystal [3,37,38]. **Figure 10** shows the $[233]_{B2}$//RD, $[111]_{B2}$//RD, and $[011]_{B2}$//RD texture components mapped on the EBSD-generated microstructures of the AR and HRDSR samples and **Figure 11** shows the total fractions of grains with the three texture components in the AR and HRDSR samples. The calculated fractions for the annealed and aged AR and HRDSR samples range between 0.4 and 0.6, but the HRDSR samples have lower fractions than the AR samples. This result implies that from a texture viewpoint, the HRDSR samples do not have an advantage in achieving better superelasticity compared to the AR samples.

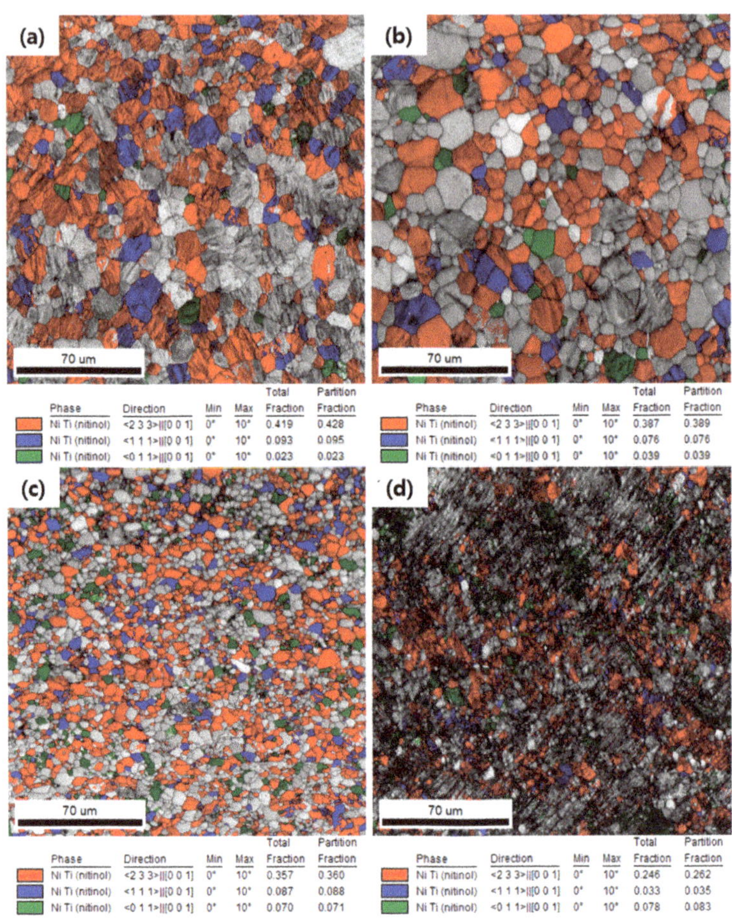

Figure 10. The $[233]_{B2}//RD$, $[111]_{B2}//RD$ and $[011]_{B2}//RD$ texture components mapped on the EBSD-generated microstructures of the AR samples (**a**) annealed for 5 min at 873 K and (**b**) annealed for 5 min at 873 K and then aged at 523 K for 16 h, and the HRDSR samples (**c**) annealed for 5 min at 873 K and (**d**) annealed for 5 min at 873 K and then aged at 523 K for 16 h.

3.2.5. Mechanical Properties

Figure 12a,b shows the Vickers hardness measurement results obtained for the AR and HRDSR samples after annealing and aging. For the AR sample, after annealing for 5 min, the hardness is slightly decreased to 259.1 Hv due to the annealing effect. However, upon subsequent aging, the hardness is increased to 280–288 Hv. The HRDSR sample shows a Hv of 354, which is significantly higher than that for the AR sample (266.8 Hv). After annealing for 5 min, the hardness is largely decreased to 251 Hv due to the occurrence of recrystallization. Upon subsequent aging, however, the hardness is increased to 327.2–342.9 Hv, which is much higher than that obtained for the AR sample aged under the same conditions. The hardness of the HRDSR sample annealed for 120 min also increases after aging, but the obtained hardness is notably lower than that of the aged HRDSR sample (after annealing for 5 min).

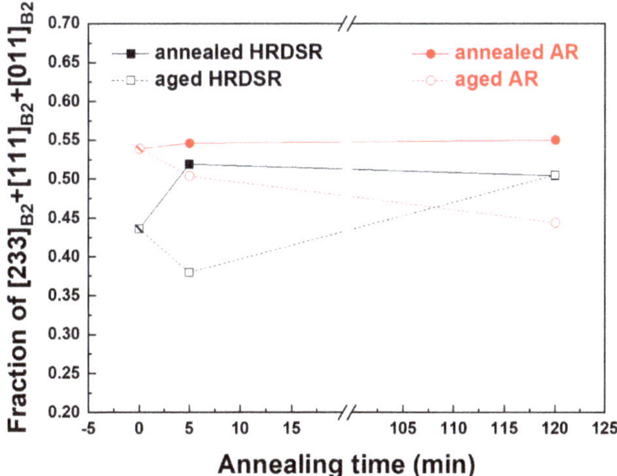

Figure 11. The total fraction of grains with the texture components of $[233]_{B2}//RD$, $[111]_{B2}//RD$ and $[011]_{B2}//RD$ texture components for the AR and HRDSR samples after annealing or annealing plus aging.

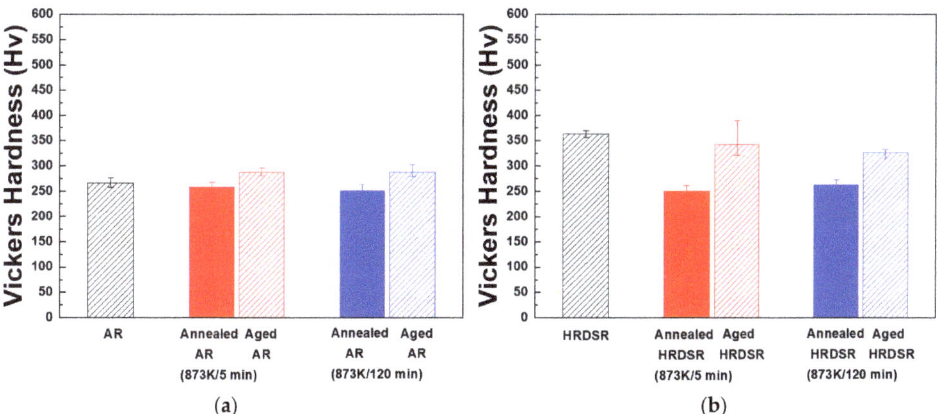

Figure 12. The Vickers hardness of the (**a**) AR and (**b**) HRDSR samples after annealing or annealing plus aging.

Figure 13a–f shows the cyclic tensile test results for the annealed AR and HRDSR samples up to 6%. The measured residual strains are plotted as a function of the number of cycles in **Figure 14**. For both the annealed AR and HRDSR materials, B19′ is expected to be directly induced from B2 during loading. For the AR sample, annealing treatment for 5 min reduces the residual strain from 3.25% to 1.4% in the first cycle, indicating a positive effect of annealing on superelasticity. As the number of cycles increases, the residual strain increases and then tends to become saturated after many cycles. For the HRDSR sample, fracture occurs before reaching a strain of 6% due to a significant deterioration of ductility after heavy plastic deformation, which is typical in many SPD-processed metals [2]. After annealing for 5 min and 120 min, the residual strain after the first cycle is approximately 2% and fracture occurs after 4~6 cycles. There is a notable difference between the annealed AR and HRDSR samples in terms of the critical stress for stress-induced martensitic transformation: the critical stresses of the annealed AR samples are lower than those for the annealed HRDSR samples (357.0–375.1 MPa vs. 433.0–437.0 MPa). This difference can be attributed to the grain-size effect on critical stresses for stress-induced martensitic transformation because, as the grain size becomes smaller, the barrier for martensitic transformation is expected to become higher [11].

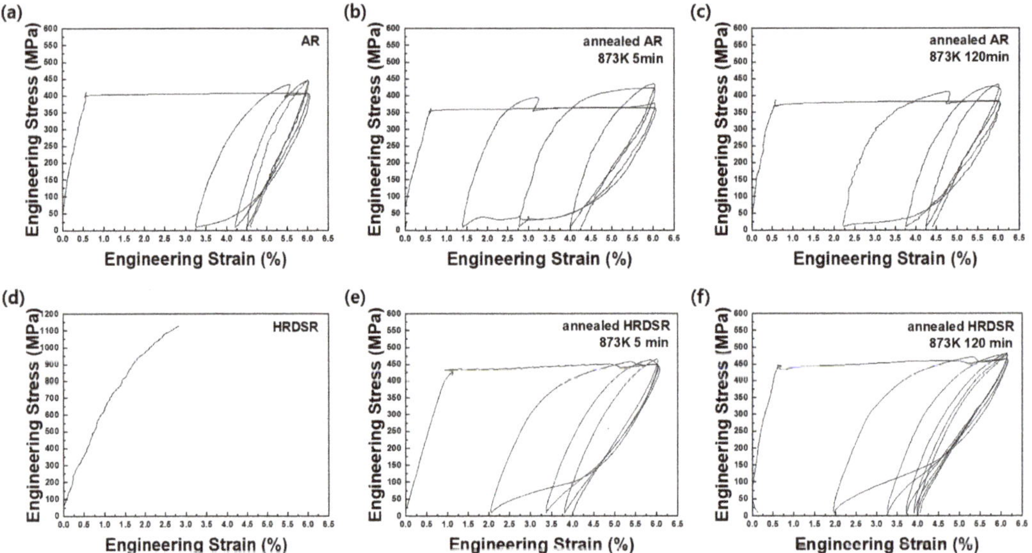

Figure 13. Superelastic cyclic tests up to the strain of 6% for (**a**) the AR sample, (**b**) the AR sample annealed at 873 K for 5 min, (**c**) the AR sample annealed at 873 K for 120 min, (**d**) the HRDSR sample, (**e**) the HRDSR sample annealed at 873 K for 5 min and (**f**) the HRDSR sample annealed at 873 K for 120 min.

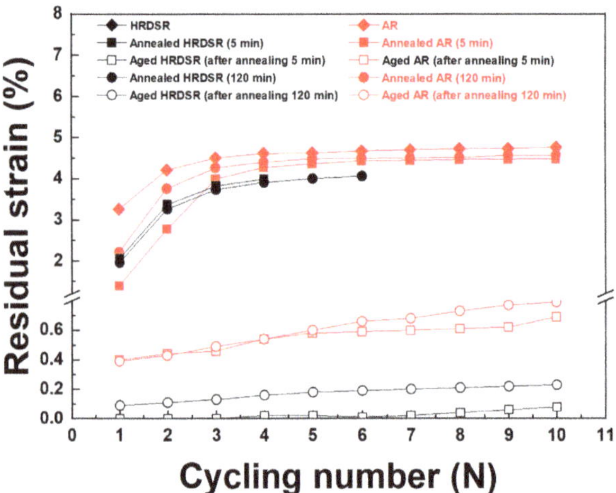

Figure 14. Residual strain as a function of cyclic number for the AR and HRDSR samples after annealing or annealing plus aging.

Figure 15a–d shows the cyclic tensile test results for the aged AR and HRDSR samples up to 6%. The measured residual strains are plotted as a function of the number of cycles in **Figure 14**. A significant improvement in superelasticity is observed in both materials compared to the cases when only annealing is applied. Unlike the annealed materials with the B2 phase, the aged materials contain a mixture of B2 and *R* phases, and the amount of *R* is expected to be largest in the aged HRDSR sample (after annealing for 5 min). The slope for elastic deformation of the aged HRDSR sample (after annealing for 5 min) is apparently lower than that of the aged HRDSR samples (after annealing for 120 min) as well as the aged AR samples. This is because the elastic modulus of the *R* phase is lower than that of B2 (20 [16] vs. 60–70 GPa [39]) such that the elastic modulus of the aged HRDSR sample (after annealing for 5 min) with the *R*-phase as a major phase is relatively low. For the aged HRDSR sample (after annealing for 5 min), where B19′ is expected to be stress induced from *R* rather than B2 during loading and that B19′ reverts to *R* during unloading, the residual strain remains virtually zero after many cycles. This result indicates that the aged HRDSR sample (after annealing for 5 min) exhibits excellent superelasticity and cyclic stability. For the aged HRDSR sample (after annealing for 120 min), where B19′ is expected to be stress induced from B2 during loading, the residual strain is larger than that for the aged HRDSR sample (after annealing for 5 min) from the first cycle and with repeated cycling, the residual strain continues to increase and then saturates beyond five cycles. Compared to the aged HRDSR samples, the aged AR samples exhibit poorer superelasticity and cyclability. It is worthwhile to note that the aged HRDSR sample (after annealing for 5 min) exhibits incremental variation in stress at plateau during repeated superelastic loading and unloading, while the other samples show a flat stress plateau. Wang et al. [40] also observed the incremental stress variation during superelastic loading in the swaged NiTi and attributed this phenomenon to the heterogeneous microstructure of the swaged sample (mixed with high- and low-angle grain boundaries) where martensitic transformation occurs first at high-angle grain boundaries and then later at low-angle grain boundaries. The microstructure of the aged HRDSR sample (after annealing for 5 min) also consists of many high-angle grain boundaries and dislocation substructures in grain interiors.

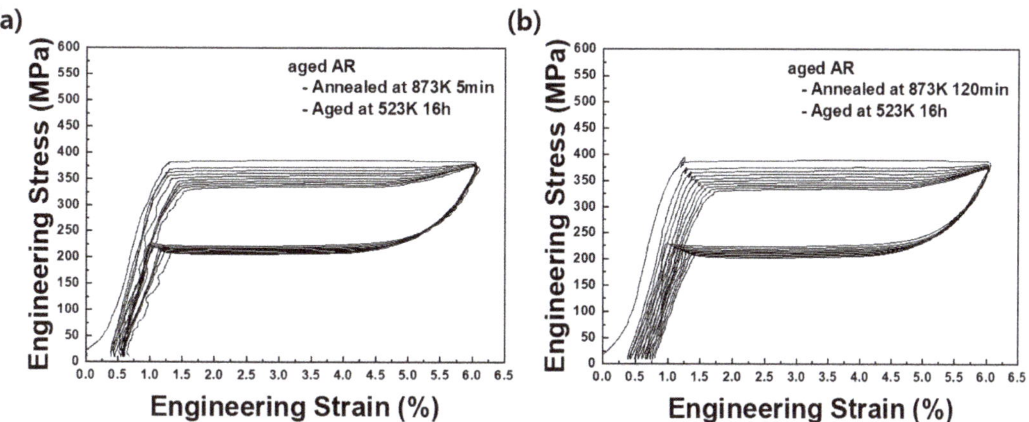

Figure 15. *Cont.*

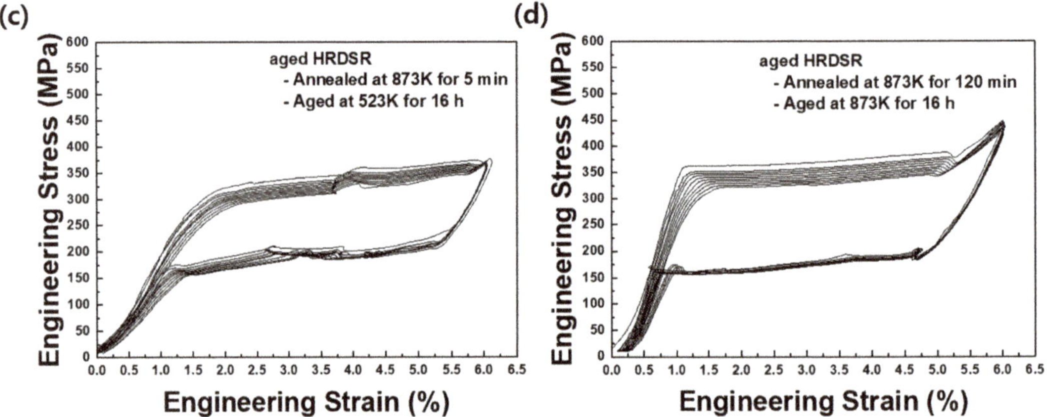

Figure 15. Superelastic cyclic tests up to the strain of 6% for (**a**) the aged AR sample (after annealing at 873 K for 5 min), (**b**) the aged AR sample (after annealing at 873 K for 120 min), (**c**) the aged HRDSR sample (after annealing at 873 K for 5 min) and (**d**) the aged HRDSR sample (after annealing at 873 K for 120 min).

From **Figure** 15**a–d**, it is also recognized that the aged HRDSR sample (after annealing for 5 min) shows the lowest critical stress for martensitic transformation among the four aged samples, even though it has the smallest grain size. This is most likely because the barrier for transformation from *R* to B19′ is smaller than that for transformation from B2 to B19′ [17]. The aged HRDSRed sample (after annealing for 5 min) also exhibits the smallest hysteresis (low dissipation energy).

4. Discussion

As the aged HRDSR samples do not have a favorable texture compared to the aged AR samples, the superior superelasticity of the former should originate from their microstructures. Tong et al. [41] proposed that the difference between the yield strength of B2 and the critical stress for the phase martensitic transformation ($\Delta\sigma$) is related to the recovery strain, which is equal to the applied strain minus the residual strain. Here, the yield stresses for the annealed and aged AR and HRDSR samples were estimated based on their Vickers hardness data using the relation of $\sigma_y = 3.03H_v$ [42], where σ_y is the yield stress (MPa) and H_v is the Vickers hardness (kg/mm^2). **Figure** 16a shows the $\Delta\sigma$ calculated using the σ_y values calculated from the H_v data and the critical stresses for the martensitic transformation measured from the tensile tests in **Figures** 13 **and** 15. The aged HRDSRed sample (after annealing for 5 min) exhibits the largest $\Delta\sigma$. This is because the critical stress for phase transformation from *R* to B19′ is lower than that from B2 to B19′ and the yield strength is high due to effective grain refinement and particle strengthening.

Figure 16**b** shows the relationship between $\Delta\sigma$ and residual strain. There is a trend that as $\Delta\sigma$ increases, the residual strain decreases. This result suggests that the HRDSR technique can greatly enhance the superelasticity of Ni-rich NiTi alloys by increasing the strength (against slip) through effective grain refinement and aging and by decreasing the critical stress for stress-induced martensite (by having the *R* phase as a major phase prior to loading). Reduction of the dislocation density by annealing is important because the presence of a high dislocation density disturbs the phase transformation from B2$\longleftrightarrow$B19′ or *R*$\longleftrightarrow$B19′ and reduces the cyclic number. However, when the dislocation density or dislocation substructure is reduced too much through long-term annealing, the nucleation sites for Ni_3Ti_4 precipitates can be greatly reduced, leading to a decrease in the yield strength and an increase in the critical stress for stress-induced martensite by decreasing the volume fraction of the *R*-phase at room temperature under unstressed condition.

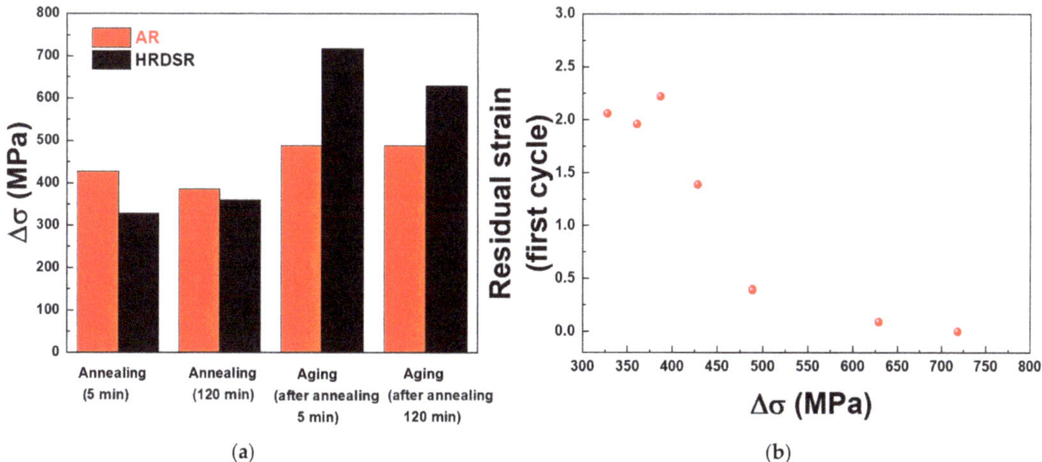

Figure 16. (**a**) The difference between the yield strength and the critical stress for phase martensitic transformation ($\Delta\sigma$) and (**b**) the relationship between $\Delta\sigma$ and residual strain.

5. Conclusions

The combined effect of grain-size reduction by SPD and aging on the superelasticity of Ni-rich NiTi alloy was studied, and the following results were obtained:

1. Severe plastic deformation by HRDSR and subsequent short-term annealing for 5 min at 873 K produces a partically recrystallized microstructure with a small grain size of 5.1 μm.
2. During the aging of the annealed HRDSR sample at 523 K for 16 h, a high density of Ni_3Ti_4 particles is densely and uniformly precipitated over the matrix, resulting in the formation of an *R* phase as the major phase at room temperature. For a long annealing time before aging, the dislocation substructure within the grain interiors is diminished, and the grain boundary area decreases, such that the precipitation of Ni_3Ti_4 during aging is small, and their distribution is inhomogeneous.
3. The difference between the yield strength and critical stress for the stress-induced martensitic transformation ($\Delta\sigma$) is found to be closely related to the superelastic strain. As $\Delta\sigma$ increases, the superelastic strain increases.
4. Superelasticity and cyclability of a Ni-rich NiTi alloy can be enhanced by increasing the strength through effective grain refinement via SPD plus annealing and aging for precipitation of Ni_3Ti_4 and by decreasing the critical stress for stress-induced martensite through incorporation of the *R*-phase as a major phase at room temperature.

Author Contributions: Conceptualization, W.-J.K.; methodology, T.-J.L.; investigation, T.-J.L.; data curation, T.-J.L.; writing—original draft preparation, T.-J.L.; writing—review and editing, W.-J.K.; funding acquisition, W.-J.K. All authors have read and agreed to the published version of the manuscript.

Funding: This research was financially supported by the National Research Foundation of Korea, funded by the Korean government (MSIT) (Project No. NRF 2020R1A4A1018826).

Data Availability Statement: The raw/processed data required to reproduce these findings cannot be shared at this time as the data also forms part of an ongoing study.

Conflicts of Interest: The authors declare no conflict of interest.

References

1. Otsuka, K.; Ren, X. Physical metallurgy of Ti–Ni-based shape memory alloys. *Prog. Mater. Sci.* **2005**, *50*, 511–678. [CrossRef]
2. Khan, M.I.; Zhou, Y. Micro-welding of nitinol shape memory alloy. *Join. Assem. Med. Mater. Devices* **2013**, 133–153.
3. Otsuka, K.; Wayman, C.M. *Shape Memory Materials*; Cambridge University Press: Cambridge, UK, 1999.

4. Baigonakova, G.; Marchenko, E.; Kovaleva, M.; Vorozhtsov, A. Influence of Wire Geometry on the Mechanical Behavior of the TiNi Design. *Metals* **2022**, *12*, 1131. [CrossRef]
5. Yasenchuk, Y.F.; Marchenko, E.S.; Gunter, S.V.; Baigonakova, G.A.; Kokorev, O.V.; Volinsky, A.A.; Topolnitsky, E.B. Softening Effects in Biological Tissues and NiTi Knitwear during Cyclic Loading. *Materials* **2021**, *14*, 6256. [CrossRef] [PubMed]
6. Khelfaoui, F.; Gue´nin, G. Influence of the recovery and recrystallization processes on the martensitic transformation of cold worked equiatomic Ti-Ni alloy. *Mater. Sci. Eng. A* **2003**, *355*, 292–298. [CrossRef]
7. Kim, J.I.; Liu, Y.; Miyazaki, S. Ageing-induced two-stage R-phase transformation in Ti -50.9 at.%Ni. *Acta Mater.* **2004**, *52*, 487–499. [CrossRef]
8. Mohamad, H.; Mahmud, A.S.; Nashrudin, M.N.; Razali, M.F. Effect of ageing temperatures on pseudoelasticity of Ni-rich NiTi shape memory alloy. *AIP Conf. Proc.* **2018**, *1958*, 020008. [CrossRef]
9. Tirry, W.; Schryvers, D. Quantitative determination of strain fields around Ni4Ti3 precipitates in NiTi. *Acta Mater.* **2005**, *53*, 1041–1049. [CrossRef]
10. Waitz, T.; Antretter, T.; Fischer, F.D.; Karnthaler, H.P. Size effects on martensitic phase transformations in nanocrystalline NiTi shape memory alloys. *Mater. Sci. Technol.* **2008**, *24*, 934–940. [CrossRef]
11. Shi, X.; Guo, F.; Zhang, J.; Ding, H.; Cui, L. Grain size effect on stress hysteresis of nanocrystalline NiTi alloys. *J. Alloy. Compd.* **2016**, *688*, 62–68. [CrossRef]
12. Sun, Q.; He, Y. A multiscale continuum model of the grain-size dependence of the stress hysteresis in shape memory alloy polycrystals. *Int. J. Solids Struct.* **2008**, *45*, 3868–3896. [CrossRef]
13. Waitz, T.; Kazykhanov, V.; Karnthaler, H. Martensitic phase transformations in nanocrystalline NiTi studied by TEM. *Acta Mater.* **2004**, *52*, 137–147. [CrossRef]
14. Wang, X.; Pu, Z.; Yang, Q.; Huang, S.; Wang, Z.; Kustov, S.; Van Humbeeck, J. Improved functional stability of a coarse-grained Ti-50.8 at.% Ni shape memory alloy achieved by precipitation on dislocation networks. *Scr. Mater.* **2019**, *163*, 57–61. [CrossRef]
15. Wang, X.; Li, K.; Schryvers, D.; Verlinden, B.; Humbeeck, J.V. R-phase transition and related mechanical properties controlled by low-temperature aging treatment in a Ti–50.8 at.% Ni thin wire. *Scr. Mater.* **2014**, *72–73*, 21–24. [CrossRef]
16. Sittner, P.; Landa, M.; Luka´s, P.; Nova´k, V. R-phase transformation phenomena in thermomechanically loaded NiTi polycrystals. *Mech. Mater.* **2006**, *38*, 475–492. [CrossRef]
17. Duerig, T.W.; Pelton, A.R.; Bhattacharya, K. The Measurement and Interpretation of Transformation Temperatures in Nitinol. *Shape Mem. Superelasticity* **2017**, *3*, 485–498. [CrossRef]
18. Kim, J.I.; Miyazaki, S. Comparison of shape memory characteristics of a Ti-50.9 At. Pct Ni alloy aged at 473 and 673 K. *Metall. Mater. Trans.* **2005**, *36A*, 3301–3310. [CrossRef]
19. Ryklina, E.; Polyakova, K.; Prokoshkin, S. Comparative Study of Shape Memory Effects in Ni-Rich Ti–Ni Alloy After Training in Various Phase States. *Shape Mem. Superelasticity* **2020**, *6*, 157–169. [CrossRef]
20. Guo, Z.; Pan, Y.; Wee, L.B.; Yu, H. Design and control of a novel compliant differential shape memoryalloy actuator. *Sens. Actuators A* **2015**, *225*, 71–80. [CrossRef]
21. Ohkata, I.; Tamura, H. The R-Phase Transformation in the Ti-Ni Shape Memory Alloy and its Application. *MRS Proc.* **1996**, *459*, 345. [CrossRef]
22. Lim, Y.G.; Kim, W.J. Characteristics and interrelation of recovery stress and recovery strain of an ultrafine-grained Ni-50.2Ti alloy processed by high-ratio differential speed rolling. *Smart Mater. Struct.* **2017**, *26*, 035005. [CrossRef]
23. Lim, Y.G.; Han, S.H.; Choi, E.; Kim, W.J. Shape memory and superelasticity of nanograined Ti-51.2 at.% Ni alloy processed by severe plastic deformation via high-ratio differential speed rolling. *Mater. Charact.* **2018**, *145*, 284–293. [CrossRef]
24. Jiang, S.; Zhang, Y.; Zhao, L.; Zheng, Y. Influence of annealing on NiTi shape memory alloy subjected to severe plastic deformation. *Intermetallics* **2013**, *32*, 344–351. [CrossRef]
25. Hu, L.; Jiang, S.; Zhang, Y. Role of Severe Plastic Deformation in Suppressing Formation of R Phase and Ni4Ti3 Precipitate of NiTi Shape Memory Alloy. *Metals* **2017**, *7*, 145. [CrossRef]
26. Prokofyev, E.; Gunderov, D.; Prokoshkin, S.; Valiev, R. Microstructure, mechanical and functional properties of NiTi alloys processed by ECAP technique. In *European Symposium on Martensitic Transformations*; EDP Sciences: Metz, France, 2009; p. 06028.
27. Shahmir, H.; Nili-Ahmadabadi, M.; Mansouri-Arani, M.; Langdon, T.G. The processing of NiTi shape memory alloys by equal-channel angular pressing at room temperature. *Mater. Sci. Eng. A* **2013**, *576*, 178–184. [CrossRef]
28. Malard, B.; Pilch, J.; Sittner, P.; Gärtnerová, V.; Delville, R.; Schryvers, D.; Curfs, C. Microstructure and functional property changes in thin Ni–Ti wires heat treated by electric current—high energy X-ray and TEM investigations. *Funct. Mater. Lett.* **2009**, *2*, 45–54. [CrossRef]
29. Delville, R.; Malard, B.; Pilch, J.; Sittner, P.; Schryvers, D. Microstructure changes during non-conventional heat treatment of thin Ni–Ti wires by pulsed electric current studied by transmission electron microscopy. *Acta Mater.* **2010**, *58*, 4503–4515. [CrossRef]
30. Tong, Y.; Hu, K.; Chen, F.; Tian, B.; Li, L.; Zheng, Y. Multiple-stage transformation behavior of Ti 49.2 Ni 50.8 alloy with different initial microstructure processed by equal channel angular pressing. *Intermetallics* **2017**, *85*, 163–169. [CrossRef]
31. Han, S.H.; Kim, W.J. Achievement of nearly fully amorphous structure from NiTi alloys via differential speed rolling at 268 K and effect of annealing on superelasticity. *Mater. Charact.* **2020**, *169*, 110584. [CrossRef]
32. Zhang, Y.; Jiang, S.; Hu, L.; Liang, Y. Deformation mechanism of NiTi shape memory alloy subjected to severe plastic deformation at low temperature. *Mater. Sci. Eng. A* **2013**, *559*, 607–614. [CrossRef]

33. Duerig, T.W.; Bhattacharya, K. The Influence of the R-Phase on the Superelastic Behavior of NiTi. *Shape Mem. Superelasticity* **2015**, *1*, 153–161. [CrossRef]
34. Badji, R.; Chauveau, T.; Bacroix, B. Texture, misorientation and mechanical anisotropy in a deformed dual phase stainless steel weld joint. *Mater. Sci. Eng. A* **2013**, *575*, 94–103. [CrossRef]
35. Hossain, R.; Pahlevani, F.; Quadir, Z.; Sahajwalla, V. Stability of retained austenite in high carbon steel under compressive stress: An investigation from macro to nano scale. *Sci. Rep.* **2016**, *6*, 34958. [CrossRef]
36. Jiang, S.-Y.; Zhang, Y.-Q.; Zhao, Y.-N.; Liu, S.-W.; Hu, L.; Zhao, C.-Z. Influence of Ni4Ti3 precipitates on phase transformation of NiTi shape memory alloy. *Trans. Nonferrous Met. Soc. China* **2015**, *25*, 4063–4071. [CrossRef]
37. Miyazaki, S.; Kimura, S.; Otsuka, K.; Suzuki, Y. The habit plane and transformation strains associated with the martensitic transformation in Ti-Ni single crystals. *Scr. Met.* **1984**, *18*, 883–888. [CrossRef]
38. Saburi, T.; Yoshida, M.; Nenno, S. Deformation behavior of shape memory Ti-Ni alloy crystals. *Scr. Met.* **1984**, *18*, 363–366. [CrossRef]
39. Šittner, P.; Heller, L.; Pilch, J.; Curfs, C.; Alonso, T.; Favier, D. Young's Modulus of Austenite and Martensite Phases in Superelastic NiTi Wires. *J. Mater. Eng. Perform.* **2014**, *23*, 2303–2314. [CrossRef]
40. Wang, Z.; Chen, J.; Kocich, R.; Tardif, S.; Dolbnya, I.P.; Kunčická, L.; Micha, J.-S.; Liogas, K.; Magdysyuk, O.V.; Szurman, I.; et al. Grain Structure Engineering of NiTi Shape Memory Alloys by Intensive Plastic Deformation. *ACS Appl. Mater. Interfaces* **2022**, *14*, 31396–31410. [CrossRef]
41. Tong, Y.; Chen, F.; Guo, B.; Tian, B.; Li, L.; Zheng, Y.; Gunderov, D.V.; Valiev, R.Z. Superelasticity and its stability of an ultrafine-grained Ti49.2Ni50.8 shape memory alloy processed by equal channel angular pressing. *Mater. Sci. Eng. A* **2013**, *587*, 61–64. [CrossRef]
42. Zhang, P.; Li, S.; Zhang, Z. General relationship between strength and hardness. *Mater. Sci. Eng. A* **2011**, *529*, 62–73. [CrossRef]

Article

Thermal–Optical Evaluation of an Optimized Trough Solar Concentrator for an Advanced Solar-Tracking Application Using Shape Memory Alloy

Nasir Ghazi Hariri *, Kamal Mohamed Nayel, Emad Khalid Alyoubi, Ibrahim Khalil Almadani, Ibrahim Sufian Osman and Badr Ahmed Al-Qahtani

Department of Mechanical and Energy Engineering, College of Engineering, Imam Abdulrahman Bin Faisal University, P.O. Box 1982, Dammam 31441, Saudi Arabia
* Correspondence: nghariri@iau.edu.sa

Abstract: One of the modern methods for enhancing the efficiency of photovoltaic (PV) systems is implementing a solar tracking mechanism in order to redirect PV modules toward the sun throughout the day. However, the use of solar trackers increases the system's electrical consumption, hindering its net generated energy. In this study, a novel self-tracking solar-driven PV system is proposed. The smart solar-driven thermomechanical actuator takes advantage of a solar heat collector (SHC) device, in the form of a parabolic trough solar concentrator (PTC), and smart shape memory alloy (SMA) to produce effective mechanical energy for solar tracking applications from sun rays. Furthermore, a thermal–optical analysis is presented to evaluate the performance of the solar concentrator for the simulated weather condition of Dammam City, Saudi Arabia. The numerical results of the thermal and optical analyses show the promising feasibility of the proposed system in which SMA springs with an activation temperature between 31.09 °C and 45.15 °C can be utilized for the self-tracking operations. The work presented adds to the body of knowledge an advanced SMA-based SHC device for solar-based self-actuation systems, which enables further expansions within modern and advanced solar thermal applications.

Keywords: thermomechanical; shape memory alloy; actuator; solar heat collector; thermal analysis; optical analysis; self-actuation; solar tracker; solar concentrator

Citation: Hariri, N.G.; Nayel, K.M.; Alyoubi, E.K.; Almadani, I.K.; Osman, I.S.; Al-Qahtani, B.A. Thermal–Optical Evaluation of an Optimized Trough Solar Concentrator for an Advanced Solar-Tracking Application Using Shape Memory Alloy. *Materials* **2022**, *15*, 7110. https://doi.org/10.3390/ma15207110

Academic Editor: Antonio Politano

Received: 31 August 2022
Accepted: 10 October 2022
Published: 13 October 2022

1. Introduction

Continuous developments within modern and efficient technology solutions for innovative solar energy applications have attracted many researchers. Similarly, energy-related companies and manufacturers are significant players in continuously exploring alternative eco-friendly energy sources through the focus on sustainability and environmental conservation [1,2]. Environmental impact is essential for sustainable development, and its influence is enhanced by the pervasive use of fossil fuels and a lack of environmental protection [1]. Solar photovoltaic (PV) has become the principal source of electricity in several economies, where PV systems contribute significantly to total power production along with other sources, such as fossil fuels and nuclear power [3,4].

Constant investigations are being evaluated to improve the productivity and efficiency of the PV industry; correlation between systems' operational structures and economic, social, and environmental aspects of renewable energy are investigated [5,6]. Solar cells are one type of modern technology that allows access to a clean and safe energy source since they convert solar radiation into electrical energy. Recently, the developments of effective PV tracking systems over fixed photovoltaic systems have improved the system's overall efficiency [7].

Teng et al. [8] studied the highest power exchange of a PV tracking system, and maximum power transmission and better solar cell efficiency were achieved with a simple control circuit to match the impedance of the system's components, enabling maximum power transmission between solar cells. Additionally, Osman, et al. [9] have studied the effect of different solar tracking systems on the performance of conventional PV modules. Their paper reveals that the dual and single trackers lead to a net energy increase compared to a fixed module—up to 29% and 19%, respectively. Moreover, the paper concluded that solar tracking systems utilize about 17% of the electrical production to run the actuators and the control system. On the other hand, Hariri et al. [10] investigated single-axis solar tracking mechanisms, in which a PV panel is moved along one axis to align it perpendicularly towards the sun's beams throughout the day using sensor and azimuth-based methods. Experimental results showed that the sensor and azimuth-based tracking systems improved overall net energy production by about 12.68% and 7.7%, respectively. In short, most current tracking systems investigate active tracking methods that depend mainly on electrical energy, so overall energy generation remains hindered due to the tracker's consumption.

Furthermore, shape memory alloy (SMA)-based actuators have been widely used in various areas, including robotic, automotive, aerospace, biomedical, and energy applications, because of their unique thermomechanical and shape memory effect (SME) properties. Whenever an external force deforms the SMA, it can return to its original state once heated above its activation temperature [11]. For instance, Catoor et al. [12] studied the relationship between temperature, displacement, and response speed. The result showed that the speed response and displacement change when there is an increase in the temperature. Moreover, this opens the door for more innovative and novel mechanism designs. For example, an SMA-based device has been used to enhance the aerodynamic performance of vehicles [13], generate constant force [14], power non-explosive reusable lock release mechanisms, and actuate light grippers [15,16]. Furthermore, SMAs have good chemical and physical properties and biocompatibility [17,18]. Hence, engineers and researchers across the globe show high interest in the unique properties of SMA, especially regarding developed thermomechanical energy, which it provides at a lower cost. Therefore, further innovations, designs, and systems have been proposed to integrate SMA as a smart actuation source [19,20]. Integrating the SMA within certain disciplines, such as robotic and biomedical, allows for further simplification on the design for SMA-based systems, in which it reduces the system's volume and weight significantly, offering compact structures and systems [21].

An SMA actuator that uses temperature as a source of activation could be activated via joule heating or an external heating/cooling source, either passive or active [22]. In this study, heat transferred from the sun is the primary heating source for activating the SMA-based actuator. A solar heat collector (SHC) will be used to improve the absorb thermal energy provided by the sun. The SHC is a device used to collect the thermal energy supplied by the sun, which is stored and carried away by the flowing fluid [23]. SHCs have been used in various applications, such as domestic heating/cooling devices, storage devices, and water desalination [24–26]. Furthermore, some criteria when designing an SHC include the environmental criteria standards, cost-effectiveness, and technical aspects [26–28]. Moreover, the efficiency of the SHC device can vary depending on the design, the fluid inside the SHC, and the material of the collector and the absorber [1,29]. Some SHC devices utilize mirrors or reflecting materials to increase the heat and the solar power for a small volume so the energy can be converted into heat and then used to generate electricity. Additionally, parabolic trough collector (PTC) technology is one of the most widely used thermal power technologies among other concentrated solar power (CSP) technologies [30]. For example, the Dubai Noor project, the most considerable solar power project established globally, contains 950 MW hybrid projects, composed of 250 MW photovoltaic and 700 MW parabolic trough power technology [31]. Due to massive land use, maintenance, and installation costs, the PTC technology has vital high and intensive capital costs. Such a complicated heating system increases the potential for the system's failure due to the massive temperature gradient on the receiver. The mirror heats the bottom side of the receiver by focusing the highly concentrated radiation intensity. In contrast, only direct solar radiation heats the top side, making the receiver's bottom side heated more than the top side; this uneven temperature on the receiver's outer surface causes high thermal stresses on the receiver side [30–32]. The temperature variance is considered the primary cause of failure, such as breakage of the receiver's glass envelope, vacuum leakage or bending within the structure, hence causing optical losses due to the misplacement of the absorber tube from the optimal focal area. Moreover, when the receiver becomes damaged frequently and affects other components, the plant's cost will also be affected. Therefore, with regards to the receiver's price and reliability, it should be able to function efficiently and work reliably and economically [30–33].

In a recent research effort presented by some of the authors, a solar-driven thermo-mechanical actuator have been proposed [1,34,35]. The proposed actuator has shown a thermal to mechanical conversion efficiency of about 19.15%, where the actuator produces about 150 N and 130 mm [34,35]. Additionally, the authors have studied the thermal profile in the studied are of Dammam, Kingdom of Saudi Arabia (KSA) details and found that the actuator would properly operates throughout the year and does not require any maintenance routine [1].

Optimizing the parabolic trough collector is a significant step; researchers simulate and model the system to investigate the system's effect and thermal performance before construction. There is a unique approach to overcoming the problem of interception rate and optical efficiency. The first method is to absorb more sun rays on the absorber tube using a larger diameter [36]. This is employed, for example, by Abengoa Company (Seville, Spain), where an extensive commercial concentrator with an absorber tube of 80 mm and 8.2 m aperture width is constructed. This system has been used in the Dubai Noor project, resulting in a lower price of 7.3 cents/kWh [31]. On the other hand, an alternative method is to add another reflector layer inside the glass cover on the receiver tube. The secondary reflector is used to reflect the sun rays to enhance the factor of interception when a large aperture is used with a small absorber tube [37]. Additionally, the third method is changing the absorber's shape; hence, by adding a secondary reflector, the study made by Gong et al. [38] displayed an optical efficiency of less than 75%. Therefore, a semicircular receiver tube with a solar radiation flux distribution is used on the circular receiver tube model to intercept more sun rays. A PTC used a semicircular receiver tube with an 80° half-rim angle and 8 m aperture width with an optical efficiency of about 79.2%.

On the other hand, the integration of thermomechanical alloy, in the shape of an SMA actuator, into self-actuation and tracking mechanisms via utilizing mainly solar energy to enhance the overall performance of PV systems has not been addressed. Therefore, in recent research efforts presented by the authors, a solar-driven thermomechanical actuator has been proposed [1,34,35]. The proposed actuator has shown a thermal-to-mechanical conversion efficiency of about 19.15%, and the actuator produces about 150 N and 130 mm [34,35]. Additionally, the authors have studied the thermal profile in the studied area of Dammam, Kingdom of Saudi Arabia (KSA), where it was found that the actuator would adequately operate throughout the year and requires no maintenance routine [1]. Therefore, this study aims to add to the body of scientific knowledge an innovative methodology for a technology solution of a novel thermomechanical SMA actuator targeting solar-based self-tracking PV applications.

2. Materials and Methods

2.1. Overview

The proposed research study focuses on a thermomechanical SMA actuator's thermal and optical performances. The thermomechanical SMA actuator has three primary elements: SMA springs, SHC, and PTC. These three elements significantly impact the actuator and, therefore, were analyzed and studied. The SHC and the PTC are placed in the actuator to increase the temperature of the SMA springs, while the activation temperature of the SMA springs is the crucial factor in the actuation mechanism. To ensure the temperature inside the SHC reaches the activation temperature of the SMA, a comprehensive study was proposed using computational fluid dynamics (CFD) simulation software version 5.6.

2.2. SMA Actuator Conceptual Design

The proposed research aimed to design and simulate an SMA solar-driven thermomechanical actuator. The configuration for the main parts of the proposed design is illustrated in Figure 1. The actuator is a piston-based linear actuator in which several SMA springs are arranged horizontally inside a SHC with a movable rod to ensure the linearity of the piston movement. The design utilizes the SME in the SMA springs by absorbing the heat and the intensive radiation reflected by PTC into the SHC. At sunrise, the SHC absorbs the solar radiation from the sun in addition to the reflected radiation from the PTC. The solar radiation collected through the sun and the PTC in the enclosed SHC leads to a greenhouse effect inside the SHC, raising its inner temperature, which allows the SMA springs to reach the activation temperature that triggers their SME. On the other hand, as the SMA springs lose thermal energy, the springs' temperature drops, and the bias load stretches the SMA springs back to their original shape.

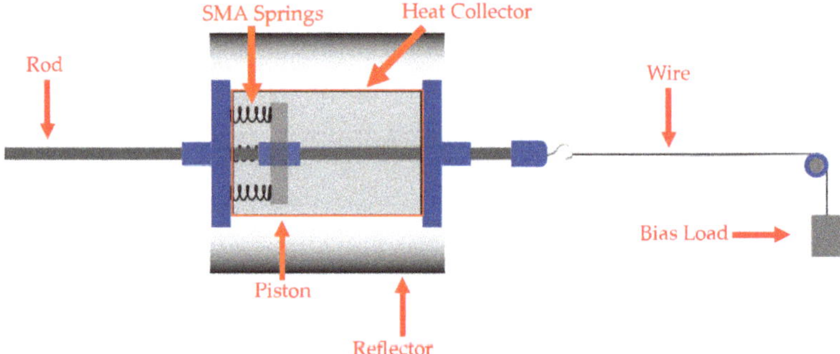

Figure 1. Conceptual design of the solar-driven SMA thermomechanical actuator.

The solar-driven thermomechanical SMA-based actuator is a passive actuator that activates using solar thermal energy given by the sun. In addition, the actuator has two actuation phases based on the SMA states—the activation and deactivation phases—as shown in Figure 2. The activation phase occurs when the SMA springs gain thermal energy that has been absorbed via the SHC, causing the movement of the piston; this movement allows the system to lift the bias load. Conversely, when the SMA springs lose thermal energy, they deactivate, and the bias load stretches the SMA springs back to their original shape.

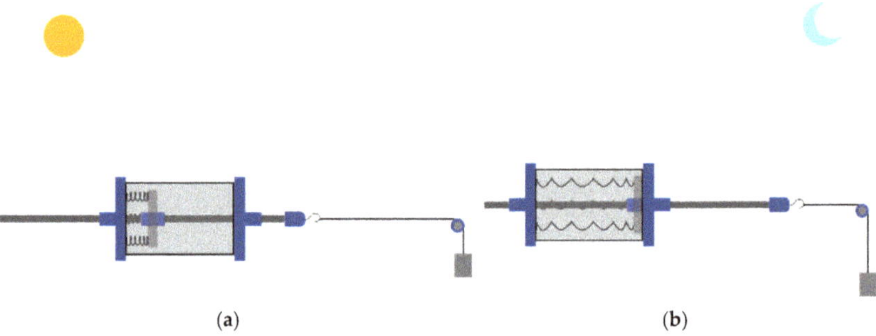

Figure 2. SMA-based actuator phases; (**a**) activated and (**b**) deactivated arrangements.

The designed actuator has three main limitations, which are environmental, mechanical, and thermal, and these factors can affect the feasibility of the proposed design. The environmental limitations are environment-dependent constraints, such as ambient temperature, cloudiness, and solar irradiance. Mechanical limitations include the mechanical properties of the available SMA, such as the allowable strain and the produced force. Regarding thermal limitations, the constraints of the SMA include activation temperature, deactivation temperature, and thermal endurance. All of the limitations mentioned have an effect on the performance and feasibility of the design.

2.3. SMA Solar Self-Tracker Conceptual Design

The proposed solution integrates two identical SMA-based, actuators allowing the system to orient the PV module into three different positions, as shown in Figure 3. During sunrise, the PV module will be oriented at an angle, casting shade on both SMA-based actuators. Both actuators will be deactivated since the internal SHC temperature is lower than the activation temperature of the SMA. During noon, one of the actuators will be exposed to the sun's radiation, allowing the internal temperature to reach the activation temperature of the SMA. The other one will be shaded, causing the PV to be fixed horizontally. Lastly, at sunset, both SMA-based actuators will be exposed to solar radiation, activating both actuators and allowing the PV to tilt at an angle.

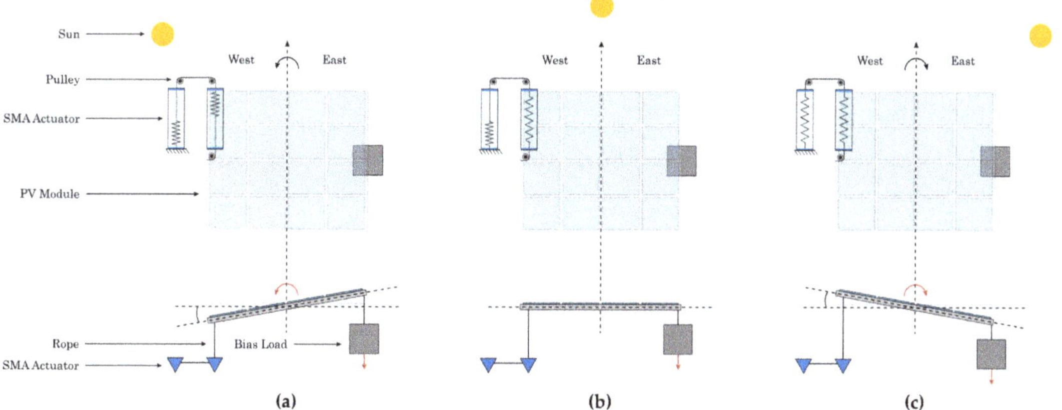

Figure 3. Thermomechanical actuator (**a**) before sunset, (**b**) at noon, and (**c**) after sunrise.

In order to ensure that the design of the thermal SMA actuator is aligned and functional, a three-dimensional (3D) model has been developed using computer-aided design (CAD) software. The thermal SMA actuator consists of seven main parts, including SMA springs, SHC, piston, PTC, bias load, rod, and stand holders, as highlighted in Figure 4. One end of the SMA springs is connected to the cover of the SHC, and the other end is connected to the piston to assure the linear movement of the actuator. The piston is attached to the rod, while the rod is set inside a linear ball bearing. Three linear ball bearings are attached inside the cover and the stand holders' holes to minimize friction and guide the linear motion along a single axis. The SHC is responsible for absorbing the thermal heat from the sun, the PTC, and the surroundings, subsequently raising the temperature inside the actuator significantly above the ambient temperature. The PTC is placed to increase the thermal energy absorbed by the bottom side of the SHC by reflecting the sun's rays.

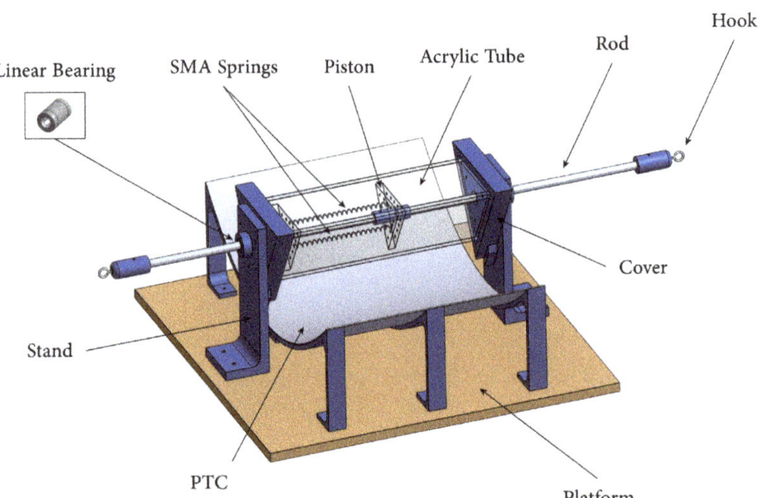

Figure 4. CAD model of the designed thermomechanical SMA actuator.

2.4. Analytical Thermal Model

The thermomechanical SMA actuator is designed to be a solar-driven actuator, and temperature is a significant key in this process. To understand the spatial variation of the temperature gradient, a simplified study for the thermal circuits is provided to highlight the heat transfer process flows under one-dimensional (1D) and steady-state conditions. As illustrated in Figure 5, the temperature inside the thermomechanical SMA actuator increases as the SHC absorbs the solar energy from the sun, the ambient weather, and the rays reflected through the PTC.

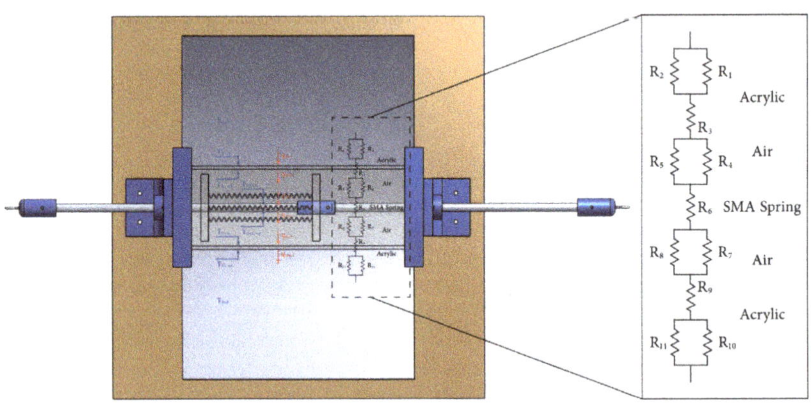

Figure 5. Thermal resistance for the thermomechanical SMA actuator.

The SMA's temperature can be obtained by analyzing the thermal circuit illustrated in Figure 5. Equation (1) is the general formula for the heat and transfer process, in which varying thermal resistance is applied depending on the heat transfer process, as shown in Equations (2)–(4) [1,34,35].

$$q = \frac{\Delta T}{R} \tag{1}$$

$$R_{cond} = \frac{L}{K} \tag{2}$$

$$R_{conv} = \frac{1}{h_{conv}} \tag{3}$$

$$R_{rad} = \frac{1}{h_{rad}} \tag{4}$$

where q represents the amount of heat transferred per unit length [W/m^2], ΔT is the temperature difference in [K], while R represents the thermal heat transfer resistance in [K/W], and the thermal resistance for conductive, convective, and radiative varies for R_{cond}, R_{conv}, and R_{rad} in [K/W]. L is the length in [m], K is the thermal conductivity in [W/m × K], h_{conv} is the convective heat transfer coefficient in [W/m^2 × K], and h_{rad} is the radiative heat transfer coefficient in [W/m^2 × K].

The acrylic plate of the SHC absorbs the thermal heat coming from the direct sun rays, the reflected rays through the PTC, and the ambient weather in the form of radiation and as a convection heat transfer process, as described in Equation (5). Furthermore, the acrylic plate will release the heat as conduction, convection, and radiation, as shown in Equation (6). In addition, the SMA springs will gain heat through radiation and convection while losing some heat throughout conduction, convection, and radiation, as shown in Equations (7) and (8). To complete the cycle, the other acrylic plate will gain the extracted heat coming from the SMA, as shown in Equation (9). At the same time, the plate will lose thermal heat in the form of conduction, convection, and radiation, as demonstrated in Equation (10).

$$q_{in,1} = \frac{T_{amb} - T_{A1_{in}}}{\frac{1}{h_{conv_{in}}} + \frac{1}{h_{rad_{in,1}}}} \tag{5}$$

$$q_{out,1} = \frac{T_{A1_{in}} - T_{A1_{out}}}{\frac{L_1}{K_1}} + \frac{T_{A1_{out}} - T_{air,1}}{\frac{1}{h_{conv_{out,1}}} + \frac{1}{h_{rad_{out,1}}}} \tag{6}$$

$$q_{in,2} = \frac{T_{air,1} - T_{SMA_{in}}}{\frac{1}{h_{conv_{in,2}}} + \frac{1}{h_{rad_{in,2}}}} \tag{7}$$

$$q_{out,2} = \frac{T_{SMA_{in}} - T_{SMA_{out}}}{\frac{L_2}{K_2}} + \frac{T_{SMA_{out}} - T_{air,2}}{\frac{1}{h_{conv_{out,2}}} + \frac{1}{h_{rad_{out,2}}}} \tag{8}$$

$$q_{in,3} = \frac{T_{air,2} - T_{A2_{in}}}{\frac{1}{h_{conv_{in,3}}} + \frac{1}{h_{rad_{in,3}}}} \tag{9}$$

$$q_{out,3} = \frac{T_{A2_{in}} - T_{A2_{out}}}{\frac{L_3}{K_3}} + \frac{T_{A2_{out}} - T_{amb}}{\frac{1}{h_{conv_{out,3}}} + \frac{1}{h_{rad_{out,3}}}} \tag{10}$$

where T is the temperature and h is the heat transfer coefficient. The subscript represents the medium and the heat transfer process based on Figure 5. The aim is to obtain the SMA temperature, which could be simplified—since the SMA springs are noticeably thin, it can be assumed that the inlet temperature of the SMA springs ($T_{SMA_{in}}$) is equal to the outer temperature of the SMA springs ($T_{SMA_{out}}$).

The outlet and the inlet heat transfer for the SMA springs have been equalized, forming Equation (11). Furthermore, the thermal process is shortened in the form of resistances, as shown throughout Equations (12)–(14), which simplify Equation (11) into (15). In order to obtain the SMA temperature, Equation (15) must be solved in terms of the T_{SMA}, as shown in Equation (16).

$$\frac{T_{air,1} - T_{SMA_{in}}}{\frac{1}{h_{conv_{in,2}}} + \frac{1}{h_{rad_{in,2}}}} = \frac{T_{SMA_{in}} - T_{SMA_{out}}}{\frac{L_2}{K_2}} + \frac{T_{SMA_{out}} - T_{air,2}}{\frac{1}{h_{conv_{out,2}}} + \frac{1}{h_{rad_{out,2}}}} \tag{11}$$

$$R_1 = \frac{1}{h_{conv_{in,2}}} + \frac{1}{h_{rad_{in,2}}} \tag{12}$$

$$R_2 = \frac{L_2}{K_2} \tag{13}$$

$$R_3 = \frac{1}{h_{conv_{out,2}}} + \frac{1}{h_{rad_{out,2}}} \tag{14}$$

$$\frac{T_{air_1} - T_{SMA_{in}}}{R_1} = \frac{T_{SMA_{in}} - T_{SMA_{out}}}{R_2} + \frac{T_{SMA_{out}} - T_{air,2}}{R_3} \tag{15}$$

$$T_{SMA} = \frac{T_{air,1} \times R_3 + T_{air,2} \times R_1}{R_1 + R_3} \tag{16}$$

Equation (16) is the simple form, and to expand it, the resistances (R_1, R_2, and R_3) could be resubstituted, while the $T_{air,1}$ and $T_{air,2}$ could be evaluated in terms of the acrylic and the ambient temperature.

2.5. D Thermal Numerical Study Setup

The offered thermomechanical SMA actuator contains complex thermal processes involving numerous inputs influencing the system's thermal behavior. Therefore, a simulation-based thermal study has been conducted to understand and optimize the thermal behavior inside the SHC, aiming to reach the activation temperature of the SMA springs inside the SHC. To start the numerical study, the CAD model of the main parts of the thermomechanical SMA actuator was imported into the CFD simulation software, as shown in Figure 6. Moreover, the simulation-based study was specified to the CFD software to be performed using the actual weather conditions of the city of Dammam (26.433° N 49.8° E), KSA.

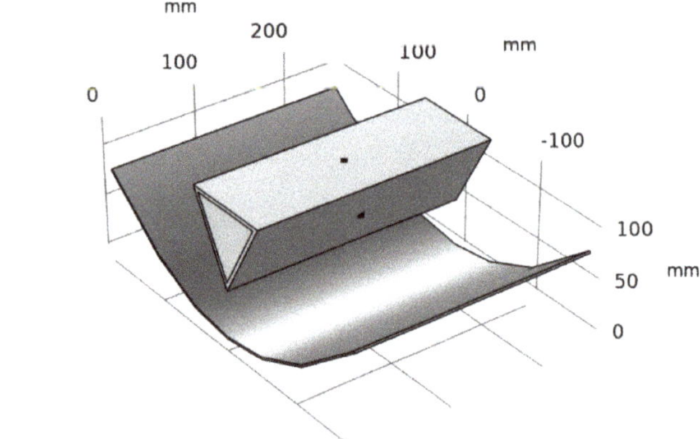

Figure 6. Imported CAD model of thermomechanical SMA actuator within the CFD software.

Two physical interfaces have been chosen to run the study, including the heat and transfer in solids and the surface-to-surface radiation physics interfaces. The study was performed in a time-dependent mode since the aim is to accurately understand the thermal behavior inside the SHC with respect to the studied area. Additionally, the material is assigned to be acrylic plastic for the side of the SHC, while the fluid inside is air. The thermodynamic properties of the actuator material, such as density, heat capacity at constant pressure, and thermal conductivity, are defined in Table 1. The ambient weather conditions are defined as being the same as the weather conditions of Dammam City, KSA using the meteorological data from ASHRAE 2017 that is available in the CFD software.

Table 1. Properties of acrylic plastic used for the SHC.

Property	Value	Unit
Heat capacity at constant pressure	1470	J/(Kg·K)
Density	1190	Kg/m^3
Thermal conductivity	0.18	W/(m·K)

To integrate the CAD model with multiphysics interfacing solution tools, a set of thermal physics must be optimized, such as the heat transfer in solids and surface-to-surface radiation. The heat transfer in solids studies the heat flux due to the convection heat transfer process on the outside surface of the SHC. The convection heat process considers the change in temperature outside the SHC surface due to the ambient temperature while considering the plate surface of the SHC. The initial value conditions for the multiphysics solver are defined as the ambient temperature given in the meteorological data in Dammam City. The fluid inside the thermomechanical SMA actuator is defined as air, and all the properties are functions of the initial value condition and the ideal gas properties. The actuator has a triangular shape, and different optimization is needed for the oriented plates. The horizontal upper plate is exposed to external natural convection, as shown in Figure 7a. The other two sides of the SHC are also exposed to external natural conviction but with an inclined surface, as shown in Figure 7b.

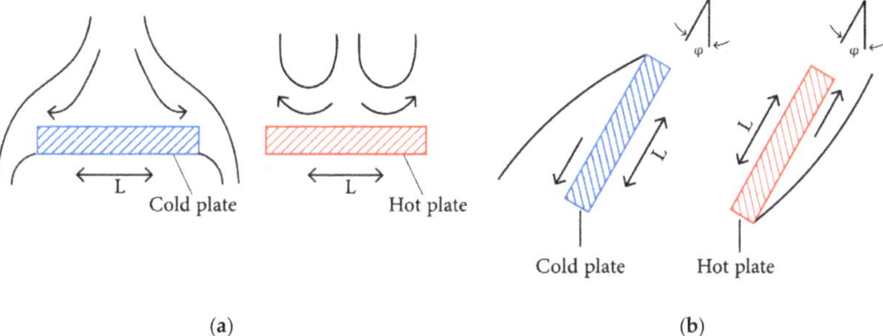

Figure 7. The external natural convection process for (**a**) upper horizontal plate and (**b**) inclined plate.

In addition, the PTC and the SHC have been mainly considered in the surface-to-surface radiation tools. The source of the radiative heat transfer process is mainly the sun, while the SHC side is defined as a blackbody. The initial values for the weather conditions for the Surface-to-surface radiation are set to be the ambient weather conditions. The external radiation source is the sun, with solar irradiance defined as the clear sky at noon. To resemble the mirror effect to the PTC, a diffusion mirror which will work as a mirror absorbing all the radiative coming and then radiating back in all directions has been defined. Both heat transfer in solids and surface-to-surface will be coupled in the multiphysics interface tools with a time-dependent study to investigate the thermal behavior of the actuator with the effect of the PTC. To obtain an accurate result, a physics-controlled fine mesh sequence has been applied to the thermomechanical SMA actuator, as shown in Figure 8.

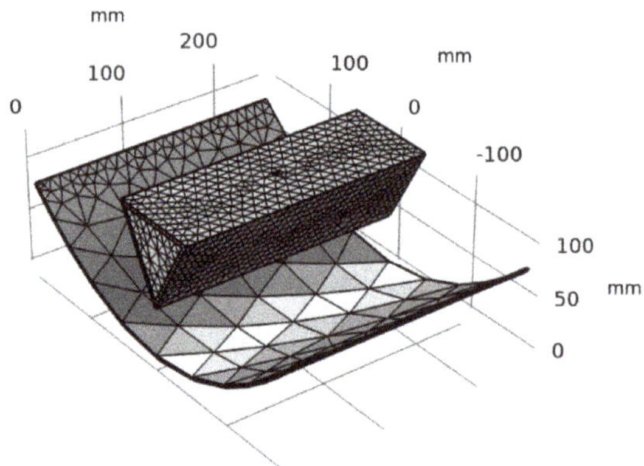

Figure 8. Physics-controlled mesh sequence in the multiphysics interface solution tool for the thermomechanical SMA actuator.

3. Results

The results in this section highlight the thermal analysis of the SHC, including the temperature gradient, midpoint temperature, and temperature profile for different shapes. In addition, the thermal performance of the SHC has been improved based on multiple comparisons of several parameters, such as reflector presence, the shape of the SHC, dimensional optimization, and orientation. The comparisons carried out help to enhance the overall thermal performance of the SHC to ensure its compatibility with the intended application. It should be noted that all comparison simulations were run in the same period (1 July 2022).

3.1. Effect of Introducing a PTC on the Thermomechanical Actuator's Thermal Behavior

In this division, the outcomes of adding a PTC to the system on the performance of the SHC are explored. Multiple comparisons are carried out to explore the feasibility of using a reflector, including inward heat flux variation on the lower surface of the SHC in different conditions. The amount of inward heat flux through the lower surface of the triangular SHC is shown in Figure 9. In the morning, when the sun rises from the east, the eastern side of the SHC surface will gain high heat flux, while the western side will be shaded, causing significantly lower inward heat flux, as illustrated in Figure 9a. At noon, the sun is perpendicular to the SHC; the upper surface casts shade, preventing the heat flux from reaching the bottom side of the SHC, as shown in Figure 9b. In Figure 9c, we can note an increase in the inward heat flux on the western side while shadows are projected on the eastern side, causing a lower inward heat flux.

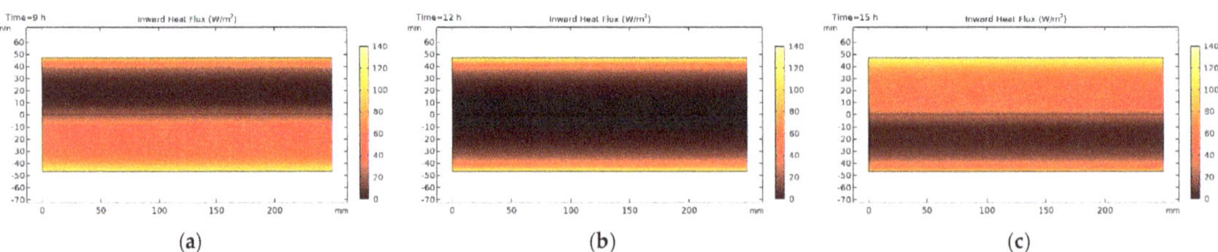

Figure 9. Surface plot of the inward heat flux through the projected area of the lower surface of the triangular SHC without a reflector at (**a**) 9:00; (**b**) 12:00; and (**c**) 15:00 on 1 July 2022.

To navigate the enhancement by adding the PTC, the same simulation has been conducted to study the inward heat flux on the lower surface of the SHC when adding the PTC. Figure 10 is a 2D surface plot illustrating the inward heat flux distribution on the projected lower surface area of the triangular SHC during the day. The surface plot facilitates investigation into how adding a reflector affects the performance of the SHC by redirecting the sun rays to the lower surface of the SHC during the day. For example, in the morning, the eastern side projects shadow onto the western side of the SHC, but the PTC compensates for this by reflecting the sun radiation, allowing the lower surface to receive higher heat flux and increase the coverage area, as shown in Figure 10a. By comparing Figures 9b and 10b, we can see the significant improvement the PTC provides.

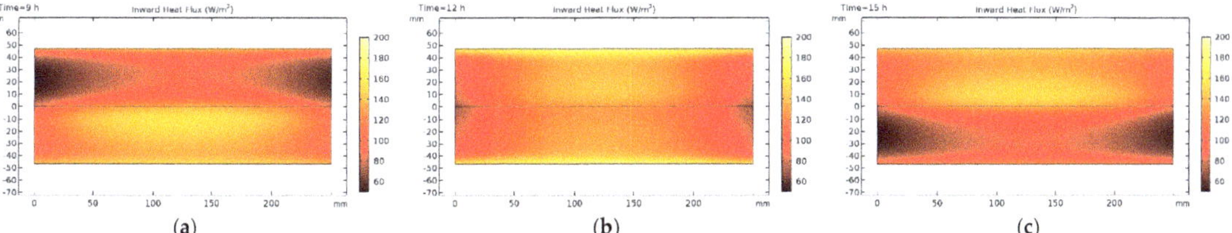

Figure 10. Surface plot of the inward heat flux through the projected area of the lower surface of the triangular SHC with a reflector at (**a**) 9:00; (**b**) 12:00; and (**c**) 15:00 on 1 July 2022.

Figure 11 demonstrates the variation in the amount of inward heat flux through the lower surface of the triangular SHC throughout the day and the effect of using a PTC. We can note that the contribution of the PTC not only increases the heat flux coverage but also highly increases the intensive heat flux on the lower surface throughout the day, as illustrated in Figure 11. In the morning, the body of the PTC casts shadows, reducing the solar radiation on the lower surface of the SHC for a period. Furthermore, at noon, the SHC without the PTC does not received any heat flux; this is incomparable to the SHC with the PTC, which almost reaches 130 [W/m^2]. It is evident in the plot that using a reflector increases the inward heat flux, which means increasing the temperature inside the SHC, which enables the SMA springs to reach this specified activation temperature. It can be concluded by the analysis of the revealed data that the addition of a reflector increases SHC compatibility with the application aimed.

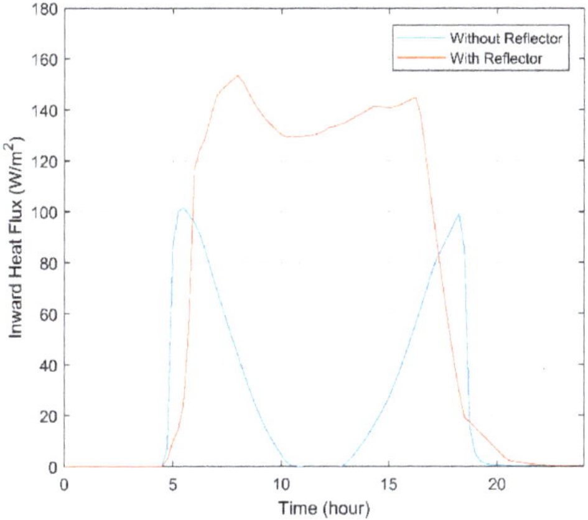

Figure 11. Variation in the amount of inward heat flux through the lower surface of the triangular SHC during a whole day with the effect of using the reflector.

3.2. Thermal Behavior of a Triangular SHC in Comparison to a Circular SHC

This section evaluates the thermal behavior of different SHC to compare circular and triangular shapes. This evaluation enables better judgment of which shape to optimize the SHC functionality and feasibility for the application of a passive solar tracking system.

One of the parameters from which the shaping capability of the SHC can be determined is the temperature gradient across the SHC. The temperature gradient across the SHC works as an indicator that the temperature is distributed evenly, allowing the temperature to reach the activation temperature along the SMA springs. Figure 12 shows a 3D plot of the different shapes' temperature gradients. The outcome of the figure showed little to no difference between the different shapes in both temperature gradient and thermal behavior.

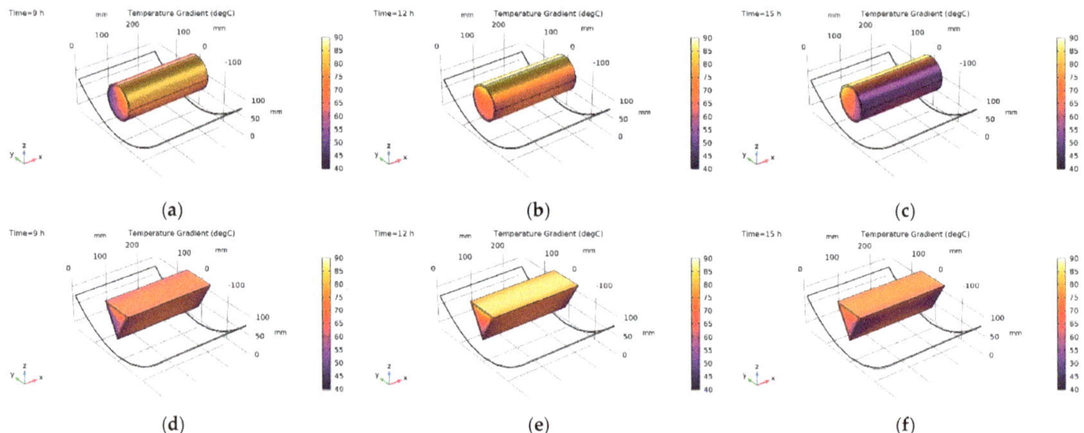

Figure 12. Three-dimensional plot of the temperature gradient of the circular SHC with a reflector at (**a**) 9:00, (**b**) 12:00, and (**c**) 15:00 and the temperature gradient of the triangular SHC with a reflector at (**d**) 9:00, (**e**) 12:00, and (**f**) 15:00.

Another parameter that must be evaluated is the ability to increase the temperature inside the SHC. Moreover, the temperature inside must be maintained above the SMA activation temperature during the daytime. Therefore, the distributed SMA springs (shown previously in Figure 4) have been integrated into a single point (midpoint). For simplicity, the midpoint is assumed to be the SMA temperature. In Figure 13 the midpoint temperature of both the circular and triangular-shaped SHCs are plotted. Both shapes increase the inside temperature 30 °C above the ambient temperature, reaching 74 °C. Furthermore, the two shapes maintained a midpoint temperature above the SMA activation temperature for longer. Moreover, the plot showed a similar outcome to Figure 12, in which minimum to no difference was indicated, except in Figure 13, where the outcome was dependent purely on the temperature of a single point (midpoint). The maximum, minimum, and average temperatures of the circular SHC were 74.01 °C, 29.77 °C, and 50.58 °C, respectively. Moreover, the maximum, minimum, and average temperatures of the triangular SHC were 74.69 °C, 29.84 °C, and 50.69 °C, respectively. The minimum differences in the temperatures were slightly inclined toward the triangular-shaped SHC; therefore, the triangular SHC is preferred due to the simplicity of manufacturing.

3.3. *Optimization of the Geometrical Parameters of the Reflector*

To ensure the optimum performance of the utilized reflector for increasing the temperature of the solar-driven thermomechanical SMA actuator, two main geometrical parameters are considered. The parameters conserved to optimize the designed reflector are the aperture width and the focal length. These parameters affect the performance of the PTC, which affects the system's performance as a whole. The first parameter is the aperture width of the reflector, where a variety of aperture widths were simulated to discover how the aperture width affects the midpoint temperature of the SHC. The second parameter was the focal length between the reflector and the SHC, where multiple focal lengths were simulated.

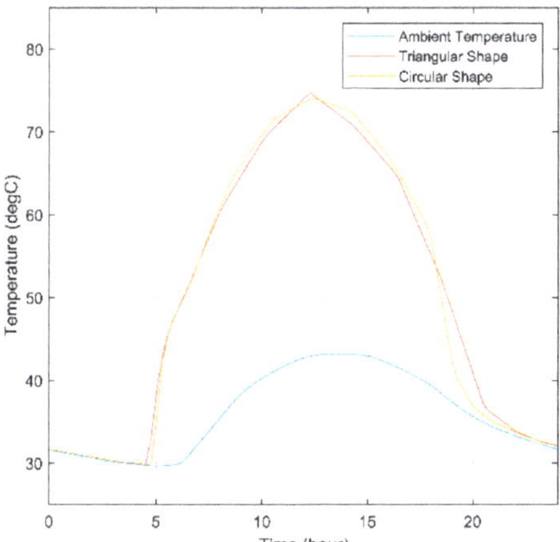

Figure 13. Midpoint temperature for different shapes of SHC.

The simulation was conducted by fixing the focal length and varying the aperture widths ranging from 300 mm to 400 mm to ensure the optimum from which the SHC reaches the highest temperature. The temperature variation of the SHC depending on aperture widths is demonstrated in Figure 14. Although the different aperture widths have the advantage of increasing the temperature depending on the time of the day, the temperature difference is minor, concluding that the aperture width can be neglected in the range mentioned.

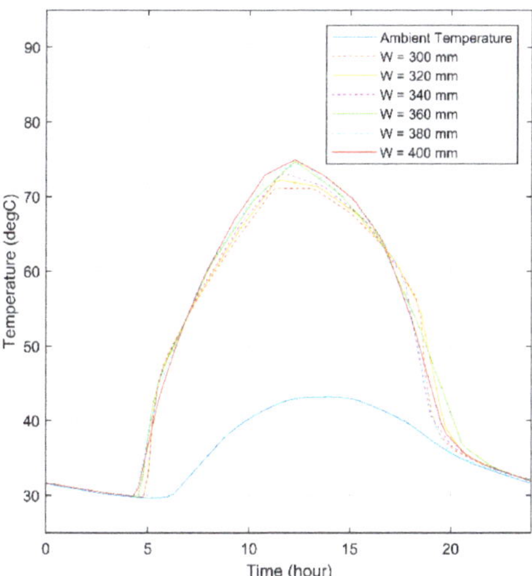

Figure 14. Midpoint temperature for triangular SHC for different values of aperture width.

In order to optimize the focal length, the aperture width was initially fixed, while multiple focal lengths were evaluated to ensure the optimized design of the system. The midpoint temperature inside the SHC is plotted in Figure 15 under different focal lengths ranging from 80 mm to 180 mm. The results of the plotted data showed that the focal length significantly affects the midpoint temperature inside the SHC. For example, the maximum midpoint temperature at noon was 76.25 °C at a focal length of 80 mm, while the minimum midpoint temperature at noon was 65.44 °C at a focal length of 180 mm. Although the maximum midpoint temperature at noon was at a focal length of 80 mm, it was noted that a smoother temperature curve with the second highest temperature at noon was at a focal length of 100 mm with a temperature of 74.17 °C, which makes it the optimal option.

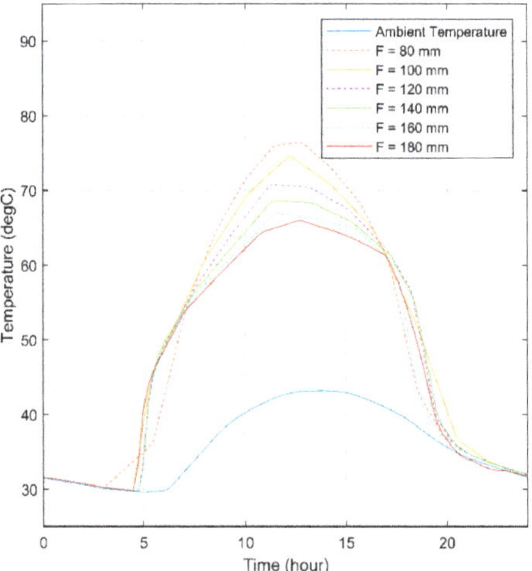

Figure 15. Midpoint temperature for triangular SHC for different values of focal length.

3.4. SHC Orientation's Effect on the Thermal Behavior

The orientation of the SHC is essential to its temperature profile, which is why a simulation of the temperature profile under different orientations is carried out. In Figure 16 the temperature profile at the center line of the SHC is plotted for both the north–south axis and the east–west axis. It can be noted that the temperature profile inside the SHC varies depending on the orientation at 09:00 and 15:00; however, the temperature profile is similar at noon. The difference in the temperature profile depending on the orientation is due to the sun's direction at different times of the day. Although the east–west-oriented SHC has a higher maximum temperature in Figure 16a,c, the maximum temperature at noon is higher in the north–south-oriented SHC as seen in Figure 16b. In addition, the temperature profile is more evenly distributed for the north–south orientation. Therefore, it could be concluded from the outcome of the simulation that the north–south is more fitted for the application intended since it provides a well-distributed temperature profile inside the SHC throughout the day.

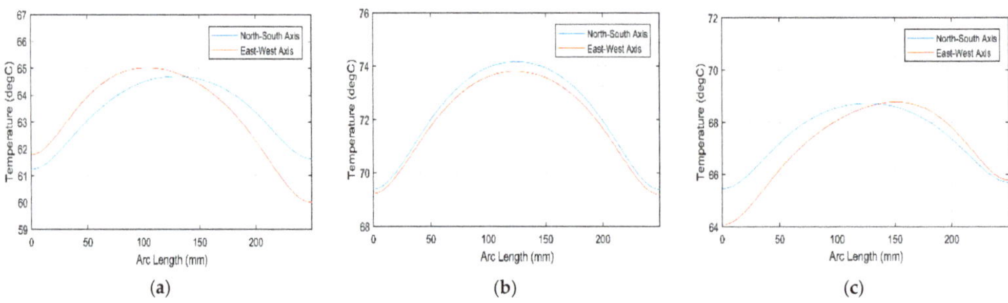

Figure 16. Temperature profile of the triangular SHC with a reflector at (**a**) 9:00; (**b**) 12:00; (**c**) and 15:00 on 1 July 2022.

3.5. One-Year Midpoint Temperature Variation in the Optimized SHC

In order to ensure the capability of the designed system through a single year, a simulation to compute the midpoint temperature inside the SHC was carried out. The computed midpoint temperature results of the simulation performed for a year are mapped in Figure 17. According to the simulation, the highest and lowest points of the maximum midpoint temperature line were 75.00 °C and 45.15 °C, respectively. In contrast, the highest and lowest points of the minimum midpoint temperature line were 31.09 °C and 10.89 °C, respectively. Moreover, the average midpoint temperature inside the SHC throughout the year was 39.46 °C. Therefore, it was concluded that any SMA springs with an activation temperature varying between 31.09 °C and 45.15 °C—in addition to being able to withstand a temperature of 75.00 °C without becoming plastically deformed or damaged—can be used in the designed SHC under Dammam City's weather conditions. Additionally, the yearly temperature profile has shown a close similarity to previous efforts by Hariri, et al. [34]; however, the temperature profile of the proposed system is higher due to the utilization of such reflectors.

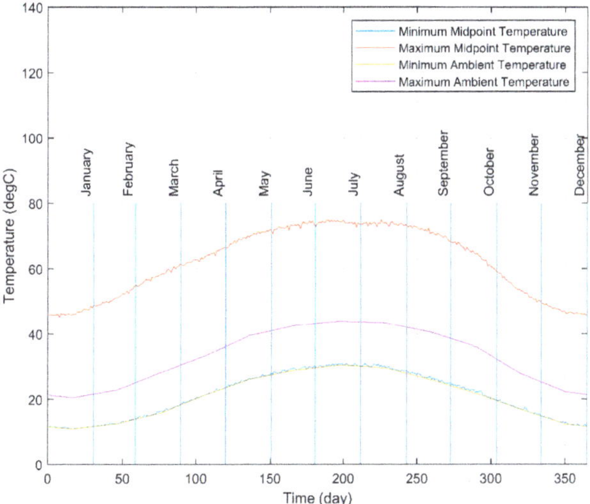

Figure 17. Maximum and minimum daily midpoint temperature and ambient temperature variation throughout 2022G.

4. Conclusions

The presented article is a thermal–optical evaluation of an optimized trough solar concentrator for advanced solar-tracking applications using shape memory alloy. The main element determining the feasibility of the solar concentrator is the temperature of the SHC, from which the thermal energy would be transferred into the SMA, which would convert the heat to mechanical energy. Multiple numerical simulations were carried out to optimize the design of the SHC. The outcomes of the tests performed proved the capability of the optimized design of the SHC, and the outcomes of the simulations were summarized as follows:

- The performance of the SHC is best with the presence of a reflector with an aperture width between 300 and 400 mm and a focal length of 100 mm.
- The SHC is better oriented on the north–south axis since this orientation provides a well-distributed temperature profile inside the SHC.
- SMA springs with an activation temperature varying between 31.09 °C and 45.15 °C, in addition to being able to withstand a temperature of 75.00 °C without becoming plastically deformed or damaged, are applicable.
- The outcomes from the numerical simulation prove the feasibility of the design for the intended application.

The presented outcomes show the feasibility of innovative technology of an actuation method that drives a passive advanced solar tracker. The proposed solar tracking system is solar-driven via the utilization of an SMA-based actuator, and it is the first of its kind. Future work on the proposed system may include thermal evaluation of the system under weather conditions in different areas to test the system's feasibility accordingly. In addition, future work may include further experimental analysis of the system and comparisons between the experimental and numerical data. Additional future work may include the shadow analysis of the PV module to ensure the optimal arrangement of the actuation mechanism.

Author Contributions: Conceptualization, N.G.H., K.M.N. and E.K.A.; methodology, K.M.N., B.A.A.-Q. and E.K.A.; software, K.M.N., E.K.A. and I.S.O.; validation, N.G.H., K.M.N., I.S.O. and I.K.A.; formal analysis, K.M.N., B.A.A.-Q. and E.K.A.; investigation, K.M.N., B.A.A.-Q. and E.K.A.; resources, N.G.H., K.M.N., E.K.A., I.K.A. and I.S.O.; data curation, K.M.N., E.K.A., I.K.A. and I.S.O.; writing—original draft preparation, K.M.N., E.K.A., I.K.A., I.S.O. and B.A.A.-Q.; writing—review and editing, N.G.H.; visualization, N.G.H., K.M.N., E.K.A., I.K.A., I.S.O. and B.A.A.-Q.; supervision, N.G.H.; project administration, N.G.H.; funding acquisition, N.G.H. All authors have read and agreed to the published version of the manuscript.

Funding: The authors extend their appreciation to the Deputyship for Research& Innovation, Ministry of Education in Saudi Arabia for funding this research work through the project number 2020-085-ENG at Imam Abdulrahman bin Faisal University/College of Engineering.

Institutional Review Board Statement: Not applicable.

Informed Consent Statement: Not applicable.

Acknowledgments: The authors are grateful to the Deputyship for Research and Innovation in Ministry of Education as well as the Deanship of Scientific Research (DSR) at Imam Abdulrahman bin Faisal University, Kingdom of Saudi Arabia for their continued guidance and support of this research.

Conflicts of Interest: The authors declare no conflict of interest.

Abbreviations

PV	Photovoltaic
SMA	Shape memory alloy
SME	Shape memory effect
SHC	Solar heat collector
PTC	Parabolic trough collector
CSP	Concentrated solar power
KSA	Kingdom of Saudi Arabia
3D	Three-dimensional
1D	One-dimensional
CFD	Computational fluid dynamics
Symbols	
q	Amount of heat transferred [W/m^2]
T	Temperature [°C]
R	Thermal resistance [°C/W]
L	Length [m]
K	Thermal conductivity [W/m × K]
h	Heat transfer coefficient [W/m^2 × K]
Subscript	
cond	Conduction heat transfer process
conv	Convection heat transfer process
rad	Radiation heat transfer process
in	Input
out	Output
A	Acrylic
amb	Ambient

References

1. Osman, I.S.; Hariri, N.G. Thermal Investigation and Optimized Design of a Novel Solar Self-Driven Thermomechanical Actuator. *Sustainability* **2022**, *14*, 5078. [CrossRef]
2. Kolosok, S.; Bilan, Y.; Vasylieva, T.; Wojciechowski, A.; Morawski, M. A Scoping Review of Renewable Energy, Sustainability and the Environment. *Energies* **2021**, *14*, 4490. [CrossRef]
3. Alkholidi, A.G.; Hamam, H. Solar Energy Potentials in Southeastern European Countries: A Case Study. *Int. J. Smart Grid* **2019**, *3*, 108–119. [CrossRef]
4. Al-Ahmed, A.; Inamuddin; Al-Sulaiman, F.A.; Khan, F. (Eds.) *The Effects of Dust and Heat on Photovoltaic Modules: Impacts and Solutions*; Springer International Publishing: Cham, Switzerland, 2022. [CrossRef]
5. Riad, A.; Zohra, M.B.; Alhamany, A.; Mansouri, M. Bio-sun tracker engineering self-driven by thermo-mechanical actuator for photovoltaic solar systems. *Case Stud. Therm. Eng.* **2020**, *21*, 100709. [CrossRef]
6. Lyulyov, O.; Pimonenko, T.; Kwilinski, A.; Dzwigol, H.; Dzwigol-Barosz, M.; Pavlyk, V.; Barosz, P. The Impact of the Government Policy on the Energy Efficient Gap: The Evidence from Ukraine. *Energies* **2021**, *14*, 373. [CrossRef]
7. Eldin, S.A.S.; Abd-Elhady, M.S.; Kandil, H.A. Feasibility of solar tracking systems for PV panels in hot and cold regions. *Renew. Energy* **2016**, *85*, 228–233. [CrossRef]
8. Teng, T.-P.; Nieh, H.-M.; Chen, J.-J.; Lu, Y.-C. Research and development of maximum power transfer tracking system for solar cell unit by matching impedance. *Renew. Energy* **2010**, *35*, 845–851. [CrossRef]
9. Osman, I.S.; Almadani, I.K.; Hariri, N.G.; Maatallah, T.S. Experimental Investigation and Comparison of the Net Energy Yield Using Control-Based Solar Tracking Systems. *Int. J. Photoenergy* **2022**, *2022*, 7715214. [CrossRef]
10. Hariri, N.G.; AlMutawa, M.A.; Osman, I.S.; AlMadani, I.K.; Almahdi, A.M.; Ali, S. Experimental Investigation of Azimuth- and Sensor-Based Control Strategies for a PV Solar Tracking Application. *Appl. Sci.* **2022**, *12*, 4758. [CrossRef]
11. Bengisu, M.; Ferrara, M. *Materials That Move*; Springer International Publishing: Cham, Switzerland, 2018. [CrossRef]
12. Catoor, D.; Ma, Z.; Kumar, S. Cyclic response and fatigue failure of Nitinol under tension–tension loading. *J. Mater. Res.* **2019**, *34*, 3504–3522. [CrossRef]
13. Sellitto, A.; Riccio, A. Overview and Future Advanced Engineering Applications for Morphing Surfaces by Shape Memory Alloy Materials. *Materials* **2019**, *12*, 708. [CrossRef] [PubMed]
14. Jani, J.M.; Leary, M.; Subic, A.; Gibson, M.A. A review of shape memory alloy research, applications and opportunities. *Mater. Des.* **2014**, *56*, 1078–1113. [CrossRef]
15. Costanza, G.; Tata, M.E.; Calisti, C. Nitinol one-way shape memory springs: Thermomechanical characterization and actuator design. *Sens. Actuators Phys.* **2010**, *157*, 113–117. [CrossRef]
16. Costanza, G.; Paoloni, S.; Tata, M.E. IR Thermography and Resistivity Investigations on Ni-Ti Shape Memory Alloy. *Key Eng. Mater.* **2014**, *605*, 23–26. [CrossRef]
17. Hartl, D.J.; Lagoudas, D.C. Aerospace applications of shape memory alloys. *Proc. Inst. Mech. Eng. Part G J. Aerosp. Eng.* **2007**, *221*, 535–552. [CrossRef]
18. Braun, D.; Weik, D.; Elsner, S.; Hunger, S.; Werner, M.; Drossel, W.-G. Position Control and Force Estimation Method for Surgical Forceps Using SMA Actuators and Sensors. *Materials* **2021**, *14*, 5111. [CrossRef]
19. Wang, M.; Yu, H.; Shi, P.; Meng, Q. Design Method for Constant Force Components Based on Superelastic SMA. *Materials* **2019**, *12*, 2842. [CrossRef] [PubMed]
20. Pan, X.; Zhang, Y.; Lu, Y.; Yang, F.; Yue, H. A reusable SMA actuated non-explosive lock-release mechanism for space application. *Int. J. Smart Nano Mater.* **2020**, *11*, 65–77. [CrossRef]
21. Lu, Y.; Xie, Z.; Wang, J.; Yue, H.; Wu, M.; Liu, Y. A novel design of a parallel gripper actuated by a large-stroke shape memory alloy actuator. *Int. J. Mech. Sci.* **2019**, *159*, 74–80. [CrossRef]
22. Hu, K.; Rabenorosoa, K.; Ouisse, M. A Review of SMA-Based Actuators for Bidirectional Rotational Motion: Application to Origami Robots. *Front. Robot. AI* **2021**, *8*, 678486. [CrossRef]
23. Suman, S.; Khan, M.K.; Pathak, M. Performance enhancement of solar collectors—A review. *Renew. Sustain. Energy Rev.* **2015**, *49*, 192–210. [CrossRef]
24. Marcic, S.; Kovacic-Lukman, R.; Virtic, P. Hybrid system solar collectors—Heat pumps for domestic water heating. *Therm. Sci.* **2019**, *23*, 3675–3685. [CrossRef]
25. Oztop, H.F.; Bayrak, F.; Hepbasli, A. Energetic and exergetic aspects of solar air heating (solar collector) systems. *Renew. Sustain. Energy Rev.* **2013**, *21*, 59–83. [CrossRef]
26. Tian, Y.; Zhao, C.Y. A review of solar collectors and thermal energy storage in solar thermal applications. *Appl. Energy* **2013**, *104*, 538–553. [CrossRef]
27. Lamrani, B.; Kuznik, F.; Draoui, A. Thermal performance of a coupled solar parabolic trough collector latent heat storage unit for solar water heating in large buildings. *Renew. Energy* **2020**, *162*, 411–426. [CrossRef]
28. Ahmad, S.H.A.; Saidur, R.; Mahbubul, I.M.; Al-Sulaiman, F.A. Optical properties of various nanofluids used in solar collector: A review. *Renew. Sustain. Energy Rev.* **2017**, *73*, 1014–1030. [CrossRef]
29. Chen, C.Q.; Diao, Y.H.; Zhao, Y.H.; Wang, Z.Y.; Liang, L.; Wang, T.Y.; Ma, C. Thermal performance of a closed collector–storage solar air heating system with latent thermal storage: An experimental study. *Energy* **2020**, *202*, 117764. [CrossRef]

30. Wu, Z.; Li, S.; Yuan, G.; Lei, D.; Wang, Z. Three-dimensional numerical study of heat transfer characteristics of parabolic trough receiver. *Appl. Energy* **2014**, *113*, 902–911. [CrossRef]
31. NOOR ENERGY 1. Acwapower. Available online: https://acwapower.com/en/projects/noor-energy-1/ (accessed on 5 March 2022).
32. Maatallah, T.; Ammar, R. Design, modeling, and optimization of a dual reflector parabolic trough concentration system. *Int. J. Energy Res.* **2020**, *44*, 3711–3723. [CrossRef]
33. Aldali, Y.; Muneer, T.; Henderson, D. Solar absorber tube analysis: Thermal simulation using CFD. *Int. J. Low-Carbon Technol.* **2013**, *8*, 14–19. [CrossRef]
34. Hariri, N.G.; Almadani, I.K.; Osman, I.S. A State-of-the-Art Self-Cleaning System Using Thermomechanical Effect in Shape Memory Alloy for Smart Photovoltaic Applications. *Materials* **2022**, *15*, 5704. [CrossRef] [PubMed]
35. Almadani, I.K.; Osman, I.S.; Hariri, N.G. In-Depth Assessment and Optimized Actuation Method of a Novel Solar-Driven Thermomechanical Actuator via Shape Memory Alloy. *Energies* **2022**, *15*, 3807. [CrossRef]
36. Gong, J.; Huang, J.; Hu, X.; Wang, J.; Lund, P.D.; Gao, C. Optimizing research on large-aperture parabolic trough condenser using two kinds of absorber tubes with reflector at 500 °C. *Renew. Energy* **2021**, *179*, 2187–2197. [CrossRef]
37. Widyolar, B.; Jiang, L.; Ferry, J.; Winston, R.; Cygan, D.; Abbasi, H. Experimental performance of a two-stage (50×) parabolic trough collector tested to 650 °C using a suspended particulate heat transfer fluid. *Appl. Energy* **2019**, *240*, 436–445. [CrossRef]
38. Gong, J.; Wang, J.; Lund, P.D.; Zhao, D.; Xu, J.; Jin, Y. Comparative study of heat transfer enhancement using different fins in semi-circular absorber tube for large-aperture trough solar concentrator. *Renew. Energy* **2021**, *169*, 1229–1241. [CrossRef]

Article

A State-of-the-Art Self-Cleaning System Using Thermomechanical Effect in Shape Memory Alloy for Smart Photovoltaic Applications

Nasir Ghazi Hariri *, Ibrahim Khalil Almadani and Ibrahim Sufian Osman

Department of Mechanical and Energy Engineering, College of Engineering, Imam Abdulrahman Bin Faisal University, P.O. Box 1982, Dammam 31441, Saudi Arabia
* Correspondence: nghariri@iau.edu.sa

Abstract: This research aims to present a state-of-the-art cleaning technology solution that effectively overcomes the dust accumulation issue for conventional photovoltaic systems. Although continuous innovations and advanced developments within renewable energy technologies have shown steady improvements over the past years, the dust accumulation issue remains one of the main factors hindering their efficiency and degradation rate. By harvesting abundant solar thermal energy, the presented self-cleaning system uses a unique thermomechanical property of Shape Memory Alloys to operate a solar-based thermomechanical actuator. Therefore, this study carries out different numerical and experimental validation tests to highlight the promising practicability of the developed self-cleaning system from thermal and mechanical perspectives. The results showed that the system has a life expectancy of over 20 years, which is closely equivalent to the life expectancy of conventional photovoltaic modules while operating under actual weather conditions in Dammam city. Additionally, the thermal to mechanical energy conversion efficiency reached 19.15% while providing average cleaning effectiveness of about 95%. The presented outcomes of this study add to the body of knowledge an innovative methodology for a unique solar-based self-cleaning system aimed toward smart and modern photovoltaic applications.

Keywords: photovoltaic energy; dust accumulation; PV efficiency; thermomechanical; shape memory alloys; PV cleaning system; passive system; soiling effect

Citation: Hariri, N.G.; Almadani, I.K.; Osman, I.S. A State-of-the-Art Self-Cleaning System Using Thermomechanical Effect in Shape Memory Alloy for Smart Photovoltaic Applications. *Materials* **2022**, *15*, 5704. https://doi.org/10.3390/ma15165704

Academic Editor: Salvatore Saputo

Received: 27 July 2022
Accepted: 16 August 2022
Published: 18 August 2022

1. Introduction

The world's awareness of renewable energy's importance and advantages arises as the destructive impact of fossil fuels becomes a severe environmental issue. Harming fossil fuels to the environment includes global warming and greenhouse emissions, affecting the overall environmental quality. The Kingdom of Saudi Arabia (KSA) has taken serious steps intending to shift toward renewable energy as an alternative to conventional energy production methods, including fossil fuels, and aspires to achieve a 50% dependence on renewable energy by 2050 [1–3]. Although many renewable energy sources are available, solar PV energy shines the most. The abundant amount of energy provided via the sun is estimated to be 8000 times higher than the global energy consumption [4–6]. However, the efficiency of PV modules is a critical concern since the conventional PV modules' efficiency can be affected by many factors; one of the main factors is the dust accumulation issue (DAI) [4,5]. The DAI is especially critical in areas with high dust intensity, such as the Middle East and North Africa (MENA).

The soiling effect is one of the main obstacles for PV systems, especially for areas with excessive dust accumulation rates, such as the MENA region, since such regions have high dust intensity. There are many factors affecting dust accumulation; a study by GHOLAMI et al. [7] showed that humidity, adhesion force, rain rate, the cover glass of the panel, wind speed and direction, and gravity affect the dust accumulation over a solar panel. Many research efforts have highlighted the significant reduction in efficiency due to soiling on PV modules [8,9] since soiling develops a layer on the outer surface of the PV module's glass, reducing the light transmissivity [10]. Furthermore, the soiling layer developed on the glass' outer surface leads to degradation effects which shorten the life span of the solar panels [10,11]. Kazem et al. [12] have demonstrated the effect of dust accumulation on the power of PV modules, where power loss due to soiling effects of up to 80% per month and 1% per day were recorded. In addition, the study suggested using some cleaning methods to eliminate the soiling effect.

In Farrokhi Derakhshandeh et al. [13] review, different cleaning systems for solar panels were compared. Although the review studied different active and passive solar cleaning systems, none of the studied passive cleaning systems can be achieved without human intersection. Deb and Brahmbhatt [14] reviewed and compared multiple studies on cleaning solar panels and proposed an automated water-free cleaning method using cleaning brushes. The cleaning system proposed showed a cleaning efficiency of about 9.05% and an effective cleaning efficiency considering the cost of the cleaning system of 6.31%. Alghamdi et al. [15] made three mechanical cleaning systems platforms to study the feasibility of each system in the Middle East region, where rainfall is lacking. The three platforms were an air-jet, vibrator, combination of the previous, and waterjet based. The results showed that waterjet cleaning increased the output power by about 27% and was considered the most efficient cleaning method in the experiment, while module vibration and air-jet reduced the soling in such a less practical matter. Alnaser et al. [16] experimented by comparing the performance of artificial cleaning with natural cleaning. Two sets of eight PV modules were used in the experiment, where the first set represented artificial cleaning while the second set represented natural cleaning. The experiment found that the first set produced an electrical power of about 1.06 KWh more than the producer's expected power, representing a 16% increase from the expected power. On the other hand, the second set produced an electrical power of 0.97 KWh, which is 5% more than the producer's anticipation. In 15 months, the production of dirty PV modules is expected to be 9% less than that of clean PV modules. Yadav et al. [17] proposed an automated cleaning system using rubber wipers powered by a direct current servo motor. The experimental outcome of the proposed system's cleaning efficiency was recorded as 97.8%.

Cleaning frequency can differ in each country depending on the climate and energy cost [18]. The optimal cleaning frequency is about 25 days; for tropical regions, the cleaning frequency is less than 10 days [19]. In order to determine the optimal cleaning frequency, the overall cleaning cost must be investigated, which depends on the climate conditions, water consumption, and electrical usage [20].

In 1932, Shape Memory Alloys (SMA) was discovered, which is a type of smart material with the property to regain a programmed shape after exposure to heat called the Shape Memory Effect (SME) [21,22]. Several alloys have SME, each with distinctive characteristics, including activation temperature and yield strength. Various materials are used to make different SMAs, such as Cu-Zn-Al, Cu-Al-Ni, Au-Cd, and TiNi-Ag [23,24]. The most used type of SMA is Nickel-Titanium alloy, also known as NiTiNOL, due to its wide range of activation temperature and low price. SMA has multiple applications in various fields, including aerospace, biomedical, robotics, aeronautics, and engineering [25–27]. The utilization of SMA in multiple applications requires the ability to control the alloys as needed, which is possible through various techniques [28,29]. The wide range of characteristics and ability to manipulate the SMA has led to various innovative applications. QADER et al. [30] reviewed how different parameters such as stress, strain, and temperature affect SMA behavior. The review compared typical material and different SMA deformation under different stresses and strains. Additionally, the review compares steal and NiTiNOL SMA strain recovery percent and the differences in the materials' elasticity. In addition, the review reveals how SMA behavior changes under various temperatures. Another review by Farber et al. [31] discusses the manufacturing of NiTiNOL using 3D printing techniques. The review shows how the Nickel/Titanium ratio affects the SMA's activation temperature and shape memory effect. Such reviews facilitate a better understanding of SMA, helping further studies to better implement SMA in various applications spatially as actuators. In addition, a recent exciting study has discussed advanced technology to obtain shape memory alloy thin layer utilizing pulsed laser deposition (PLD) process [32].

Multiple studies and reviews discuss the utilization of SMA in actuation techniques for different applications. A review by Yuan et al. [33] shows different SMA-based rotary actuators' designs. The actuators mentioned in the review were categorized depending on two features movement continuity and whether the motion is single rotation direction or bi-directional rotation. Depending on the freedom and continuity of motion and other actuator attributes, the actuation mechanism's application differs. A review by Costanza and Tata [34] discusses the recent utilization of SMA in aerospace and aeronautics. The review shows that SMA reduces noise, increases thrust, and optimizes the efficiency of wing morphing and propulsion system. As for SMA applications in aerospace, the review shows that SMA is used for "isolating the micro-vibrations, for low-shock release devices, and self-deployable solar sails." All of these studies demonstrate how SMA is being recognized for actuation applications in various fields.

SMA actuators can be activated via joule heating, external/internal passive, or forced heating [35]. Likewise, the offered method to activate the SMA-based actuator in this paper is via solar heat collector (SHC). SHC is a unique heat exchanger device that absorbs the natural sun rays as thermal energy input, converting it into another useful form of energy [36,37]. Solar thermal energy is more efficient than solar photovoltaic energy by about 77% [2,38,39], and many researchers have suggested different technologies to enhance its thermal efficiency [40]. SHC applications are varied, including industrial heating or cooling, energy storage systems, and water desalination systems [41,42]. SHCs have different types based on various categories, where they can be divided based on their designed shape, derived method, and the utilized fluid [36,43]. Additionally, multiple materials are used to construct the SHC absorber, such as aluminum and copper [36,40,44]. At the same time, multiple fluids are also used inside the SHC, such as water, nanofluids, and air [1,45]. Designing the SHC is a critical step in any thermal energy system, where three main criteria should be considered: technical, cost, and environmental consecrations [46]. The technical aspect focuses on the efficiency of the SHC. In contrast, the cost aspect estimates the system's overall cost, while the environmental aspect predicts the effect of the SHC on the environment [2,47,48].

In the literature and previous studies, minimum research discussed the utilization of the natural sun rays to run thermomechanical or SMA-based actuators; however, previous research efforts by the authors have highlighted its applicability. Osman and Hariri [2] presented a detailed thermal study for a novel thermomechanical actuator. The results conclude that the system can operate under the actual weather condition of the studied area. Similarly, Almadani et al. [1] offered a mechanical study for a similar thermomechanical actuator. The output force and displacement of the actuator showed significant advantages, which revealed its great feasibility as a solar-based actuator.

This research effort aims to offer a novel design of a PV cleaning system for DAI that utilizes the unique property of SMA. Additionally, the paper presents different feasibility tests for the suggested technology solution, including numerical and practical experiments. The study is unique in its approach since the proposed PV cleaning system uses the available sun rays to operate in an innovative mechanism while eliminating human intervention and electrical storage configurations. Furthermore, the study offers a unique technology solution to overcome a critical issue within the renewable energy field, which increases the encouragement toward the transition to green, smart, and clean cities across the globe.

2. Materials and Methods

The proposed PV cleaning system contains three main parts: the cleaning spindle, gear mechanism, and thermomechanical SMA actuator assembly, as shown in Figure 1. For the working principle of the proposed design, the thermomechanical SMA actuator transforms thermal energy into mechanical energy in the form of a linear motion. Then, linear movement is converted into rotational movement through a gear mechanism, thus, allowing the cleaning spindle to slide over the PV module's surface, which helps remove the accumulated dust particles. Therefore, this section discusses the conceptual design of the thermomechanical actuator and highlights the evaluated numerical thermal study applied. After that, it presents the development of the thermomechanical cleaning system and the evaluated mechanical assessment tests.

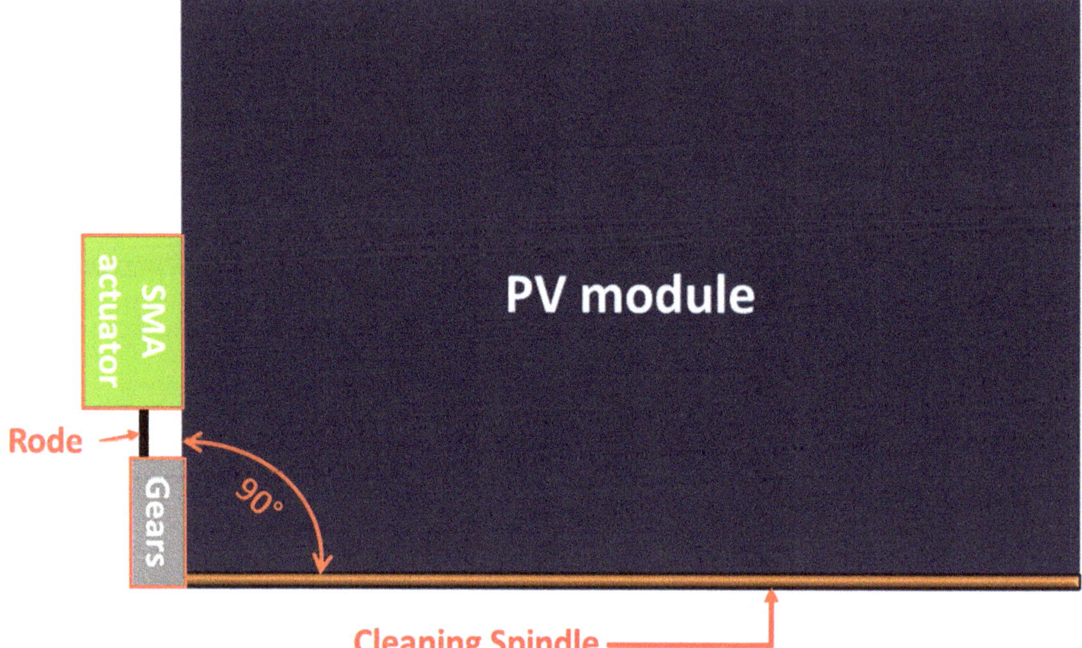

Figure 1. Conceptual design model of the cleaning system.

2.1. Conceptual Designs of the Thermomechanical Actuators

The SMA-based actuator uses the sun's thermal energy to give a smart movement and produces considerable force via the actuator. The actuator is typically deactivated at night-time while activated during the daytime using the sun's rays and when it gains sufficient thermal energy from the sun, as shown in Figure 2. Therefore, the SHC supplied heat to the actuator continuously throughout the day in order to give the spring-based NiTiNOL actuator enough thermal activation energy. The actuator continues receiving solar radiation even after reaching the desired activation temperature, which becomes essential to ensure adequate activation temperature throughout the year since temperatures of ambient and actuator vary significantly. On the contrary, the actuator is typically deactivated at night-time due to the absence of the sun's rays. The primary purpose of using a bias-load is to retain the NiTiNOL springs to their initial shape at night-time, thus, completing a single cycle daily.

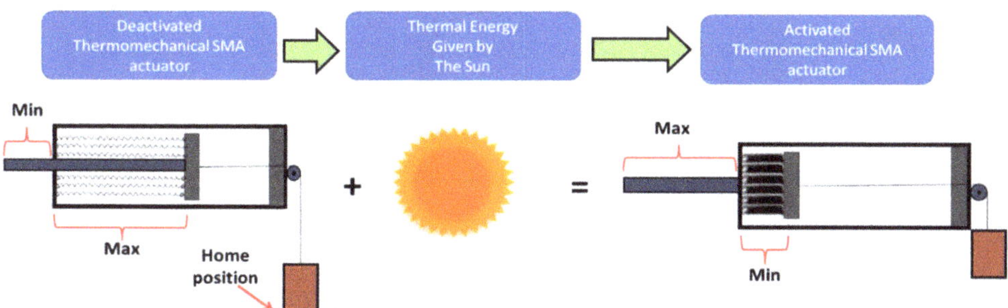

Figure 2. Basic conceptual procedures of the thermomechanical solar-based linear actuator.

The proposed thermomechanical SMA piston-based actuator consists of five main components: SHC, piston, rod, bias-load, and NiTiNOL springs. On the other hand, the actuation mechanism includes five main phases, wherein the first phase, the NiTiNOL springs, are under the austenite starting temperature (As) and extended while the bias load is at the home position. Secondly, the NiTiNOL springs absorb adequate heat to reach (As) and start contracting to pull the bias load. In the third phase, the NiTiNOL springs arrive at the austenite final temperature (Af), allowing them to reach the maximum deflection under the applied load. Fourthly, the NiTiNOL springs dissipate the heat causing the spring to reach martensite start temperature (Ms) and extend as the bias load pulls the piston. In the final phase, the NiTiNOL springs are in the (Mf), letting the springs return to the original extended shape. Multiple Computer-Aided Designs (CAD) of SHCs devices were designed and evaluated to optimize the design with the aim of completing one actuation cycle per day throughout the year. As the heat collectors change in size and shape, all components inside the heat collector change accordingly. Therefore, the first actuator design (actuator A) has a glass cylindrical heat collector, as shown in Figure 3a. The Second actuator design (actuator B) has a glass-covered aluminum isosceles trapezoid prism heat collector, as shown in Figure 3b. The third actuator design (actuator C) combines glass and aluminum for the equilateral triangle prism heat collectors, as shown in Figure 3c.

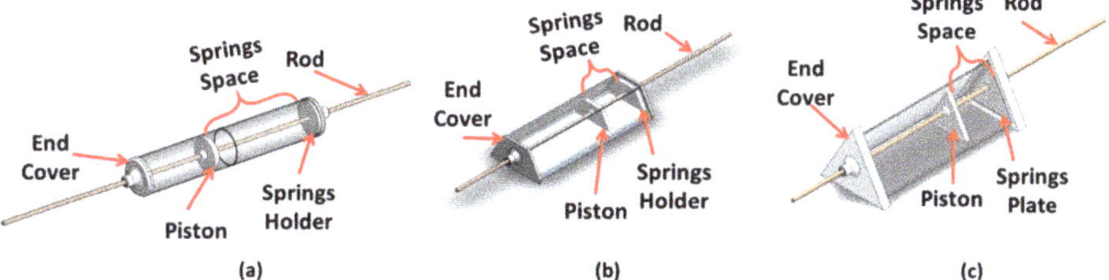

Figure 3. Overview of developed CAD models for (**a**) actuator A, (**b**) actuator B, and (**c**) actuator C.

2.2. Numerical Study Setup of the Thermomechanical Actuators

Since the proposed system is entirely dependent on thermal energy to operate, a thermal investigation of the system becomes crucial. Therefore, a time-dependent three-dimensional (3D) thermal simulation study has been conducted for the proposed designs under the weather conditions of Dammam city, KSA using Computational Fluid Dynamics (CFD) software version 5.6. The three developed designs were imported to CFD software, where different studies have been conducted, and their results are presented in upcoming Section 3. Various boundary conditions were applied, knowing that the SHC gains thermal energy via the sun's radiation and convection heat transfer processes. Due to the SHC design, the applied amount of heat is transferred to the NiTiNOL springs by the radiation and convection heat transfer processes. Therefore, the stated boundary conditions were applied as heat flux, located at different locations depending on the state heat transfer process. In addition, an external heat source that represents the actual sun rays of Dammam city was also included in the thermal studies as boundary conditions. Additionally, the initial conditions have been chosen to simulate the actual weather conditions of the studied area. Eventually, fine mesh and time steps for each designed model and iteration were adopted.

2.3. Development of a State-of-the-Art PV Cleaning System

The proposed solar-driven smart PV modules cleaning system offers a novel solution to the dust accumulation dilemma, which varies depending on the weather and the location in which PV modules are installed. The design criteria of the smart cleaning system are meant to be portable, concise, and easy to install in order to facilitate the integration into existing solar arrays. The novel smart cleaning system consists of four main components: the thermomechanical SMA actuator, power transmission mechanism, cleaning spindle, and PV connectors. The mechanism used to transmit power from the thermomechanical SMA actuator into the cleaning spindle is a rack and pinion mechanism, where the rack is connected to the thermomechanical SMA actuator's rod mounted into the PV module's frame by a PV connector. In addition, the pinion is attached to a PV connector and mounted into the PV module's frame. As for the cleaning spindle, the spindle is mounted into the pinion using the cleaning spindle holder. Figure 4 represents all components within the smart cleaning system.

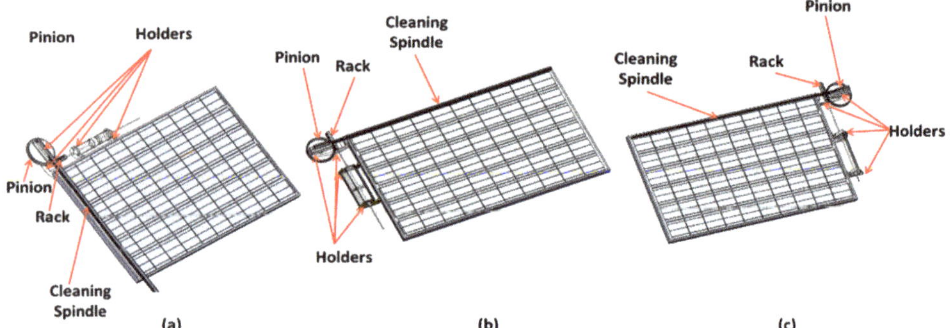

Figure 4. Detailed CAD model of the PV cleaning system based on (**a**) Actuator A, (**b**) Actuator B, and (**c**) Actuator C.

The force transmission unit has been designed according to the desired performance of the smart cleaning system, where a 150 mm linear motion is converted into a 90° rotational motion. In order to do so, a rack and pinion force transmission mechanism is designed and fabricated, where the linear travel distance of the rack is known and is used to calculate the required pinion circular pitch diameter (Dp) as in Equation (1).

$$S = \theta * r \tag{1}$$

where, S is the travel distance of the linear actuator, θ is the rotation angle of the cleaning spindle, and r is the radius of the pinion gear that is also equal to Dp. Multiple design iterations were computed to determine the optimized design and number of teeth (T_1) required for the pinion gear to achieve the desired motion for the intended cleaning system. For that, T_1 is computed as shown in Equations (2)–(5), respectively, where P_d is the pitch diameter, m is the number of module, a_w is the addendum, and d_w is the dedendum of the gear mechanism.

$$P_d = \frac{T_1}{D_p} \tag{2}$$

$$m = \frac{D_p}{T_1} \tag{3}$$

$$a_w = m \tag{4}$$

$$d_w = 1.2m \tag{5}$$

The following design condition shown in Equation (6) was considered for selecting T_1 of the pinion gear to avoid potential interference and validate the design [49].

$$T_1 \geq \frac{2a_w \frac{1}{T_2} P_d}{\sqrt{1 + \frac{1}{T_2}\left(\frac{1}{T_2} + 2\right)\sin^2 \varphi - 1}} \tag{6}$$

where, T_2 value is set to be quarter of T_1 since the required gear rotation is 90°. In addition, φ is the pressure angle with a value of 20°. Furthermore, the coverage area of the cleaning system is another critical design element for the feasibility of the cleaning system. Since most solar systems consist of multiple PV solar arrays, the cleaning system was designed to meet the needs of more than a single PV module. This was achieved by adequate arrangements of the actuator's installed location, as seen in Figure 5. Since the designed actuation mechanism is installed on the outer frame of the PV module, this adds several advantages from a practical perspective. This includes improved accessibility and placement of the PV cleaning assembly as well as implementations of actuator arrangements for multiple PV arrays, even for already existing PV farms.

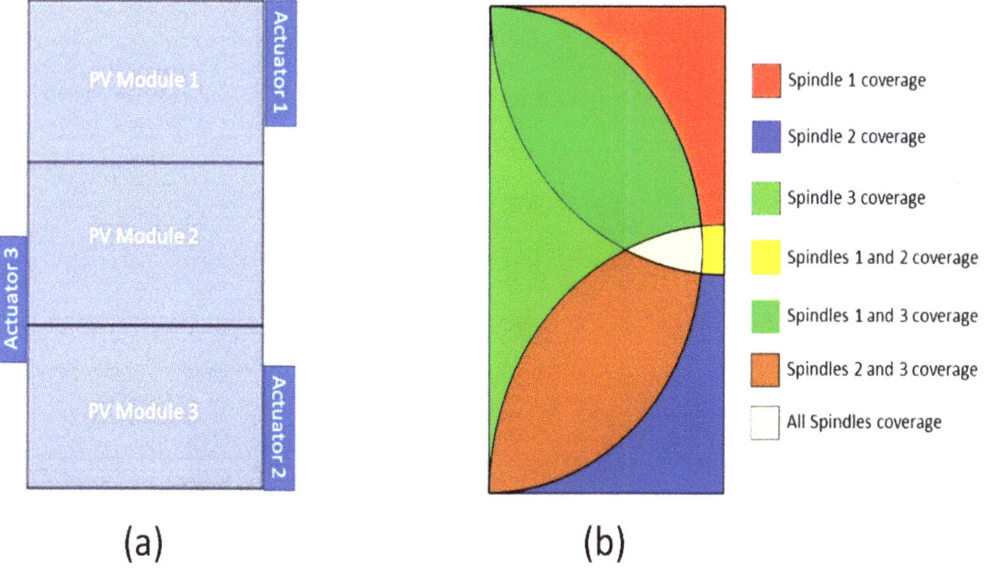

Figure 5. (**a**) Arbitrary CAD example of dusty PV modules string, and (**b**) coverage area of multiple actuator arrangements used for cleaning.

The fabrication of the thermomechanical actuator involves various steps, as shown in Figure 6. It consists of two aluminum surfaces, an acrylic sheet, a piston, a springs' plate, and a rod. In order to fabricate all different parts, several rapid manufacturing techniques were used, including laser cutting and 3D printing. In addition, an aluminum sheet was utilized to support the actuator's structure against the axial forces that can cause bending or buckling of the actuator's structure. It also acts as a reflecting element of the applied solar rays, which helps increase the temperature within the actuator.

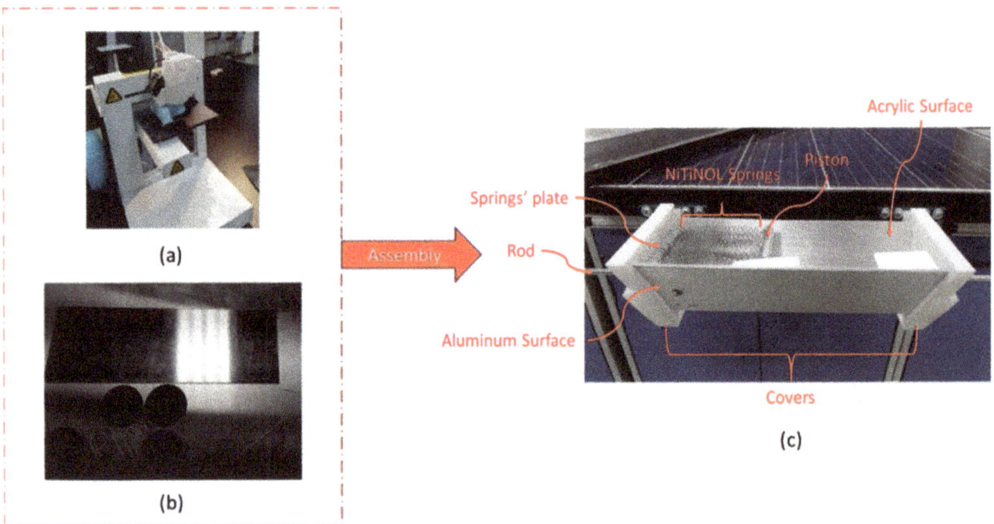

Figure 6. Implemented rapid prototyping techniques with (**a**) 3D printer, (**b**) laser cutting machine, and (**c**) fully assembled actuator.

In order to test the cleaning effectiveness of the smart cleaning system, indoor experiments have been carried out, where a heat source, in the form of a hot air gun, was applied to activate the actuation mechanism efficiently. Thus, it allows rotating the cleaning spindle over the PV module's top surface, which removes the accumulated dust particles. Different amounts of dust were placed over the PV module, as seen in Figure 7, to investigate the cleaning effectiveness versus dust densities.

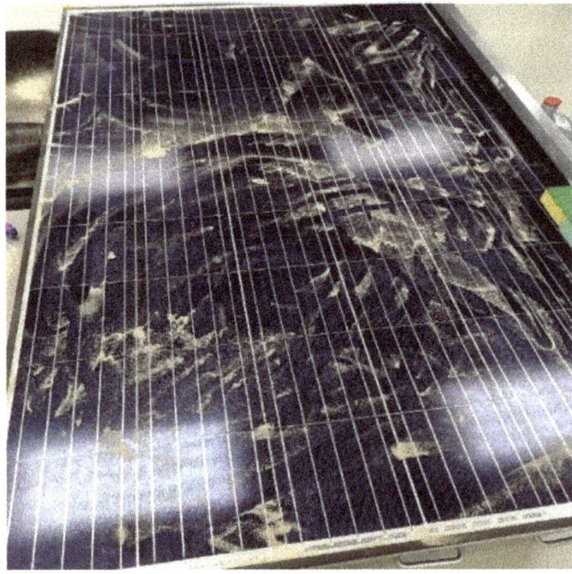

Figure 7. A prepared uncleaned PV module for the cleaning effectiveness test.

2.4. Mechanical Assessment Setup of the Thermomechanical Actuator

This section discusses elements of all assessment platforms, including the developed mechanical model, electrical components, software setup, and experimental criteria. In order to carry out the tests for both the NiTiNOL springs and the actuator, two assessment platforms were built. The first test platform, seen in Figure 8a, was designed to test the generated force and displacement of the NiTiNOL springs. An active-controlled system, in the form of a gain scheduling proportional–integral–derivative (PID) controller, was implemented to quantitatively study the SMA spring-based actuator's life expectancy. In addition, assessment tests of both displacement and force produced by the actuator's assembly versus temperature inside the actuator were studied, as seen in Figure 8b.

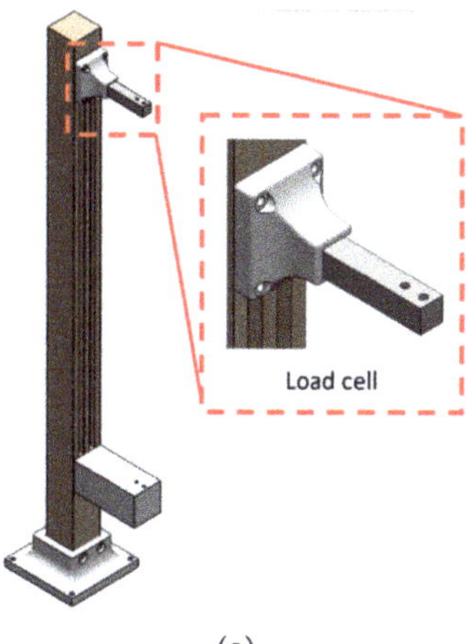

(a)

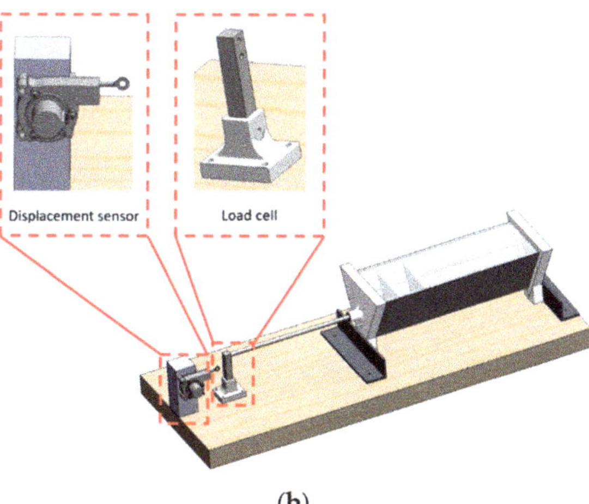

(b)

Figure 8. Mechanical models of the assessment platforms for the (**a**) SMA spring, and (**b**) actuator's assembly tests.

The two designed assessment platforms utilized various sensors to measure all parameters connected to a microcontroller for data collection. An accurate rotary encoder of 2000 pulses per revolution was utilized to measure the displacement produced by the actuator. In addition, a precise load cell was used for measuring the generated force, while a temperature, model LM35, was used for the temperature data. All these sensors were calibrated and programmed; a designed gain-scheduling PID controller was integrated to test the NiTiNOL springs' life expectancy. Figure 9a shows the overall block diagram of the PID-based controller, while Figure 9b represents the flow chart diagram of the developed system.

It can be observed that each experimental test has its experimental procedure. For instance, the first assessment platform of the NiTiNOL spring force test has one side of the spring connected to the load cell while the other side was connected to a fixed hook. As the applied current passes through the spring via the Joule heating effect, heat is developed within the SMA spring, producing a recovery force. Similarly, the load cell was replaced with the rotary encoder to measure the produced motion for the displacement test. On the other hand, the applied tests for the actuation mechanism of the complete actuator assembly were performed with the replacement of the NiTiNOL spring with the actuator while reading the temperature inside the actuator using a temperature sensor.

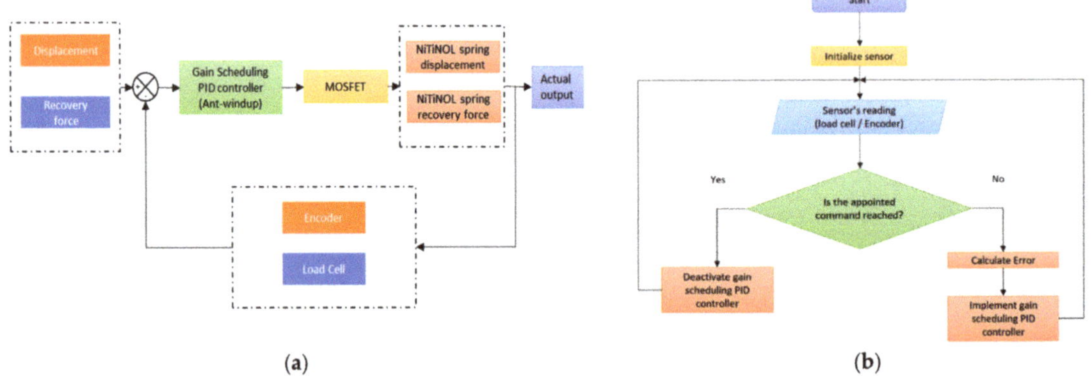

Figure 9. (**a**) Block diagram of the process gain scheduling PID controlled system, and (**b**) flow chart diagram of the controller working principle.

3. Results and Discussion

The proposed self-operating solar-driven PV cleaning system offers a novel solution to the dust accumulation dilemma, which varies depending on the weather and the location in which PV modules are installed. Therefore, different tests were performed to validate its functionality, and their results are presented in this section. First, Section 3.1 discusses the structural analysis of the designed solar-based actuator, while Section 3.2 highlights the thermal analysis of the thermomechanical system. In addition, Section 3.3 presents the dynamic analysis of the generated forces and displacement of the actuator, while Section 3.4 highlights the overall performance of the self-cleaning PV system.

3.1. Structural Analysis of the Thermomechanical Actuator

All main parts were simulated under the expected applied load to ensure the mechanical feasibility of the designed smart cleaning system. Several structural Finite Element Analysis (FEA) tests were made to avoid failure or fracture within the system's components. The FEA results have predicted locations of critical regions where expected maximum deformations or stresses might occur. For instance, Figure 10 demonstrates an example of the maximum stresses expected under an applied load of 150 N, where the yield strength of the chosen materials (acrylic: 4.5×10^7 N/m^2, ABS: 3.9×10^7 N/m^2). Therefore, calculations for the factor of safety of all parts within the actuator assembly were computed as seen in Table 1. It can be seen that the safety factor for all parts is above one for a maximum expected load of 150 N.

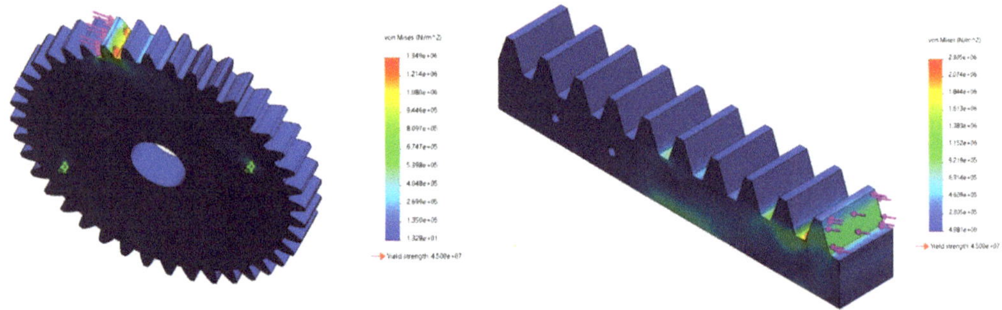

Figure 10. Maximum stresses developed within rack and pinion gear arrangement.

Table 1. FEA results of all developed parts within the mechanical system.

Part	Displacement (mm)	Strain	Stress (N/m^2)	Factor of Safety
Actuator A Holder	2.92	5.67×10^{-3}	1.7×10^{7}	2.29
Actuator B Holder	4.12	8.12×10^{-3}	2.73×10^{7}	1.43
Actuator C Holder	5.1	1.02×10^{-2}	3.09×10^{7}	1.26
Pinion	2.08×10^{-2}	3.77×10^{-4}	1.35×10^{6}	33.33
Rack	5.52×10^{-2}	6.47×10^{-4}	2.31×10^{6}	19.5
Spindle holder	1.09×10^{-1}	7.88×10^{-4}	2.76×10^{6}	1.41

3.2. *Thermal Behavior of the Thermomechanical Actuator*

Another fundamental analysis for the design of the thermomechanical actuator is the thermal performance of the designed SHC, which includes the thermal distribution and profile throughout the year. Figure 11 presents an example of a 1-day temperature distribution of the designed SHC of actuator B at different day times. It is evident in Figure 11a that the temperature on the east side (− y-axis) has received the highest temperature of about 52 °C due to the sun's location in the morning. Alternatively, the temperature distribution inside the SHC is identical through the SHC length at the zenith time and when the sun is vertically toward the system, as shown in Figure 11b. It can be noticed that the maximum temperature of about 81 °C within the SHC was obtained during the zenith period. On the other hand, the west direction (+ y-axis) absorbs the most heat after noontime with a maximum temperature above 65 °C, as seen in Figure 11c. Lastly, the temperature distribution at late night times is shown in Figure 11d, reaching its minimum values during the day as expected.

Furthermore, an extended quantitative analysis of temperature profiles for the designed SHC devices under actual weather conditions has been studied for an entire year. This allows verifying the functionality of the proposed actuators as standalone smart thermomechanical devices. As a result, the yearly temperature profiles for the three designed actuators mentioned earlier were computed, as shown in Figure 12. It can be seen that the optimized design was found to be actuator C, which shows the highest temperature profile and, most importantly, variation throughout the year. Overall, the obtained results show the ability of each SHC to increase the temperature, where actuators C, B, and A were the most efficient, respectively. The optimized design of actuator C increases the internal temperature by almost double the ambient temperature of the studied region. However, actuator B increases the temperature by about 70%, while Actuator A only increases the temperature by about 5%. The effect of the ambient temperature profile can also be seen in governing the temperature inside all designed SHCs; additionally, obtained results from this graph show that the deactivation levels are almost similar throughout the year, which is bounded by the ambient temperature during the evening time.

In addition, Table 2 shows statistically averaged data for the temperature profile inside the SHC over one year. These results have also concluded that actuator C has the highest maximum, range, and standard deviation, while actuator B has the highest mean and median values; additionally, actuator A has the lower average temperature. Hence, the highlighted statistical outcome supports that actuator C is the most efficient and adequate actuator for the studied area since it has the highest range and standard deviation values.

The daily temperature variation is the mathematical difference between the highest and the lowest temperature of the SHC in a single day. This analytical data is crucial as it directly infers about the possible activating and deactivating ranges for the SMA-based actuator. It indeed permits daily activation and deactivation of the actuation mechanism due to the extensive range of temperature spans for substantial daily temperature variation. Moreover, Figure 13 shows the histogram plot of the daily temperature variation for the three proposed actuators. Actuators A, B, and C recorded an average daily temperature variation of 14.62, 19.83, and 33.65 °C, respectively. Therefore, it has been found through the conducted analysis that the optimized actuator, actuator C, is the most suitable actuator for the intended self-cleaning application, where it recorded an average temperature variation of 130.16% and 69.69% compared to actuators A and B, respectively.

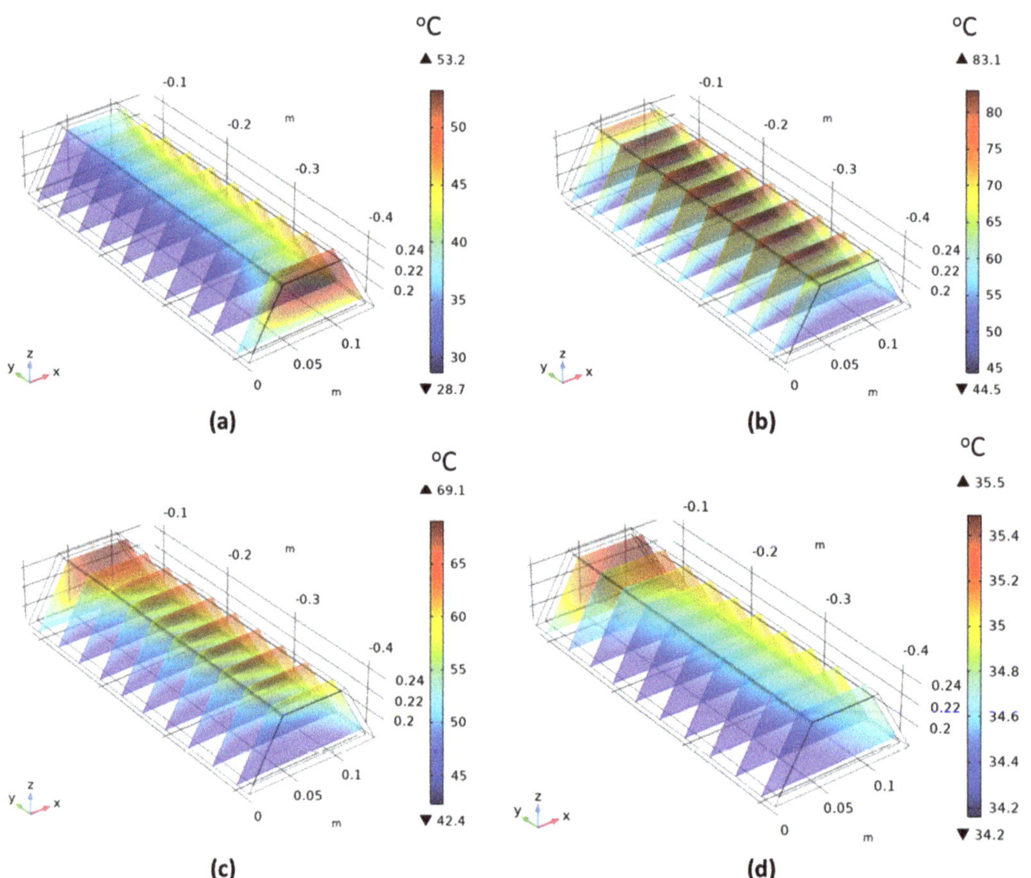

Figure 11. Example of 1-day temperature distributions for the designed SHC through (**a**) morning, (**b**) zenith, (**c**) afternoon, and (**d**) night periods.

The proposed PV cleaning system depends entirely on the thermal energy delivered to the SMA springs by solar radiation. An activation temperature lines term has been introduced here, which shows the temperature at which the SMA springs are anticipated to activate in particular months. Figure 14 presents these activation temperature lines for the proposed three actuators. The graph indicates that actuators A and B require a minimum of two activation temperature lines; hence, two sets of SMA springs are needed to operate independently throughout the year. In addition, actuators A and B have dedicated temperature activation lines for warm weather and another for cold weather. The fact that both actuators require two activation temperature lines led to the understanding that each of these actuators would need at least two maintenance routines per year for practical implementation. On the other hand, actuator C requires a single activation temperature line throughout the year in which minimum yearly maintenance routine is needed for its operation.

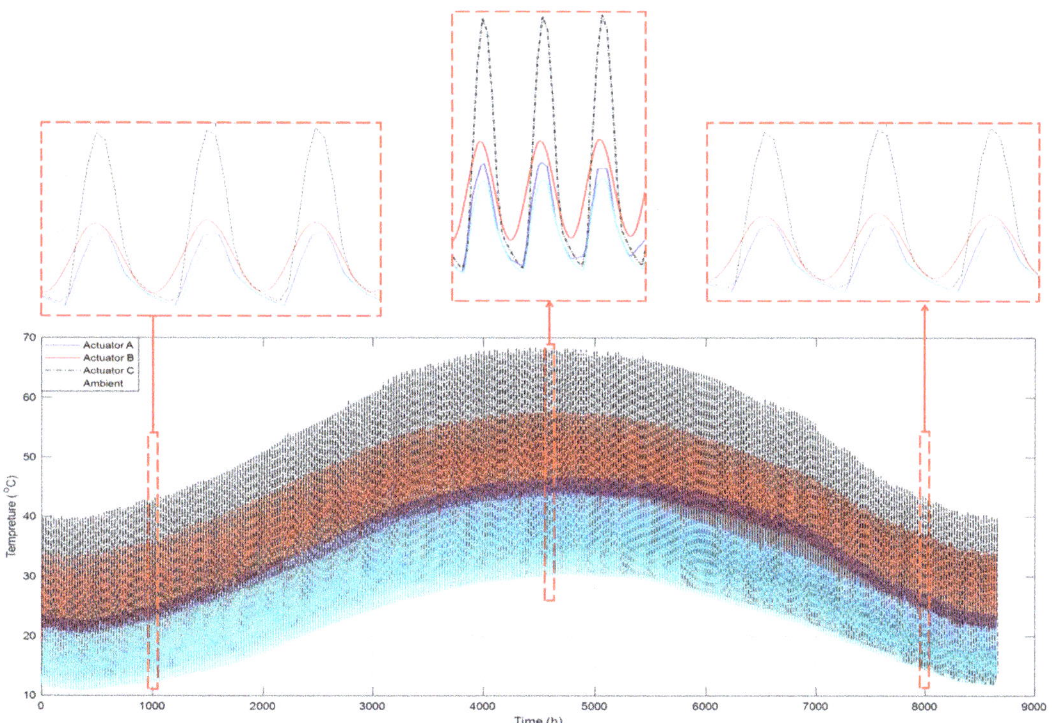

Figure 12. Temperature profiles inside the SHC over an entire year.

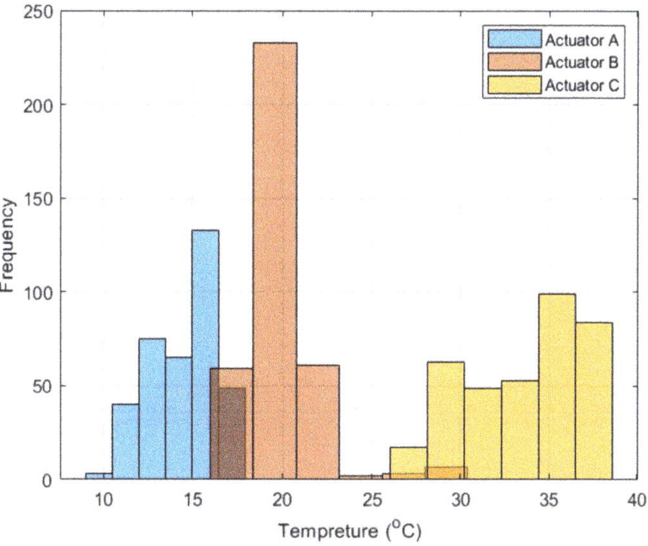

Figure 13. Histogram plot for the daily temperature variation of the three actuator designs.

Table 2. Statistical data for the temperature inside the SHC over one year.

	Actuator A (°C)	Actuator B (°C)	Actuator C (°C)
Max	49.61	62.1	68.35
Min	13.58	13.5	11.2
Mean	31.03	39.7	35.05
Median	31.2	40	32.2
Range	36.03	48.5	57.15
Standard Deviation	9.34	10.8	15.16

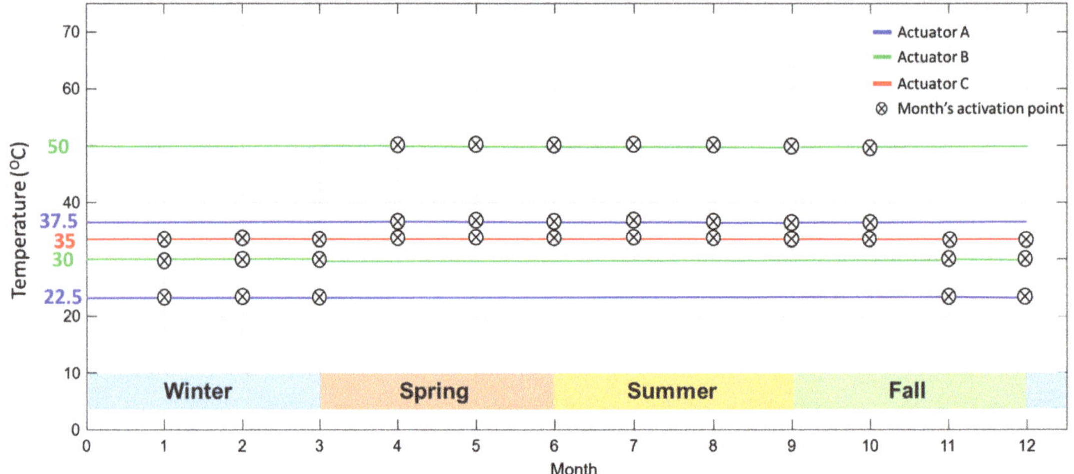

Figure 14. Activation temperature lines for the proposed three actuators.

The comprehensive thermal studies for the different actuator designs concluded that actuator C is effective for the self-cleaning application. This valuable finding is supported by the outcomes of the conducted numerical study, where this actuator design has the most extensive daily temperature variation and can function throughout the year without needing maintenance.

3.3. *Dynamic Appraisal of the Thermomechanical Actuator*

Since the thermal investigation successfully identified an adequate design, the mechanical performance of this novel thermomechanical actuator is also considered. This section highlights multiple tests conducted to ensure the feasibility of utilizing the thermomechanical SMA solar-driven actuator as the actuation mechanism for the cleaning system from a mechanical energy perspective. The experimental tests evaluated the life expectancy of the NiTiNOL springs integrated into the actuation mechanism. They also assessed the actuation mechanism's force and displacement.

In order to carry out an expedited life expectancy test for the NiTiNOL spring actuator, a gain-scheduling PID controller with an anti-windup mechanism was implemented for the active actuation mechanism. In addition, multiple commands were used to ensure the controller's usability, such as step, square, staircase, and sinusoidal command signals, as shown in Figure 15. the responses show the great accuracy of the controlled system, which enables further investigation of the expected life expectancy test.

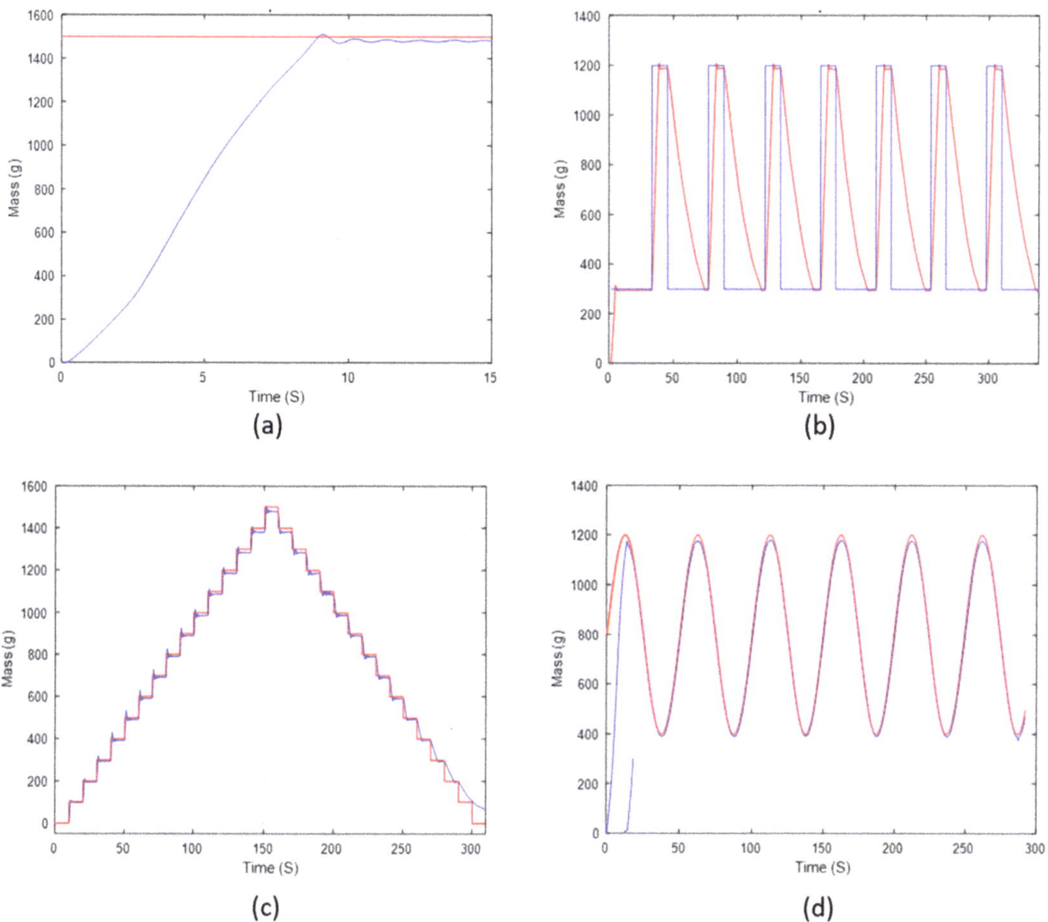

Figure 15. Gain scheduling PID controller's Response to (**a**) step, (**b**) square-wave, (**c**) staircase, and (**d**) sinusoidal-wave command signals.

For the life expectancy test, a long-duration square wave command signal was selected to rapidly simulate the force and displacement responses of the NiTiNOL spring in response to an applied heat source. Figure 16 shows, for instance, an example of the square wave used in both the displacement and force life expectancy tests. The NiTiNOL spring actuator used in the test handled over 7900 cycles of varying force between 2.95 N and 11.8 N, corresponding to about 20% and 80% of the maximum expected force. As for the displacement test, the total number of cycles ran was over 9200 cycles of varying displacement of 30 mm and 50 mm, corresponding to about 40% and 70% of the maximum displacement of the NiTiNOL spring. It can be concluded from the results of the life expectancy tests that the operation of the self-cleaning system daily can last over 20 years in service, which is approximately the life expectancy of conventional PV modules. The fact that the life expectancy of the actuator is closely similar to that of a conventional PV module shows the compatibility of the introduced self-cleaning mechanism and the PV module for the actual implementation of smart solar applications.

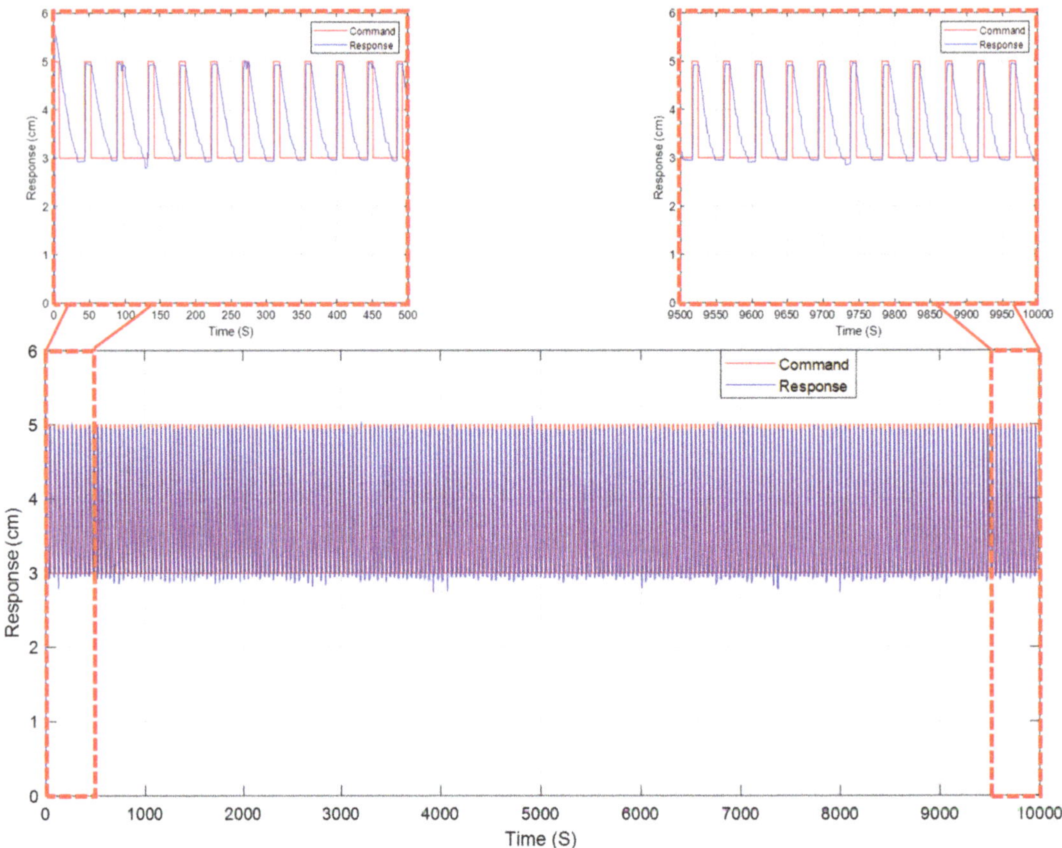

Figure 16. Long-duration displacement response of a square wave command signal.

Consequently, the force and displacement of the full-scale actuation mechanism were also tested to ensure that it can act as the actuation mechanism for the cleaning system. Figure 17 shows the hysteresis behavior of force, displacement, and a combined-case plots for the actuator. The carried-out results of the force test showed that the maximum pushing force was about 152.3 N at a temperature of 70.5 °C, while the maximum pulling force was 151.1 N. It can be found that a maximum bidirectional force of about 150 N can be produced. As for the maximum displacement, a stroke of about 127 mm at a temperature of 70.4 °C was recorded. Both force and displacement responses were plotted vs. temperature to extract the hysteresis behavior of the actuator, in which the mechanical work of the actuation mechanism was estimated of about 22.5 J.

The overall results of the dynamic appraisal have shown that the NiTiNOL springs used to convert the thermal energy into mechanical energy have a life expectancy of over 20 years, matching the life expectance of conventional PV modules. Additionally, experimental tests for the dynamic performance of the actuator showed that a bidirectional force over 150 N and a displacement of 127 mm could be produced, which achieves the desired design criteria for cleaning PV system.

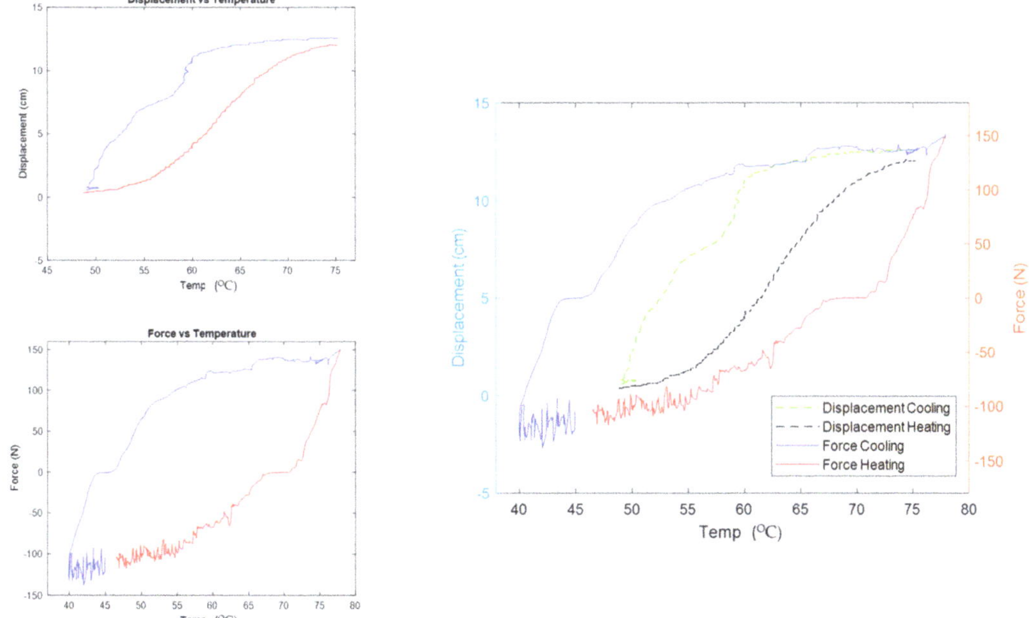

Figure 17. Displacement and force vs. temperature (hysteresis behavior).

3.4. Testing of the PV Cleaning System

The smart cleaning system has been manufactured and evaluated for a single full-scale PV module of 250 W, as seen in Figure 18. Multiple indoor experiments were carried out to test the effectiveness of the smart cleaning system, where a heat source of an applied hot air device was applied to the fabricated thermomechanical actuator. As a result, it was found that the generated movement of the actuation mechanism enabled the rotation of the cleaning spindle over the PV module surface to remove the accumulated dust particles as designed successfully.

Figure 18. Actual setup for the smart PV cleaning system.

Furthermore, different amounts of dust were placed over the PV module to study the cleaning effectiveness based on different dust densities. The cleaning effectiveness test showed a promising result, as seen in Figure 19a, compared to the cleaning effectiveness of various cleaning methods mentioned in the literature. The average cleaning effectiveness of the cleaning system under different dust densities was computed from the data highlighted in Figure 19b to be 95.14%. As presented in the figure, the dust density over a PV module increases and the cleaning effectiveness increases, showing a direct correlation. The direct relation between the dust density and the cleaning effectiveness led to the conclusion that the cleaning method presented is more effective after sandstorms because additional dust particles accumulate over the surface of PV modules during such environmental and harsh weather conditions. Thus, periodic cleaning of PV modules leads to continuous power generation and minimum power losses due to dust particles accumulating over the surface, which is dominant in high dust density regions such as the MENA region.

(a)

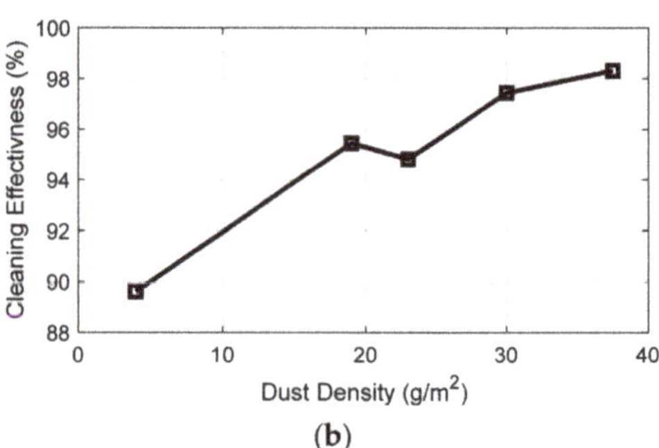

(b)

Figure 19. The cleaning effectiveness of the presented SMA-driven cleaning method under different dust densities, where (**a**) the actual performance of the self-cleaning operations, and (**b**) the quantitative outcome of the cleaning effectiveness percentages versus dust densities.

The outcomes show that the developed actuation mechanism can function throughout the year with minimum maintenance process for its operation. Additionally, the similar expected life expectancy of the SMA-based cleaning actuator compared to conventional PV modules adds enormous benefits for its real-life implementation with modern PV applications.

4. Conclusions

To conclude, the presented paper discusses the design and fabrication of a novel passive autonomous smart cleaning system to solve the dust accumulation issue facing PV systems around the globe, particularly in the MENA region. Multiple tests were conducted, including thermal analysis of different designs of a thermomechanical SMA actuator, its mechanical assessment, and cleaning effectiveness of the proposed innovative self-cleaning solution, which assure the promising feasibility and real-life applicability. It can be briefly summarized that this study presents:

- Solar-powered and effective PV cleaning systems are a first-of-its-kind novel approach to overcome the dust accumulation issue facing PV systems.
- Thermal analysis results prove that the standalone cleaning system can operate under actual weather conditions in the MENA region, Dammam city.
- The optimized design of the developed SHC shows a minimum maintenance routine required for a continuous full-year operation.
- A solar-based bidirectional actuator with a stroke of about 126 mm and maximum push and pull forces of about 152.3 N and 151.1 N, respectively, were successfully achieved.
- The actuator's thermal to mechanical conversion efficiency was recorded to be 19.15%, whereas the self-cleaning system has average cleaning effectiveness of 95%.

Future work could consider further thermal analyses of the novel cleaning system for several areas with different weather conditions to assess the feasibility of the design in different environments and worldwide. Furthermore, outdoor experiments are recommended to evaluate the designed cleaning system under real-world conditions. Lastly, a comprehensive energy comparison between a naturally cleaned PV module and a PV module equipped with the self-cleaning system is encouraged for further assessment and cost analysis.

Author Contributions: Conceptualization, N.G.H., I.K.A. and I.S.O.; methodology, I.S.O. and I.K.A.; software, I.K.A. and I.S.O.; validation, I.S.O., I.K.A. and N.G.H.; formal analysis, I.S.O., I.K.A. and N.G.H.; investigation, I.K.A., I.S.O. and N.G.H.; resources, I.K.A., I.S.O. and N.G.H.; data curation, I.K.A., I.S.O. and N.G.H.; writing—original draft preparation, I.K.A. and I.S.O.; writing—review and editing, N.G.H.; visualization, I.S.O., I.K.A. and N.G.H.; supervision, N.G.H.; project administration, N.G.H.; funding acquisition, I.S.O. All authors have read and agreed to the published version of the manuscript.

Funding: The publication is part of the project funded by the Deanship of Scientific Research (DSR) at Imam Abdulrahman Bin Faisal University under the project ID: 2022-003-Eng.

Acknowledgments: The authors are grateful to the Deanship of Scientific Research (DSR) at Imam Abdulrahman bin Faisal University, Kingdom of Saudi Arabia, for their continued guidance and financial support of this research project under the project ID (2022-003-Eng).

Conflicts of Interest: The authors declare no conflict of interest.

References

1. Almadani, I.K.; Osman, I.S.; Hariri, N.G. In-Depth Assessment and Optimized Actuation Method of a Novel Solar-Driven Thermomechanical Actuator via Shape Memory Alloy. *Energies* **2022**, *15*, 3807. [CrossRef]
2. Osman, I.S.; Hariri, N.G. Thermal Investigation and Optimized Design of a Novel Solar Self-Driven Thermomechanical Actuator. *Sustainability* **2022**, *14*, 5078. [CrossRef]
3. Amran, Y.A.; Amran, Y.M.; Alyousef, R.; Alabduljabbar, H. Renewable and sustainable energy production in Saudi Arabia according to Saudi Vision 2030; Current status and future prospects. *J. Clean. Prod.* **2020**, *247*, 119602. [CrossRef]
4. Osman, I.S.; Almadani, I.K.; Hariri, N.G.; Maatallah, T.S. Experimental Investigation and Comparison of the Net Energy Yield Using Control-Based Solar Tracking Systems. *Int. J. Photoenergy* **2022**, *2022*, 7715214. [CrossRef]
5. Hariri, N.G.; AlMutawa, M.A.; Osman, I.S.; AlMadani, I.K.; Almahdi, A.M.; Ali, S. Experimental Investigation of Azimuth- and Sensor-Based Control Strategies for a PV Solar Tracking Application. *Appl. Sci.* **2022**, *12*, 4758. [CrossRef]
6. Zsiborács, H.; Bai, A.; Popp, J.; Gabnai, Z.; Pályi, B.; Farkas, I.; Hegedűsné Baranyai, N.; Veszelka, M.; Zentkó, L.; Pintér, G.J.S. Change of real and simulated energy production of certain photovoltaic technologies in relation to orientation, tilt angle and dual-axis sun-tracking. A case study in Hungary. *Sustainability* **2018**, *10*, 1394. [CrossRef]
7. Gholami, A.; Mohammad, A.; Zandi, M.; Ghoachani, R.G. Dust accumulation on photovoltaic modules: A review on the effective parameters. *Sigma J. Eng. Nat. Sci.* **2021**, *39*, 45–57.
8. Syafiq, A.; Pandey, A.; Adzman, N.; Abd Rahim, N.J.S.E. Advances in approaches and methods for self-cleaning of solar photovoltaic panels. *Sol. Energy* **2018**, *162*, 597–619. [CrossRef]
9. Gupta, V.; Sharma, M.; Pachauri, R.K.; Dinesh Babu, K.N. Comprehensive review on effect of dust on solar photovoltaic system and mitigation techniques. *Sol. Energy* **2019**, *191*, 596–622. [CrossRef]
10. Said, S.A.; Hassan, G.; Walwil, H.M.; Al-Aqeeli, N.J.R.; Reviews, S.E. The effect of environmental factors and dust accumulation on photovoltaic modules and dust-accumulation mitigation strategies. *Renew. Sustain. Energy Rev.* **2018**, *82*, 743–760. [CrossRef]
11. Altıntaş, M.; Arslan, S. The Study of Dust Removal Using Electrostatic Cleaning System for Solar Panels. *Sustainability* **2021**, *13*, 9454. [CrossRef]
12. Kazem, H.A.; Chaichan, M.T.; Al-Waeli, A.H.; Sopian, K. A review of dust accumulation and cleaning methods for solar photovoltaic systems. *J. Clean. Prod.* **2020**, *276*, 123187. [CrossRef]
13. Farrokhi Derakhshandeh, J.; AlLuqman, R.; Mohammad, S.; AlHussain, H.; AlHendi, G.; AlEid, D.; Ahmad, Z. A comprehensive review of automatic cleaning systems of solar panels. *Sustain. Energy Technol. Assess.* **2021**, *47*, 101518. [CrossRef]
14. Deb, D.; Brahmbhatt, N.L. Review of yield increase of solar panels through soiling prevention, and a proposed water-free automated cleaning solution. *Renew. Sustain. Energy Rev.* **2018**, *82*, 3306–3313. [CrossRef]
15. Alghamdi, A.S.; Bahaj, A.S.; Blunden, L.S.; Wu, Y. Dust Removal from Solar PV Modules by Automated Cleaning Systems. *Energies* **2019**, *12*, 2923. [CrossRef]
16. Alnaser, N.W.; Al Othman, M.J.; Dakhel, A.A.; Batarseh, I.; Lee, J.K.; Najmaii, S.; Alothman, A.; Al Shawaikh, H.; Alnaser, W.E. Comparison between performance of man-made and naturally cleaned PV panels in a middle of a desert. *Renew. Sustain. Energy Rev.* **2018**, *82*, 1048–1055. [CrossRef]

17. Yadav, V.; Suthar, P.; Mukhopadhyay, I.; Ray, A. Cutting edge cleaning solution for PV modules. *Mater. Today Proc.* **2021**, *39*, 2005–2008. [CrossRef]
18. Chiteka, K.; Arora, R.; Sridhara, S.N.; Enweremadu, C.C. A novel approach to Solar PV cleaning frequency optimization for soiling mitigation. *Sci. Afr.* **2020**, *8*, e00459. [CrossRef]
19. Alvarez, D.L.; Al-Sumaiti, A.S.; Rivera, S.R. Estimation of an Optimal PV Panel Cleaning Strategy Based on Both Annual Radiation Profile and Module Degradation. *IEEE Access* **2020**, *8*, 63832–63839. [CrossRef]
20. Shah, A.H.; Hassan, A.; Laghari, M.S.; Alraeesi, A. The Influence of Cleaning Frequency of Photovoltaic Modules on Power Losses in the Desert Climate. *Sustainability* **2020**, *12*, 9750. [CrossRef]
21. Ölander, A. An electrochemical investigation of solid cadmium-gold alloys. *J. Am. Chem. Soc.* **1932**, *54*, 3819–3833. [CrossRef]
22. Vernon, L.B.; Vernon, H.M. Process of Manufacturing Articles of Thermoplastic Synthetic Resins. U.S. Patent 2,234,993, 18 March 1941.
23. Ruth, D.J.S.; Sohn, J.-W.; Dhanalakshmi, K.; Choi, S.-B.J. Control Aspects of Shape Memory Alloys in Robotics Applications: A Review over the Last Decade. *Sensors* **2022**, *22*, 4860. [CrossRef]
24. Baigonakova, G.; Marchenko, E.; Chekalkin, T.; Kang, J.-h.; Weiss, S.; Obrosov, A. Influence of Silver Addition on Structure, Martensite Transformations and Mechanical Properties of TiNi–Ag Alloy Wires for Biomedical Application. *Materials* **2020**, *13*, 4721. [CrossRef]
25. Braga, R.; Rodrigues, P.F.; Cordeiro, H.; Carreira, P.; Vieira, M.T. The Study of New NiTi Actuators to Reinforce the Wing Movement of Aircraft Systems. *Materials* **2022**, *15*, 4787. [CrossRef]
26. Benafan, O.; Moholt, M.; Bass, M.; Mabe, J.; Nicholson, D.; Calkins, F.J. Recent advancements in rotary shape memory alloy actuators for aeronautics. *Shape Mem. Superelast.* **2019**, *5*, 415–428. [CrossRef]
27. Chaudhari, R.; Vora, J.J.; Parikh, D.M. A review on applications of nitinol shape memory alloy. *Recent Adv. Mech. Infrastruct.* **2021**, 123–132.
28. Song, G.; Ma, N. Control of shape memory alloy actuators using pulse-width pulse-frequency (PWPF) modulation. *J. Intell. Mater. Syst. Struct.* **2003**, *14*, 15–22. [CrossRef]
29. Minorowicz, B.; Milecki, A. Design and Control of Magnetic Shape Memory Alloy Actuators. *Materials* **2022**, *15*, 4400. [CrossRef]
30. QADER, I.N.; Mediha, K.; DAĞDELEN, F.; ABDULLAH, S.S. The effect of different parameters on shape memory alloys. *Sak. Univ. J. Sci.* **2020**, *24*, 892–913. [CrossRef]
31. Farber, E.; Zhu, J.-N.; Popovich, A.; Popovich, V. A review of NiTi shape memory alloy as a smart material produced by additive manufacturing. *Mater. Today Proc.* **2020**, *30*, 761–767. [CrossRef]
32. Cimpoeşu, N.; Stanciu, S.; Vizureanu, P.; Cimpoeşu, R.; Achiţei, C.; Ioniţă, I. Obtaining shape memory alloy thin layer using PLD technique. *J. Min. Metall. Sect. B Metall.* **2014**, *50*, 69–76. [CrossRef]
33. Yuan, H.; Fauroux, J.C.; Chapelle, F.; Balandraud, X. A review of rotary actuators based on shape memory alloys. *J. Intell. Mater. Syst. Struct.* **2017**, *28*, 1863–1885. [CrossRef]
34. Costanza, G.; Tata, M.E. Shape memory alloys for aerospace, recent developments, and new applications: A short review. *Materials* **2020**, *13*, 1856. [CrossRef]
35. Hu, K.; Rabenorosoa, K.; Ouisse, M. A Review of SMA-Based Actuators for Bidirectional Rotational Motion: Application to Origami Robots. *Front. Robot. AI* **2021**, *8*, 678486. [CrossRef] [PubMed]
36. Suman, S.; Khan, M.K.; Pathak, M. Performance enhancement of solar collectors—A review. *Renew. Sustain. Energy Rev.* **2015**, *49*, 192–210. [CrossRef]
37. Shafieian, A.; Khiadani, M.; Nosrati, A. Strategies to improve the thermal performance of heat pipe solar collectors in solar systems: A review. *Energy Convers. Manag.* **2019**, *183*, 307–331. [CrossRef]
38. Lupu, A.; Homutescu, V.; Balanescu, D.; Popescu, A. Efficiency of solar collectors—A review. *IOP Conf. Ser. Mater. Sci. Eng.* **2018**, *444*, 082015. [CrossRef]
39. Tian, Y.; Zhao, C.Y. A review of solar collectors and thermal energy storage in solar thermal applications. *Appl. Energy* **2013**, *104*, 538–553. [CrossRef]
40. Zayed, M.E.; Zhao, J.; Du, Y.; Kabeel, A.E.; Shalaby, S.M. Factors affecting the thermal performance of the flat plate solar collector using nanofluids: A review. *Sol. Energy* **2019**, *182*, 382–396. [CrossRef]
41. Marčič, S.; Kovačič-Lukman, R.; Virtič, P. Hybrid system solar collectors-heat pumps for domestic water heating. *Therm. Sci.* **2019**, *23*, 3675–3685. [CrossRef]
42. Oztop, H.F.; Bayrak, F.; Hepbasli, A. Energetic and exergetic aspects of solar air heating (solar collector) systems. *Renew. Sustain. Energy Rev.* **2013**, *21*, 59–83. [CrossRef]
43. Ghritlahre, H.K.; Prasad, R.K. Application of ANN technique to predict the performance of solar collector systems—A review. *Renew. Sustain. Energy Rev.* **2018**, *84*, 75–88. [CrossRef]
44. Imtiaz Hussain, M.; Ménézo, C.; Kim, J.-T. Advances in solar thermal harvesting technology based on surface solar absorption collectors: A review. *Sol. Energy Mater. Sol. Cells* **2018**, *187*, 123–139. [CrossRef]
45. Bakari, R.; Minja, R.J.A.; Njau, K.N. Effect of Glass Thickness on Performance of Flat Plate Solar Collectors for Fruits Drying. *J. Energy* **2014**, *2014*, 247287. [CrossRef]
46. Al Hanai, T.; Hashim, R.B.; El Chaar, L.; Lamont, L.A. Environmental effects on a grid connected 900 W photovoltaic thin-film amorphous silicon system. *Renew. Energy* **2011**, *36*, 2615–2622. [CrossRef]

47. Chen, C.; Diao, Y.; Zhao, Y.; Wang, Z.; Liang, L.; Wang, T.; Zhu, T.; Ma, C. Thermal performance of a closed collector–storage solar air heating system with latent thermal storage: An experimental study. *Energy* **2020**, *202*, 117764. [CrossRef]
48. Mehrpooya, M.; Hemmatabady, H.; Ahmadi, M.H. Optimization of performance of Combined Solar Collector-Geothermal Heat Pump Systems to supply thermal load needed for heating greenhouses. *Energy Convers. Manag.* **2015**, *97*, 382–392. [CrossRef]
49. Budynas, R.G.; Nisbett, J.K. Gear-general. *Shigley's Mech. Eng. Des.* **2011**, 674–714.

Article

Microwave versus Conventional Sintering of NiTi Alloys Processed by Mechanical Alloying

Rodolfo da Silva Teixeira [1,*], Rebeca Vieira de Oliveira [2], Patrícia Freitas Rodrigues [3], João Mascarenhas [4], Filipe Carlos Figueiredo Pereira Neves [4] and Andersan dos Santos Paula [2]

1 Departamento de Engenharia de Materiais, Escola de Engenharia de Lorena, Universidade de São Paulo, Polo Urbo Industrial, Gleba AI-6, Lorena 12602-810, Brazil
2 Seção de Engenharia de Materiais, Instituto Militar de Engenharia (IME), Praça General Tibúrcio 80, Urca, Rio de Janeiro 22290-270, Brazil
3 University of Coimbra, Department of Mechanical Engineering, CEMMPRE, R. Luís Reis Santos, 3030-790 Coimbra, Portugal
4 Laboratório Nacional de Energia e Geologia (LNEG), Estrada do Paço do Lumiar, 22, 1649-038 Lisboa, Portugal
* Correspondence: rodolfoteixeira@live.com

Abstract: The present study shows a comparison between two sintering processes, microwave and conventional sintering, for the manufacture of NiTi porous specimens starting from powder mixtures of nickel and titanium hydrogenation–dehydrogenation (HDH) milled by mechanical alloying for a short time (25 min). The samples were sintered at 850 °C for 15 min and 120 min, respectively. Both samples exhibited porosity, and the pore size results are within the range of the human bone. The NiTi intermetallic compound (B2, R-phase, and B19′) was detected in both sintered samples through X-ray diffraction (XRD) and electron backscattering diffraction (EBSD) on scanning electron microscopic (SEM). Two-step phase transformation occurred in both sintering processes with cooling and heating, the latter occurring with an overlap of the peaks, according to the differential scanning calorimetry (DSC) results. From scanning electron microscopy/electron backscatter diffraction, the R-phase and B2/B19′ were detected in microwave and conventional sintering, respectively. The instrumented ultramicrohardness results show the highest elastic work values for the conventionally sintered sample. It was observed throughout this investigation that using mechanical alloying (MA) powders enabled, in both sintering processes, good results, such as intermetallic formation and densification in the range for biomedical applications.

Keywords: shape memory alloys; mechanical alloying and milling; microwave processing; mechanical properties; powder metallurgy

Citation: Teixeira, R.d.S.; Oliveira, R.V.d.; Rodrigues, P.F.; Mascarenhas, J.; Neves, F.C.F.P.; Paula, A.d.S. Microwave versus Conventional Sintering of NiTi Alloys Processed by Mechanical Alloying. *Materials* **2022**, *15*, 5506. https://doi.org/10.3390/ma15165506

Academic Editor: Salvatore Saputo

Received: 9 June 2022
Accepted: 4 July 2022
Published: 11 August 2022

1. Introduction

Technological advancements push research in the search for new materials and properties. So-called "intelligent" or "smart" materials are of great industrial interest because of their wide application due to their properties. In the class of intelligent materials, NiTi alloys stand out for their properties, of which their shape memory effect (SME) and superelasticity (SE) confer a singular value for these alloys [1,2]. NiTi alloys are most widely applied in the aerospace and automotive industries and in the manufacture of biomedical equipment and microdevices [3–6]. These alloys exhibit three phases in the matrix: the B2 austenite phase, the B19′ martensite phase, and the intermediate premartensitic R-phase. The austenitic phase has a polygonal/equiaxed grain microstructural appearance. In contrast, the martensitic phases, B19′ and R-phase, show a triangular morphology (composed of primary and secondary plates) and "herring-bone" appearance, respectively, where several martensitic phase nuclei originate and grow within each austenitic phase grain [7]. The phases present at room temperature depend on the composition of the alloy. The phase transformations may occur from B2↔ B19′ or with the presence of the intermediary R-phase (B2 ↔ R ↔ B19′) [7,8].

Bulk NiTi alloy production can occur via different routes that can be divided into two large groups: melting and powder metallurgy. Contamination issues during the melting processes can arise because of titanium's high reactivity with oxygen and carbon, resulting in the formation of oxides (Ti_4Ni_2O) and carbides (TiC) in the liquid state, and the formation of oxides (Ti_2O) in the solid state. This can be considered a relevant problem because additional phases change the functional properties of NiTi alloys [9,10].

In the case of the powder metallurgy route, there are many advantages compared to the melting processes, namely the possibility of large-scale association within dimension control, cost reduction, impurity reduction, and the feasibility of achieving mechanical properties in ranges of values such as those resulting from conventional metallurgical processes [11]. Mechanical alloying (MA) is an interesting powder metallurgy technique, involving repeated welding, fracturing, and rewelding of the powder particles. This technique allows homogeneous materials from the mixture of two or more elements or compounds [11–13]. Conventional sintering (CS) is the simplest form of sintering NiTi alloys. It is relatively inexpensive but usually requires a very long time. However, the main disadvantages of this method are associated with the formation of oxides and other secondary phases in the solid state, which alters the stoichiometric composition of the alloy, affecting the phase transformation temperature [7,14]. Microwave sintering (MW) is a recent method used for the densification of powders, resulting in bulk samples. Its advantages include high heating rates coupled with a low sintering time and reduced energy consumption compared to conventional sintering [15–18]. The synthesis of porous NiTi alloys is relevant for the manufacture of biomaterials for bone-replacement implants due to their properties, including low density, tissue in-growth ability (securing a firm fixation to the implants), and SME and SE [3,15,16]. It is complicated to evaluate the mechanical behavior of NiTi shape memory alloys (SMAs) due to their asymmetric transformation behavior [19,20]. Usually, the SME and SE are evaluated by conventional tests, for example, the uniaxial tensile, point bending, and compression tests [19–22]. However, asymmetric behavior is influenced by the grain size (nanoscale, microscale, and macroscale) and the nature of the stress–strain applied [19,20]. The authors of [23] propose that martensitic variants (MVs) are directly associated with asymmetric behavior.

The prime novelty of this research is the short milling time (25 min) in the mechanical alloying process of NiTi alloys, which provided the formation of a lamellar structure. The formation of this lamellar structure in such a short time has never been reported in the open literature for NiTi alloys processed by mechanical alloying. Additionally, in the two sintering processes subsequently evaluated, microwave and conventional sintering, the structure made the formation of a NiTi matrix possible, which was evaluated for its densification, microstructural effect, and micromechanical behavior under ultramicrohardness.

2. Materials and Methods

The raw materials used in this study were as follows: The nickel powder (median particle size ≈ 2 μm), supplied by JB Química (São Paulo/SP, Brazil), was obtained by oxi-reduction and had a near-spherical morphology. The titanium powder (particle size < 150.0 μm), donated by BRATS (São Paulo/SP, Brazil), was produced by the hydrogenation–dehydrogenation process (HDH) and had an irregular morphology. Figure 1 shows the schematic diagram of the process adopted in the present study.

The powders were mixed to obtain equiatomic NiTi single-phase alloy and milled in a planetary ball mill PM400 (Retsch, Haan, Germany) at 300 rpm for 25 min (after each 5 min of milling, the process was interrupted for 3 min to cool the vials). These experiments were carried out under an argon atmosphere at room temperature and using stainless steel vials (250 mL) and balls (balls with a diameter of 15 mm). The powder milling charge was 18 g, and the ball/powder weight ratio was 20:1.

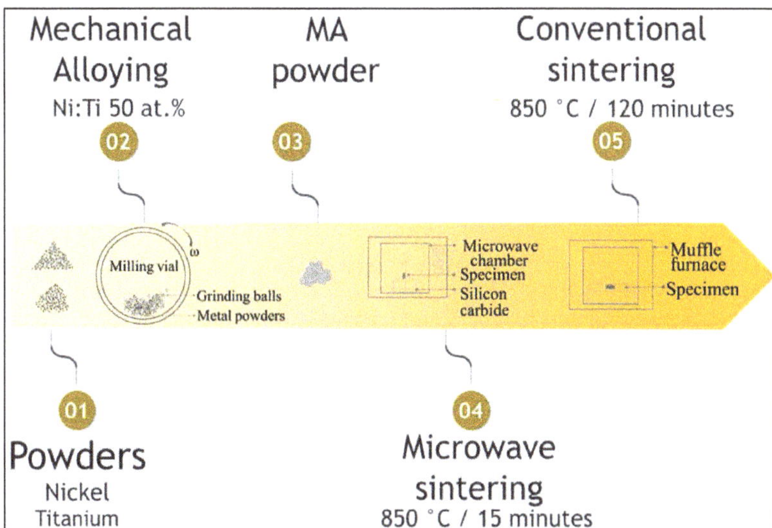

Figure 1. Schematic of the adopted mechanical alloying and sintering processes.

After milling, the samples were cold-pressed into green pellets (∅ 8 mm × 14 mm), using a uniaxial pressure of 53 MPa for 60 s, and then vacuum encapsulated in a quartz tube for the realization of the sintering. Microwave (MW) sintering was performed at a target temperature of 850 °C for 15 min and a rate of 50 °C/min in an MW oven from Microwave Research & Applications, Inc. (Carol Stream, IL, USA), operating at 2.45 GHz with a power of 1 kW. The temperature was measured by a platinum-shielded S-type thermocouple. Conventional sintering (CS) was performed in a muffle furnace, a Quimis model, for 120 min at the same temperature and heating rate. The conventional sintered samples were then quenched in water at room temperature.

The specimen's density and porosity were measured by the Archimedes method, based on standard ABNT NBR 16661:2017. The pore size examinations were carried out using optical microscopy (Olympus, Tokyo, Japan). The average porosity was measured using five optical micrographs taken from the cross-sections of the sintered samples.

Differential thermal analysis (DTA) was performed on the MA-processed powders for 25 min (SETARAM, Provence-Alpes-Cote d'Azur, France). The heating/cooling cycles comprised a temperature range from 22 to 1050 °C (heating/cooling rates of 10 K·min^{-1}) under an argon atmosphere.

Differential scanning calorimetry (DSC) was carried out under a DSC-60 (Shimadzu, Kyoto, Japan) fitted with a LN_2 cooling attachment and inert gas (N_2) atmosphere. The heating/cooling cycles had temperatures of 150 °C and −150 °C (heating/cooling rate of 10 K·min^{-1}), and the phase transformation temperatures extracted from the DSC curves were obtained by the tangent method.

X-ray diffraction (XRD) patterns were recorded in a diffractometer (PANalytical, Worcestershire, UK) with a cobalt anode, operated at 40 kV and 40 mA. Scanning electron microscopy (SEM) analysis was performed using a scanning electronic microscope (SEM) (FEI, Hillsboro, OR, USA) with secondary electron (ETD), backscattered electron (BSE), and electron backscattered diffraction (EBSD) detectors.

Instrumented ultramicrohardness analysis was conducted with a Berkovich indenter (Shimadzu DUH-211S, Kyoto, Japan) using the load–unload cycle method and two different maximum load levels: 1.0 gf/9.81 mN and 20.0 gf/196.1 mN. In each sample, ten tests were performed from the longitudinal cross-section of the sintered samples along the ½ diameter.

The XRD, SEM/EBSD, and instrumented ultramicrohardness specimens, after being cut by a precision cutting machine, were prepared by conventional mechanical grinding followed by electrochemical polishing at room temperature using an electrolyte of H_2SO_4 (20%) and CH_3OH (80%) at 30 V. To reveal the phases, the samples were immersed for 60 s in the following etching solution: 50 mL glycerin ($C_3H_8O_3$), 18 mL concentrated acetic acid ($C_2H_4O_2$), 20 mL concentrated nitric acid (HNO_3), and 16 mL concentrated hydrofluoric acid (HF).

3. Results

3.1. Mechanical Alloying

The starting morphology of the powders is shown in Figure 2. After the MA process, a lamellar structure was formed, as shown in Figure 2c. The presence of this structure can promote increases in the diffusion processes [24–26]. This was a remarkable result; lamellar structure has only been observed in the literature for milling times ranging from 4 to 180 h [27–29]. However, even with long times, it is not guaranteed to obtain a lamellar structure because there are other process parameters that must be controlled, as has been observed in [24,28,30].

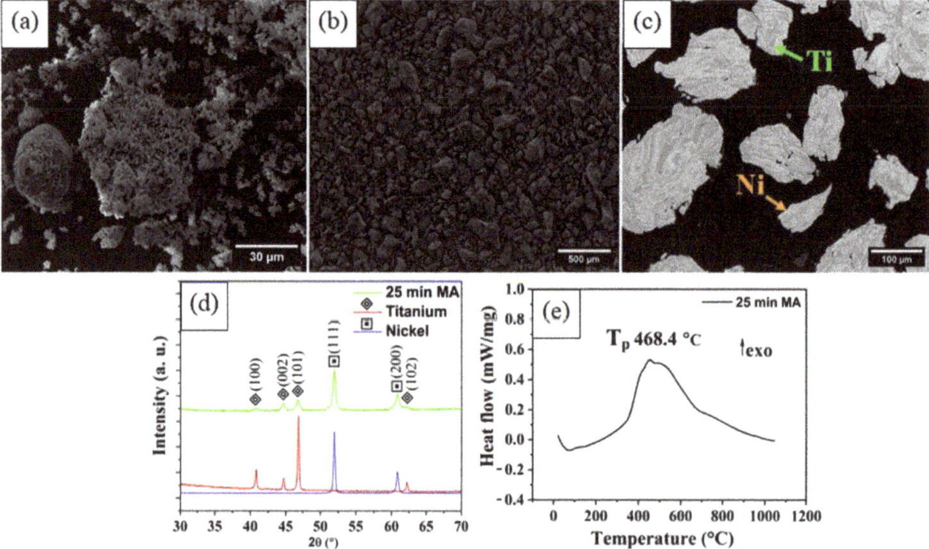

Figure 2. SEM morphology of the elementary powders: (**a**) nickel and (**b**) titanium. (**c**) Micrography of the MA-processed powders. (**d**) XRD patterns of the starting powders and the MA-processed powders. (**e**) DTA heating curve of MA-processed powders.

Figure 2d presents the XRD patterns of the starting powders and MA-processed powders. This analysis reveals that no intermetallic phase was detected and only peaks from the nickel and titanium initial powders were recorded. It is worth mentioning that XRD has a detection limit that depends on the equipment [31–33], meaning that if any intermetallic phase is formed below the detection limit, it will not be significant in the face of the detected elements.

The DTA curve is shown in Figure 2e, where the effect of the MA-process was analyzed. The curve shows only one peak, corresponding to the reactions between the nickel and titanium to the formation of the intermetallic compound of the Ni-Ti system. As each powder particle contained nickel and titanium in the form of a lamellar structure, the reaction was slow and had low energy. As a consequence, the DTA peak has low intensity and is very broad [34]. The temperature of the peak (T_p) was determined to be 468.4 °C (Figure 2e), it should be noted that in a mixture as it is (without MA), this reaction occurs at a higher temperature (±950 °C) and with much higher intensity [35]. This temperature differs significantly from the previous results reported in the literature, as it is lower than those obtained for milling times of 4–20 h [29,35,36].

3.2. Sintering Characterization

Table 1 summarizes the density, porosity, and pore size values obtained for the MW and CS samples. For comparison purposes, Table 1 also shows some results found in the literature [16,37].

Table 1. Density, porosity, and pore size values of the porous NiTi alloys prepared by MW and CS, and from the literature results.

Sample	Density (g/cm^3)	Porosity (%)	Pore Size (μm)
MW	3.31 ± 0.02	51.3 ± 0.29	120 ± 13.84
CS	3.76 ± 0.06	58.2 ± 0.98	155 ± 13.01
* −200 mesh	—	51.9 ± 1.21	97 ± 3.51
* −120 mesh	—	51.5 ± 1.08	149 ± 7.30
* −60 mesh	—	50.1 ± 0.95	238 ± 7.10
* −45 mesh	—	52.9 ± 1.13	294 ± 9.95
** 0 wt.%	5.10 ± 0.20	22.0 ± 0.31	≈26 ± 2.76
** 10 wt.%	3.64 ± 0.02	42.0 ± 0.59	≈120 ± 5.52
** 20 wt.%	3.20 ± 0.02	51.0 ± 0.59	≈151 ± 5.10
** 30 wt.%	2.40 ± 0.60	64.0 ± 1.18	≈178 ± 2.50

*, ** Results from [16] and [37], respectively.

According to the literature, human bone shows a density of 1.8–2.1 g/cm^3, the ideal porosity is in the range of 30–90% and the optimal pore size is between 100 μm and 500 μm [16,38,39]. Thus, the MW and CS samples prepared in the present study exhibited porosity and pore size values within the range of the human bone but higher densities. However, the values obtained for the CS sample are higher than the ones obtained for the MW sample.

The MW sample showed values close to those found in the literature for porous NiTi alloys also prepared by MW sintering [16,37] (Table 1). However, ref. [16] used a pore size controller (sieved pure NH_4HCO_3) and tested different mesh pore sizes. In that study, the porosity results were similar for all samples independently of the initial mesh of the pore size controller. The pore size values obtained in the present study are higher than the ones obtained for the sample sieved to 200 mesh by [16] but smaller than the others.

The authors of [37] used different concentrations of NH_4HCO_3 while keeping the mesh size at 250 (Table 1). When comparing the MW sample sintered in the present study, similar density and pore size values were observed for the sample with 20 wt.% NH_4HCO_3 (Table 1). However, when compared to the sample without pore size controllers (0 wt.%) of reference [16], it is seen that the density is smaller (3.31 ± 0.02 g/cm^3 vs. 5.1 ± 0.20 g/cm^3) while the porosity (51.3 ± 0.29% vs. 22 ± 0.31%) and the pore size (120 ± 13.84 μm vs. 26 ± 2.76 μm) are higher. It is evident that the MA processing promoted alterations in these properties. The formation of the lamellar structure (Figure 2c) provided the acceleration of the diffusion process during sintering and more uniform porous specimens, as has been previously investigated in literature [40,41].

The DSC curves are shown in Figure 3. The dashed line represents the phase present at room temperature (20.0 °C). In the DSC curve of the MW sample (Figure 3a), during heating, a two-stage martensitic transformation (B2 → R → B19′) was observed. During the cooling, as has been noted, an overlap of the DSC peaks was observed with two-step transformations (B2 → R → B19′) [7,13,42–46]. Regarding the CS sample of the DSC curve, (Figure 3b), equal behavior was also noted.

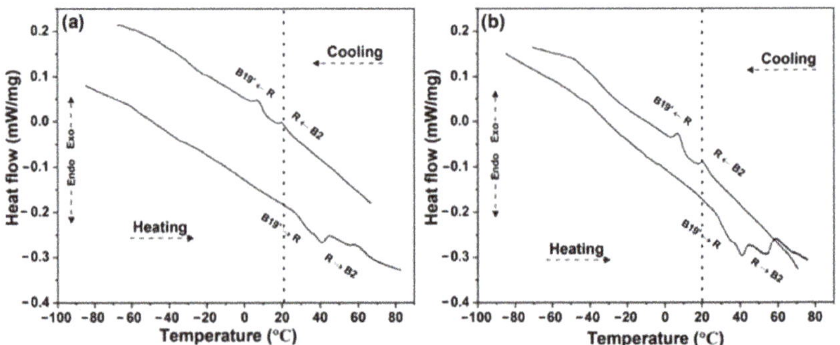

Figure 3. DSC curves showing the phase transformations for (**a**) MW and (**b**) CS samples. Dot line: room temperature (20 °C).

The transformation temperatures and enthalpy of the MW and CS samples are listed in Table 2. The R_s and A_f temperatures were above room temperature (20 °C) in both sintering processes. During the cooling and heating, a significant variation in the transformation temperatures was not identified, as they did not surpass one degree Celsius during both sintering processes.

Table 2. Phase transformation temperatures in degrees Celsius and enthalpy of MW and CS samples. "-" Not detected; "*" undefined.

Sample	Cooling								
	B2 → R			B2 → B19′			R → B19′		
	R_s	R_p	R_f	M'_s	M'_p	M'_f	M_s	M_p	M_f
CS	25.6	20.7	17.8	-	-	-	11.7	7.4	2.9
MW	24.9	19.9	17.7	-	-	-	11.5	7.4	3.4
	Heating								
Sample	**B19′→ R**			**B19′→ B2**			**R→ B2**		
	R'_s	R'_p	R'_f	A'_s	A'_p	A'_f	A_s	A_p	A_f
CS	25.8	*	*	*	*	*	*	*	58.3
MW	27.4	*	*	*	*	*	*	*	57.6
	Enthalpy (mJ/g)								
Sample	**First Peak on Cooling**		**Second Peak on Cooling**		**First Peak on Heating**		**Second Peak on Heating**		
CS	1.329		0.856				11.891		
MW	0.279		0.769				11.164		

* It was not possible to accurately determine the transformation temperatures due to overlapping peaks.

Few studies have observed similar behavior. For example, in [37], two-step martensitic transformations in microwave sintering were identified, and compared to this study, the variations in the R_s and A_f temperatures were approximately 8 °C lower. However, ref. [47] observed a multi-step transformation during the heating and a two-step transformation during the cooling, in the temperature range of 850 to 1000 °C/1 h (variation of 50 °C). Those authors mixed powders by MA for 3 h.

An increase in enthalpy in the CS sample was observed when compared to the MW sample at both peaks of the cooling and heating curves. As is known, the enthalpy changes as a function of the nickel atomic content [48], and the presence of $TiNi_3$ in the MW sample is the reason for the low values of enthalpy.

The XRD patterns of the MW and CS samples are shown in Figure 4. The NiTi phases (B2, R-phase, and B19′) were identified in both samples (Figure 4a), although the R-phase in the CS sample only had one peak with overlap for B2 (1 0 0). In addition to the NiTi phases, $TiNi_3$ was also recorded in the MW sample (Figure 4a), and Ti_2Ni and Ti_3Ni_4 in the CS sample (Figure 4b). The XRD results show that the MW sample had more peaks for the R-phase and B19′. In contrast, the CS sample had more for B2 and B19′. These results are not exactly what was expected from the DSC results.

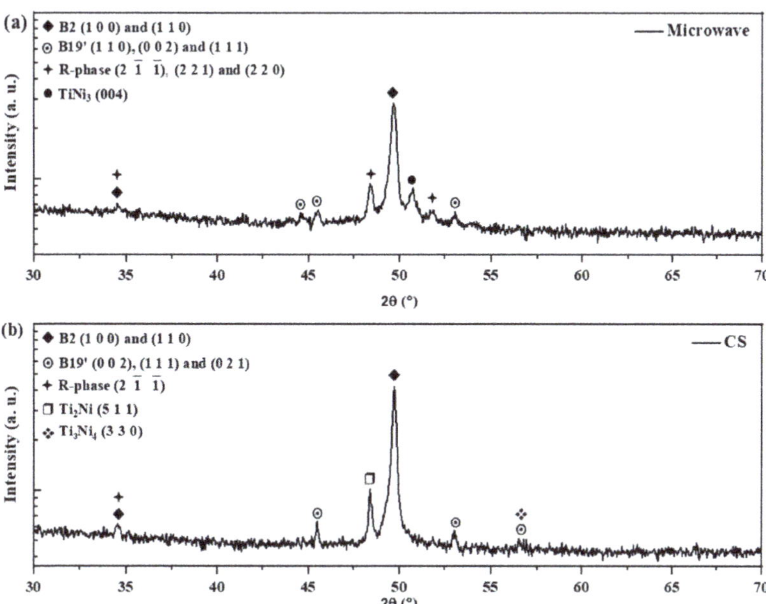

Figure 4. XRD patterns (**a**) MW and (**b**) CS sample.

Nevertheless, these results are an improvement when compared to those of other studies related to the MW sintering of NiTi alloys. For example, in reference [15], the formation of NiTi intermetallic phases was not observed after MW sintering at 850 °C and 950 °C of the cold-pressed powders blended in a rotating mixer. The justification for such behavior can possibly be associated with the MA process that, as mentioned previously, led to the formation of a lamellar structure, (Figure 2c), which can promote an increase in the diffusion processes. In the CS sample, in addition to the NiTi phase, only Ti_2Ni and precipitated Ti_3Ni_4 were recorded as secondary phases. This result could be associated with the high sintering time of CS (120 min vs. 15 min).

MA enabled the modification of the diffusion process that occurs in the MW and CS processes by promoting the formation of the B2, R-phase, and B19′ phases and by avoiding the formation of a second phase in the CS sample ($TiNi_3$) and MW sample (Ti_2Ni and Ti_4Ni_3). In the literature, it is reported that intermetallic $TiNi_3$ and Ti_2Ni form when pure nickel and titanium powder mixtures are submitted to conventionally sintered samples [7,10].

SEM micrographs of the MW and CS sintered samples before etching are shown in Figure 5a,b. It can be observed that the MW sample (Figure 5a) shows more uniform pores compared to the CS sample (Figure 5b). Considering the short milling time and the relatively low sintering temperature, the obtained results are truly relevant compared to those in the literature [15,47].

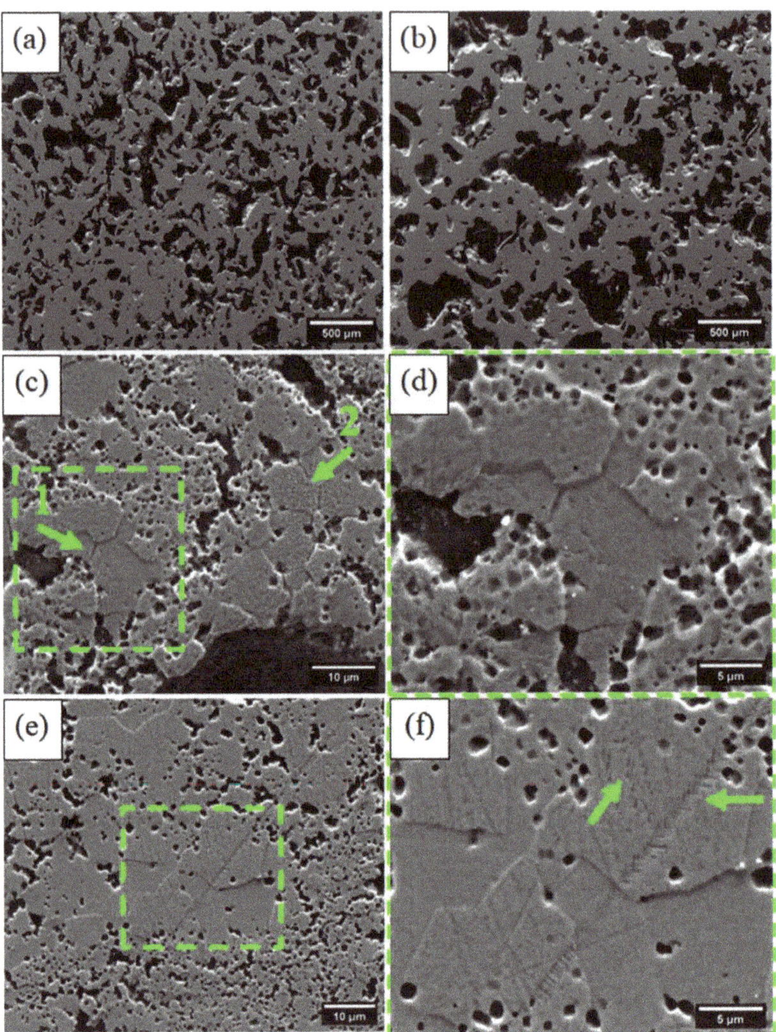

Figure 5. SEM micrographs of the sample porous NiTi: (**a**) MW sample and (**b**) CS sample before etching. (**c–f**) After etching to reveal the possible phases. MW sample indicated by arrows 1 and 2: (**c**) general aspect and (**d**) magnification from arrow 1; CS sample: (**e**) general aspect and (**f**) magnification from the square, showing the martensitic phase, indicated by the arrows.

After metallographic etching, the previous B2 boundaries were identified in both sintered samples (Figure 5c–f), as highlighted in the green squares. In a single B2 grain, a significant amount of the B19′ and R-phase could be formed, preserving the previous B2 boundary [7,49–53]. In addition, the B19′ was also identified in the CS sample (Figure 5f). These results corroborate the XRD analyses (Figure 4).

The EBSD results are shown in Figure 6. In the image quality (IQ) map, a different microstructure between the MW (Figure 6a) and CS samples (Figure 6b) can be observed. Nevertheless, for the MW sample (Figure 6a), there is a possibility of a signal mixture of the R-phase variants multiple grains formed within a given austenite grain. The phase map results show that the MW sample (Figure 6c) had a higher amount of R-phase and the CS sample (Figure 6d) had a higher amount of B2 and B19′. These results are consistent with those observed in the XRD analysis.

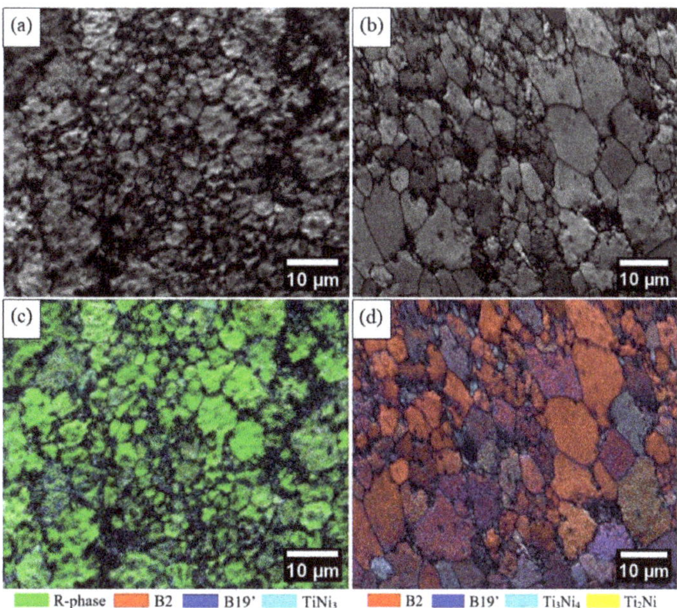

Figure 6. The image quality (IQ) map: (**a**) MW and (**b**) CS sample. EBSD phase maps + IQ map: (**c**) MW sample with the R-phase in green, B2 in red, B19′ in blue, and $TiNi_3$ in cyan; (**d**) CS sample with B2 in red, B19′ in blue, Ti_3Ni_4 in cyan, and Ti_2Ni in yellow.

The force vs. depth curves (loading and unloading) of the MW and CS sintered samples are shown in Figure 7. Generically, the curves exhibit heterogeneous behavior. However, one curve of the MW sample, highlighted in Figure 7a, displays a significant elastic return (until 0.02 μm), which indicate a possible SE effect. This result is supported by the XRD (Figure 4) and SEM (Figure 5) analyses, which showed B2 presence. However, the non-complete return possibly indicates that the pressure exerted by the load maximum reached the plastic deformation region by screw dislocations, as noted in the study of [54], who proved stress-induced martensitic transformation during a nanoindentation test of NiTi alloys (superelastic). Nevertheless, each of these measures mechanically requests a small number of grains for the indentation and deformed volume around it, and with this, each curve can translate the mechanical behavior in the function of the crystallographic orientation of these grains [54,55]. In addition, the nature of the strain–stress applied has influence on the asymmetric transformation behavior [19,20].

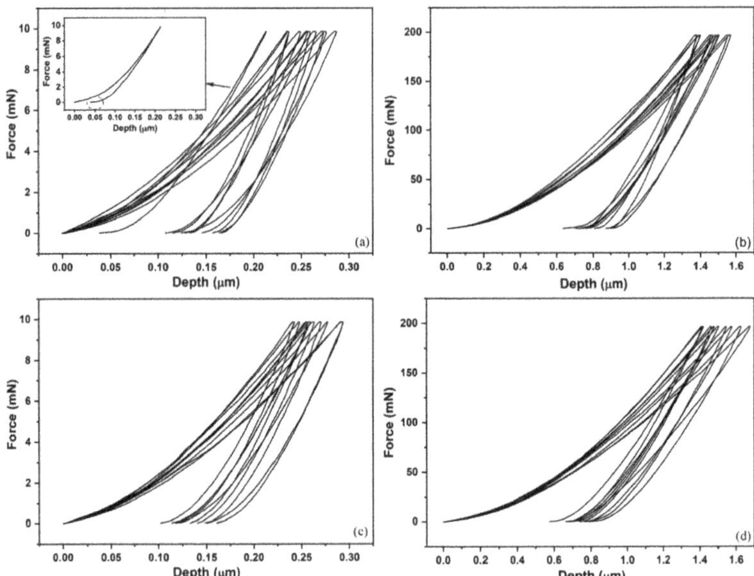

Figure 7. Force vs. depth (loading and unloading curves). (**a**,**b**) MW sample and (**c**,**d**) CS sample; maximum forces of 1.0 gf/9.81 mN and 20.0 gf/196.1 mN, respectively.

The results of the total work, elastic work, and plastic work, shown in Figure 8, indicate that the higher average and maximum force values were obtained for the CS sample. This is consistent with what was expected from the XRD and EBSD analyses, in which a B2 majority was recorded.

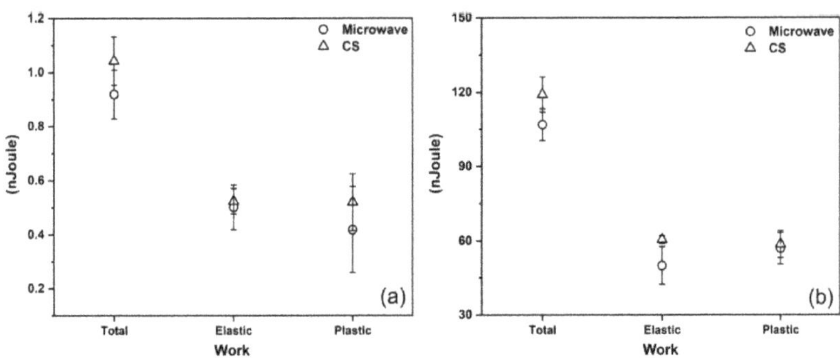

Figure 8. Total, elastic, and plastic works: maximum forces of (**a**) 1.0 gf/9.81 mN and (**b**) 20.0 gf/196.1 mN.

4. Conclusions

The comparative sintering study starting from MA powders milled for 25 min allowed us to conclude that:

- The density value obtained for the MW sintered sample (3.31 $\pm$ 0.02 g/cm^3) was closer to human bone (1.8–2.1 g/cm^3) than the density value obtained for the CS (3.76 $\pm$ 0.06 g/cm^3). The results of porosity and pore size for both processes exhibited results within the range of human bone;

- When compared to CS, the overall results obtained from combining MA and MW sintering were very promising, as it allowed for the formation of intermetallic NiTi (B2, R-phase, and B19′) in a very short time and the formation of homogeneous pores;
- Two-step transformations were detected on the heating and cooling curves for both sintering processes, with an overlapping peak on the cooling curves, probably due to the presence of a second phase and the intermediate premartensitic R-phase. No significant differences in the transformation temperatures were detected between the sintering methods;
- The EBSD results show that the MW sample had a higher amount of R-phase and the CS sample had higher amounts of B2 and B19′;
- Through instrumented ultramicrohardness analysis, heterogeneous behavior could be identified based on the curves of both samples, as expected. In MW's curves, one curve was observed with a significant elastic return. The CS sample showed the highest values of elastic work compared to the MW sample at both maximum forces.

Author Contributions: R.d.S.T.: Investigation, Visualization, Writing—original draft, Writing—review and editing. R.V.d.O.: Investigation, Writing—review and editing. P.F.R.: Investigation, Writing—review and editing. J.M.: Methodology, Validation, Writing—review and editing. F.C.F.P.N.: Supervision, Methodology, Validation, Writing—review and editing, Data curation, Visualization. A.d.S.P.: Supervision, Validation, Writing—review and editing. All authors have read and agreed to the published version of the manuscript.

Funding: This research received no external funding.

Institutional Review Board Statement: Not applicable.

Informed Consent Statement: Not applicable.

Data Availability Statement: Not applicable.

Acknowledgments: The authors acknowledge BRATS, in Cajamar/SP, Brazil, for donating the titanium HDH powders, and Ladário da Silva and José Augusto Oliveira Huguenin, who allowed the use of the Multiuser Laboratory of Materials Characterization of the Institute of Exact Sciences of the Universidade Federal Fluminense (ICEx/UFF) in Volta Redonda/RJ, Brazil, and the use of the instrumented ultramicrohardness equipment for the analysis of the samples under study. Teixeira, R.S. acknowledge the funding of Capes by PhD Scholarship Program. PFR acknowledge the funding of CEMMPRE by Project PTDC/CTM-CTM/29101/2017—POCI-01-0145-FEDER-029101 funded by FEDER funds through COMPETE2020—"Programa Operacional Competitividade e Internacionalização" (POCI) and by national funds (PIDDAC) through FCT/MCTES. This research is sponsored by FEDER funds through the program COMPETE—"Programa Operacional Factores de Competitividade"—and by national funds through FCT—Fundação para a Ciência e a Tecno-logia—Portugal, under the project UIDB/00285/2020. The funding of European Program IRSES—Marie Curie PIRSES GA-2013-612585 (under the MIDAS project—"Micro and Nanoscale Design of Thermally Actuating Systems").

Conflicts of Interest: The authors declare no conflict of interest.

References

1. Velmurugan, C.; Senthilkumar, V.; Dinesh, S.; Arulkirubakaran, D. Review on phase transformation behavior of NiTi shape memory alloys. *Mater. Today Proc.* **2018**, *5*, 14597–14606. [CrossRef]
2. Farber, E.; Zhu, J.N.; Popovich, A.; Popovich, V. A review of NiTi shape memory alloy as a smart material produced by additive manufacturing. *Mater. Today Proc.* **2019**, *30*, 761–767. [CrossRef]
3. Patel, S.K.; Behera, B.; Swain, B.; Roshan, R.; Sahoo, D.; Behera, A. A review on NiTi alloys for biomedical applications and their biocompatibility. *Mater. Today Proc.* **2020**, *33*, 5548–5551. [CrossRef]
4. Miyazaki, S.; Duerig, T.W.; Melton, K.N. *Engineering Aspects of Shape Memory Alloys*; Duerig, T.W., Melton, K.N., Stockel, D., Wayman, C.M., Eds.; Butterworth-Heinenmann: Guildford, UK, 1990; pp. 394–413. [CrossRef]
5. Hartl, D.J.; Lagoudas, D.C. Aerospace applications of shape memory alloys. *Proc. Inst. Mech. Eng. Part G J. Aerosp. Eng.* **2007**, *221*, 535–552. [CrossRef]
6. Jani, J.M.; Leary, M.; Subic, A. Shape memory alloys in automotive applications. *Appl. Mech. Mater.* **2014**, *663*, 248–253. [CrossRef]
7. Otsuka, K.; Ren, X. Physical metallurgy of Ti-Ni-based shape memory alloys. *Prog. Mater. Sci.* **2005**, *50*, 511–678. [CrossRef]
8. Otsuka, K.; Wayman, C.M. *Shape Memory Materials*; Cambridge University Press: Cambridge, UK, 1998.

9. Bram, M.; Ahmad-Khanlou, A.; Heckmann, A.; Fuchs, B.; Buchkremer, H.P.; Stöver, D. Powder metallurgical fabrication processes for NiTi shape memory alloy parts. *Mater. Sci. Eng. A* **2002**, *337*, 254–263. [CrossRef]
10. Elahinia, M.H.; Hashemi, M.; Tabesh, M.; Bhaduri, S.B. Manufacturing and processing of NiTi implants: A review. *Prog. Mater. Sci.* **2012**, *57*, 911–946. [CrossRef]
11. El-Eskandarany, M.S. The history and necessity of mechanical alloying. *Mech. Alloy.* **2015**, *2*, 13–47. [CrossRef]
12. Farvizi, M. Challenges of Using Elemental Nickel and Titanium Powders for the Fabrication of Monolithic NiTi Parts. *Arch. Metall. Mater.* **2017**, *62*, 1075–1079. [CrossRef]
13. Suryanarayana, C. Mechanical alloying and milling. *Prog. Mater. Sci.* **2001**, *46*, 1–184. [CrossRef]
14. Elahinia, M.; Shayesteh Moghaddam, N.; Taheri Andani, M.; Amerinatanzi, A.; Bimber, B.A.; Hamilton, R.F. Fabrication of NiTi through additive manufacturing: A review. *Prog. Mater. Sci.* **2016**, *83*, 630–663. [CrossRef]
15. Xu, J.L.; Jin, X.F.; Luo, J.M.; Zhong, Z.C. Fabrication and properties of porous NiTi alloys by microwave sintering for biomedical applications. *Mater. Lett.* **2014**, *124*, 110–112. [CrossRef]
16. Xu, J.L.; Bao, L.Z.; Liu, A.H.; Jin, X.F.; Luo, J.M.; Zhong, Z.C.; Zheng, Y. Effect of pore sizes on the microstructure and properties of the biomedical porous NiTi alloys prepared by microwave sintering. *J. Alloy. Compd.* **2015**, *645*, 137–142. [CrossRef]
17. Dong, B.; Xiao, Y.; Xu, F.; Hu, X.; Liu, W.; Wu, X. Quantitative tomography of pure magnetic-induced effects on metallics during microwave sintering. *J. Alloy. Compd.* **2018**, *749*, 103–112. [CrossRef]
18. Oghbaei, M.; Mirzaee, O. Microwave versus conventional sintering: A review of fundamentals, advantages and applications. *J. Alloy. Compd.* **2010**, *494*, 175–189. [CrossRef]
19. Chen, X.; Chen, W.; Ma, Y.; Zhao, Y.; Deng, C.; Peng, X.; Fu, T. Tension-Compression asymmetry of single-crystalline and nanocrystalline NiTi shape memory alloy: An atomic scale study. *Mech. Mater.* **2020**, *145*, 103402. [CrossRef]
20. Kato, H. Four-point bending tests to reveal tension-compression flow stress asymmetry in NiTi shape memory alloy thin plate. *Mater. Sci. Eng. A* **2019**, *755*, 258–266. [CrossRef]
21. Wang, J.; Pan, Z.; Carpenter, K.; Han, J.; Wang, Z.; Li, H. Comparative study on crystallographic orientation, precipitation, phase transformation and mechanical response of Ni-rich NiTi alloy fabricated by WAAM at elevated substrate heating temperatures. *Mater. Sci. Eng. A* **2021**, *800*, 140307. [CrossRef]
22. Bimber, B.A.; Hamilton, R.F.; Keist, J.; Palmer, T.A. Anisotropic microstructure and superelasticity of additive manufactured NiTi alloy bulk builds using laser directed energy deposition. *Mater. Sci. Eng. A* **2016**, *674*, 125–134. [CrossRef]
23. Šittner, P.; Novák, V. Anisotropy of martensitic transformations in modeling of shape memory alloy polycrystals. *Int. J. Plast.* **2000**, *16*, 1243–1268. [CrossRef]
24. Novák, P.; Moravec, H.; Salvetr, P.; Průša, F.; Drahokoupil, J.; Kopeček, J.; Karlík, M.; Kubatík, T.F. Preparation of nitinol by non-conventional powder metallurgy techniques. *Mater. Sci. Technol.* **2015**, *31*, 1886–1893. [CrossRef]
25. Nobuki, T.; Crivello, J.C.; Cuevas, F.; Joubert, J.M. Fast synthesis of TiNi by mechanical alloying and its hydrogenation properties. *Int. J. Hydrog. Energy* **2019**, *44*, 10770–10776. [CrossRef]
26. Nováka, P.; Průša, F.; Nová, K.; Bernatiková, A.; Salvetr, P.; Kopeček, J.; Haušild, P. Application of mechanical alloying in synthesis of intermetallics. *Acta Phys. Pol. A* **2018**, *134*, 720–723. [CrossRef]
27. Rostami, A.; Bagheri, G.A.; Sadrnezhaad, S.K. Microstructure and thermodynamic investigation of Ni–Ti system produced by mechanical alloying. *Phys. B Condens. Matter* **2019**, *552*, 214–220. [CrossRef]
28. Mousavi, T.; Karimzadeh, F.; Abbasi, M.H. Synthesis and characterization of nanocrystalline NiTi intermetallic by mechanical alloying. *Mater. Sci. Eng. A* **2008**, *487*, 46–51. [CrossRef]
29. Zhou, Y.; Li, C.J.; Yang, G.J.; Wang, H.D.; Li, G. Effect of self-propagating high-temperature combustion synthesis on the deposition of NiTi coating by cold spraying using mechanical alloying Ni/Ti powder. *Intermetallics* **2010**, *18*, 2154–2158. [CrossRef]
30. Novák, P.; Moravec, H.; Vojtìch, V.; Knaislová, A.; Školáková, A.; Kubatík, T.F.; Kopeček, J. Powder-metallurgy preparation of NiTi shape-memory alloy using mechanical alloying and spark-plasma sintering. *Mater. Tehnol.* **2017**, *51*, 141–144. [CrossRef]
31. Kang, D.H.; Kim, J.H.; Park, H.; Yoon, K.H. Characteristics of (Pb 1-x Sr x)TiO3 thin film prepared by a chemical solution processing. *Mater. Res. Bull.* **2001**, *36*, 265–276. [CrossRef]
32. Koorösy, G.; Tomolya, K.; Janovszky, D.; Sólyom, J. Evaluation of XRD analysis of amorphous alloys. *Mater. Sci. Forum* **2012**, *729*, 419–423. [CrossRef]
33. Khan, H.; Yerramilli, A.S.; D'Oliveira, A.; Alford, T.L.; Boffito, D.C.; Patience, G.S. Experimental methods in chemical engineering: X-ray diffraction spectroscopy—XRD. *Can. J. Chem. Eng.* **2020**, *98*, 1255–1266. [CrossRef]
34. Neves, F.; Martins, I.; Correia, J.B.; Oliveira, M.; Gaffet, E. Reactive extrusion synthesis of mechanically activated Ti-50Ni powders. *Intermetallics* **2007**, *15*, 1623–1631. [CrossRef]
35. Neves, F.; Braz Fernandes, F.M.; Correia, J.B. Effect of mechanical activation on Ti-50Ni powder blends reactivity. *Mater. Sci. Forum* **2010**, *636–637*, 544–549. [CrossRef]
36. Ye, L.L.; Liu, Z.G.; Raviprasad, K.; Quan, M.X.; Umemoto, M.; Hu, Z.Q. Consolidation of MA amorphous NiTi powders by spark plasma sintering. *Mater. Sci. Eng. A* **1997**, *241*, 290–293. [CrossRef]
37. Xu, J.L.; Bao, L.Z.; Liu, A.H.; Jin, X.J.; Tong, Y.X.; Luo, J.M.; Zhong, Z.; Zheng, Y. Microstructure, mechanical properties and superelasticity of biomedical porous NiTi alloy prepared by microwave sintering. *Mater. Sci. Eng. C* **2015**, *46*, 387–393. [CrossRef]
38. Kaya, M.; Orhan, N.; Tosun, G. The effect of the combustion channels on the compressive strength of porous NiTi shape memory alloy fabricated by SHS as implant material. *Curr. Opin. Solid State Mater. Sci.* **2010**, *14*, 21–25. [CrossRef]

39. Barrabés, M.; Sevilla, P.; Planell, J.A.; Gil, F.J. Mechanical properties of nickel-titanium foams for reconstructive orthopaedics. *Mater. Sci. Eng. C* **2008**, *28*, 23–27. [CrossRef]
40. Parvizi, S.; Hasannaeimi, V.; Saebnoori, E.; Shahrabi, T.; Sadrnezhaad, S.K. Fabrication of porous NiTi alloy via powder metallurgy and its mechanical characterization by shear punch method. *Russ. J. Non-Ferr. Met.* **2012**, *53*, 169–175. [CrossRef]
41. Crone, W.C.; Yahya, A.N.; Perepezko, J.H. Bulk shape memory NiTi with refined grain size synthesized by mechanical alloying. *Mater. Sci. Forum* **2002**, *386-388*, 597–602. [CrossRef]
42. Lu, H.Z.; Ma, H.W.; Luo, X.; Wang, Y.; Wang, J.; Lupoi, R.; Yin, S.; Yang, C. Microstructure, shape memory properties, and in vitro biocompatibility of porous NiTi scaffolds fabricated via selective laser melting. *J. Mater. Res. Technol.* **2021**, *15*, 6797–6812. [CrossRef]
43. Shamsolhodaei, A.; Zarei-Hanzaki, A.; Moghaddam, M. Structural and functional properties of a semi equiatomic NiTi shape memory alloy processed by multi-axial forging. *Mater. Sci. Eng. A* **2017**, *700*, 1–9. [CrossRef]
44. Chu, C.L.; Chung, C.Y.; Lin, P.H. DSC study of the effect of aging temperature on the reverse martensitic transformation in porous Ni-rich NiTi shape memory alloy fabricated by combustion synthesis. *Mater. Lett.* **2005**, *59*, 404–407. [CrossRef]
45. Jiang, S.Y.; Zhang, Y.Q.; Zhao, Y.N.; Liu, S.W.; Hu, L.; Zhao, C.Z. Influence of Ni4Ti3 precipitates on phase transformation of NiTi shape memory alloy. *Trans. Nonferrous Met. Soc. China Engl. Ed.* **2015**, *25*, 4063–4071. [CrossRef]
46. Hu, L.; Jiang, S.; Zhang, Y. Role of severe plastic deformation in suppressing formation of R phase and Ni4Ti3 precipitate of NiTi shape memory alloy. *Metals* **2017**, *7*, 145. [CrossRef]
47. Tang, C.Y.; Zhang, L.N.; Wong, C.T.; Chan, K.C.; Yue, T.M. Fabrication and characteristics of porous NiTi shape memory alloy synthesized by microwave sintering. *Mater. Sci. Eng. A* **2011**, *528*, 6006–6011. [CrossRef]
48. Khalil-Allafi, J.; Amin-Ahmadi, B. The effect of chemical composition on enthalpy and entropy changes of martensitic transformations in binary NiTi shape memory alloys. *J. Alloy. Compd.* **2009**, *487*, 363–366. [CrossRef]
49. Laplanche, G.; Birk, T.; Schneider, S.; Frenzel, J.; Eggeler, G. Effect of temperature and texture on the reorientation of martensite variants in NiTi shape memory alloys. *Acta Mater.* **2017**, *127*, 143–152. [CrossRef]
50. Mao, S.C.; Luo, J.F.; Zhang, Z.; Wu, M.H.; Liu, Y.; Han, X.D. EBSD studies of the stress-induced B2-B19′ martensitic transformation in NiTi tubes under uniaxial tension and compression. *Acta Mater.* **2010**, *58*, 3357–3366. [CrossRef]
51. Yu, H.; Qiu, Y.; Young, M.L. Influence of Ni4Ti3 precipitate on pseudoelasticity of austenitic NiTi shape memory alloys deformed at high strain rate. *Mater. Sci. Eng. A* **2021**, *804*, 140753. [CrossRef]
52. Laplanche, G.; Kazuch, A.; Eggeler, G. Processing of NiTi shape memory sheets-Microstructural heterogeneity and evolution of texture. *J. Alloy. Compd.* **2015**, *651*, 333–339. [CrossRef]
53. Weafer, F.M.; Guo, Y.; Bruzzi, M.S. The effect of crystallographic texture on stress-induced martensitic transformation in NiTi: A computational analysis. *J. Mech. Behav. Biomed. Mater.* **2016**, *53*, 210–217. [CrossRef] [PubMed]
54. Pfetzing-Micklich, J.; Wieczorek, N.; Simon, T.; Maaß, B.; Eggeler, G. Direct microstructural evidence for the stress induced formation of martensite during nanonindentation of NiTi. *Mater. Sci. Eng. A* **2014**, *591*, 33–37. [CrossRef]
55. Laplanche, G.; Pfetzing-Micklich, J.; Eggeler, G. Orientation dependence of stress-induced martensite formation during nanoindentation in NiTi shape memory alloys. *Acta Mater.* **2014**, *68*, 19–31. [CrossRef]

Article

The Study of New NiTi Actuators to Reinforce the Wing Movement of Aircraft Systems

Rafael Braga [1], Patrícia Freitas Rodrigues [1,*], Hélder Cordeiro [2], Pedro Carreira [3] and Maria Teresa Vieira [1]

1 University of Coimbra, Department of Mechanical Engineering, CEMMPRE, R. Luís Reis Santos, 3030-790 Coimbra, Portugal; rafael.braga@dem.uc.pt (R.B.); teresa.vieira@dem.uc.pt (M.T.V.)
2 Moldes RP, R. José Alves Júnior, 411, 2430-076 Marinha Grande, Portugal; helder.cordeiro@moldesrp.pt
3 CDRSP-Centre for Rapid and Sustainable Product Development, Polytechnic Institute of Leiria, General Norton de Matos Street, Apart. 4133, 2411-901 Leiria, Portugal; pedro.s.carreira@ipleiria.pt
* Correspondence: pf.rodrigues@uc.pt

Abstract: Actuators using Shape Memory Alloy (SMA) springs could operate in different mechanical systems requiring geometric flexibility and high performance. The aim of the present study is to highlight the potential of these actuators, using their dimensional variations resulting from the phase transformations of NiTi springs (SMA) to make the movements of the system's mobile components reversible. This reversibility is due to thermal-induced martensitic transformation of NiTi springs. The transformation promotes the extended and retracted of the springs as the phase changing (martensite–austenite) creates movement in part of the system. Therefore, the phase transition temperatures of NiTi, evaluated by differential scanning calorimetry (DSC), are required to control the dimensional variation of the spring. The influence of the number of springs in the system, as well as how impacts on the reaction time were evaluated. The different numbers of springs (two, four, and six) and the interspaces between them made it possible to control the time and the final angle attained in the mobile part of the system. Mechanical resistance, maximum angle, and the system's reaction time using different NiTi springs highlight the role of the actuators. Fused Deposition Modelling (FDM)/Material Extrusion (MEX) or Selective Laser Sintering (SLS) was selected for shaping the composite matrix system. A new prototype was designed and developed to conduct tests that established the relationship between the recoverable deformation of the matrix suitable for the application as well as the number and distribution of the actuators.

Keywords: actuator; aircraft systems wing; shape memory alloy; NiTi spring; additive manufacturing

Citation: Braga, R.; Rodrigues, P.F.; Cordeiro, H.; Carreira, P.; Vieira, M.T. The Study of New NiTi Actuators to Reinforce the Wing Movement of Aircraft Systems. *Materials* **2022**, *15*, 4787. https://doi.org/10.3390/ma15144787

Academic Editor: Salvatore Saputo

Received: 3 June 2022
Accepted: 23 June 2022
Published: 8 July 2022

1. Introduction

Aircraft system wing design must satisfy a specific range of flight conditions. Insertion of actuators made from smart materials such as shape memory alloys may overcome the present limitations. Researchers are currently developing new solutions to modify wing design using smart actuators [1]. Mabe J. et al. developed an aircraft system that includes devices that function depending on the performance of an alloy with shape memory [2,3]. The challenge in this approach is to design a structure strong and flexible enough to alter the angle of inclination suitable for the application envisaged. Phase transformations of shape memory alloys (SMAs) can be used to detect variations around them and respond to external stimuli instantly. SMAs have significant recoverable deformation capability and actuating function when SMA structures are submitted to cyclic loading [4,5]. Among the SMAs, NiTi is the most studied due to its specific properties—superelasticity and shape memory effect [6,7]. The change of properties in shape memory alloys (SMAs) can be detected by martensitic transformation. Stress-induced martensite (SIM) promotes superelasticity, and thermal-induced martensitic transformation (TIM) develops the shape memory effect. Therefore, martensitic transformation in NiTi alloys presents both thermal and mechanical hysteresis. Ni content can control the properties. Ni-rich and equiatomic

NiTi alloys display the superelastic effect at room temperature and above and Ti-rich NiTi alloys display the shape memory effect above room temperature [8,9]. When the material is cooled from the austenite (A) domain, the martensite (M) starts at M_s temperature. The transformation from austenite to martensite is referred to as direct transformation and finishes at martensite M_f temperature. When the material is cooled from the austenite (A) domain, the martensite starts its formation at M_s and the martensite finishes at M_f temperature. On the other hand, when phase M is heated, the austenite phase starts to form at A_s temperature. This transformation finishes when the A_f temperature is reached. When the material is deformed up to 10% in the martensitic domain, it can retain the deformation, as long as the martensitic stability temperature range is preserved. However, the material starts recovering the original shape when it is heated above A_s temperature. When A_f is achieved, the shape is recovered by the shape memory effect [5,10].

Different applications of NiTi have contributed to developing new concepts of actuators. The widespread use of NiTi is due to its fatigue behavior and mechanical strength. Depending on the customized application, NiTi is furnished commercially in several forms: spring, wire, thin films, powder particles, and nanoparticles [11]. Many research activities stem from actuator processing, involving NiTi-based materials, especially wire and spring units to construct linear and rotary actuators [12,13]. The typical straight-drawn wire of commercial NiTi has a maximum recovery strain limit of ~4%. A maximum linear strain of ~20% can be achieved using coiled NiTi wires by successively tuning their shape memory properties in the linear coil prototype [14,15], which is the driving force for the movement exerted by the spring on the prototype parts, according to the specified properties of the spring (Figure 1).

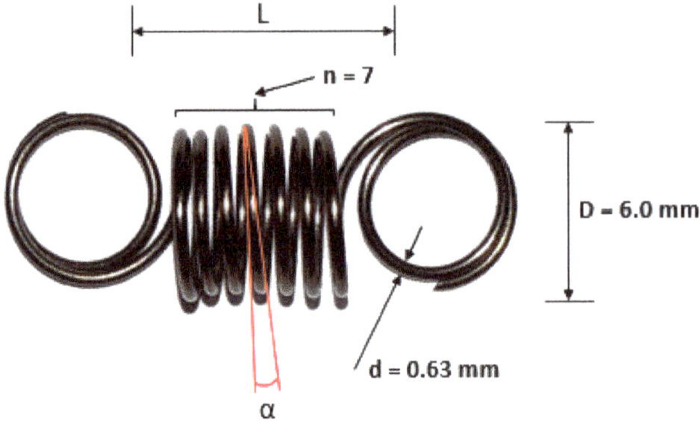

Figure 1. The geometrical design of a NiTi coil spring, thickness (d), outer diameter (D), length (L), number of active coils (n), and initial pitch angle (α).

The linear actuators activated by shape memory alloys with integrated stroke control are commonly designed and manufactured. The extended and retracted of the actuator are performed by the Joule effect, which can possibly determine the angle to perform in the prototype. One of the main functions of this kind of system applied in the prototype is the possibility to modulate the extension and retraction just by acting on the corresponding SMA springs by heat-induced from an electrical current. As the voltage applied is changed, the extended or retracted springs can be activated in order to set the desired position and overcome the limits of the standard configuration SMA spring [16–18]. The aim of this study is to analyze in detail how spring actuators (NiTi) affect the angle of inclination of the composite prototype as a function of the number and distance between the springs.

2. Experimental Procedure

A prototype of the actuator developed to control aircraft system wings constituted by support (nylon) with the incorporation of SMA actuator springs is shown in Figure 2a. The dimensions of the prototype are 130 mm in length and 60 mm in width (Figure 2b). Six SMAs springs (diameter of wire 0.63 mm) were inserted in support specially developed to accomplish the objective to accommodate each one (Figure 2c). Two tiny grooves are cut along the length of the prototype to adjust an SMA wire (on each side) used as an electrical conductor; each SMA spring has contact with the SMA wire (Figure 2c). In the electrical circuit, the current goes through the system of springs to actuate in two, four, six, or eight SMA springs simultaneously. The distance between springs is 10 mm. This system placed in the wings provides control of the structure through different angles. Figure 2d shows the aircraft wing with the actuator.

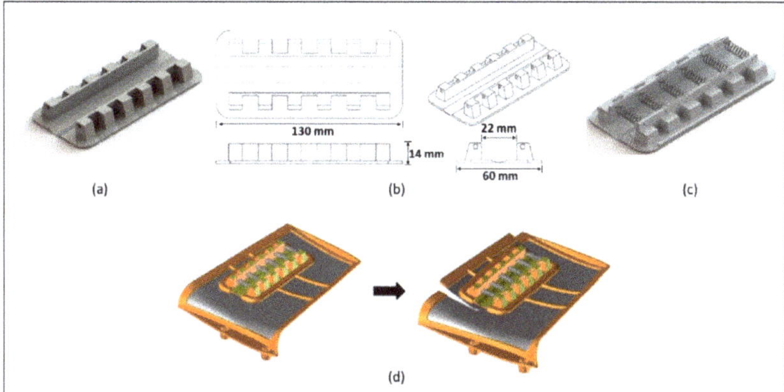

Figure 2. Geometry and assembly of prototype: (**a**) nylon support; (**b**) scheme; (**c**) support with SMA wire and springs; (**d**) aircraft component system before and after SMA actuator effect.

In this study, the support of prototype production was performed from the CAD project to Material Extrusion (MEX)* and Selective Laser Sintering (SLS). The materials used in each part (support and springs) of the prototype were tested, evaluated, and selected according to their properties.

2.1. Material

2.1.1. Spring and Wire

There are several ways to create an electromechanical actuator system. In the present study, the application of the NiTi springs was the solution selected. Therefore, an overview of the properties of springs must be provided. In this case the shape memory alloy NiTi springs (SAES Group-Tensile Spring) have shape memory effect. The dimensions of each spring are, as follows: outer diameter (D) = 6.0 mm, thickness (d) = 0.63 mm, average typical force = 2.0 N and number of turns in the coil (n) = 7 (Figure 1).

The spring has a martensitic phase at room temperature (25 °C), as shown by the Differential Scanning Calorimetry (DSC) curve (Figure 3). Tests were carried out at temperatures ranging from −100 °C to 100 °C under a controlled heating/cooling rate of 10 °C/min. Before analysis by DSC, the specimens were cut and chemically etched (10%HF + 45%HNO_3 + 45% H_2O (vol.%)) to remove oxides, as well as the layer deformed by the cutting operation (final mass: ~17 mg). The phase transformation temperatures are: A_s = 51 °C and A_f = 69 °C; R_s = 58 °C; R_f = 46 °C, M_s = −3 °C; M_f = −42 °C. A superelastic NiTi wire (Fort Wayne Metals) with a diameter of 0.5 mm was selected to be used as an electric conductor making contact between the springs, with A_f = 27.9 °C [19–22].

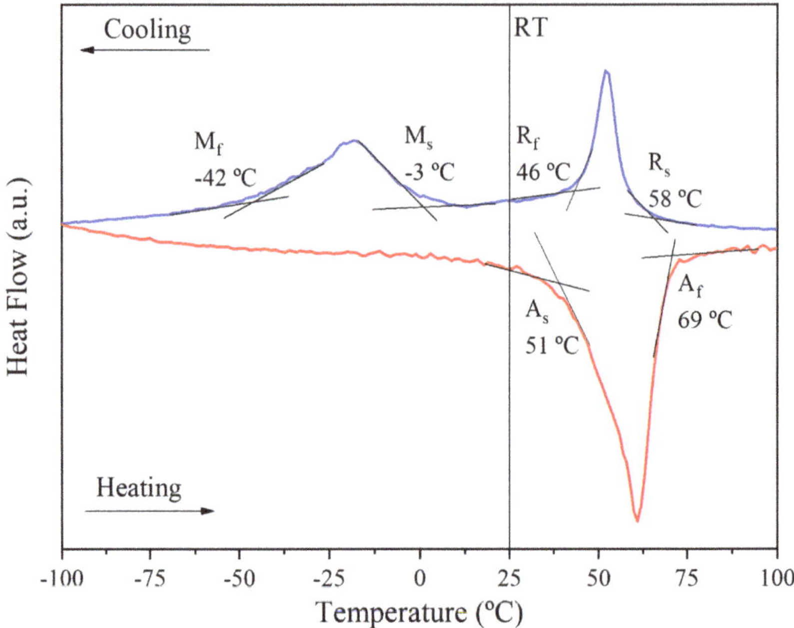

Figure 3. DSC curve of NiTi spring.

A Shimadzu Autograph AG-X Universal Tester Machine equipped with a 5 kN load cell was used to test the tensile behavior of the NiTi springs and wires. The SMA spring was connected to the machine using a hook (Figure 4). Then, tensile tests under displacement control at room temperature (25 °C) were performed. After reaching the predefined length (22.0 mm) in the design of the prototype, the displacement was stopped and complete heating/cooling cycles were carried out on the spring. Firstly, the springs were heated (>A_f temperature) and then cooled to room temperature (RT) for ten (S10), twenty (S20), and thirty (S30) seconds, in order to simulate the real action of the spring and measure the resistance force imposed by the spring in the test (15 cycles by spring test).

Figure 4. Experimental setup for tensile tests of the NiTi SMA spring.

2.1.2. Support

The selection of the material to produce the support resulted from a comparison of Polylactic Acid (PLA), Acrylonitrile Butadiene Styrene (ABS), and Polyamide (PA). The polymeric filament materials PLA (PRIMAVALUE™) and ABS (PRIMASELECT™) are from the supplier PrimaCreator and have 1.75 mm in diameter. Three-dimensional Systems furnished the polymeric powder (Nylon (PA12)) (DuraForm® ProX® PA). Table 1 summarizes the polymeric material properties selected.

Table 1. Temperature, fracture toughness, and Young's modulus of PLA, ABS, and PA12 (CES 2013 Edupack).

Material	Service Temperature (°C)	Fracture Toughness (MPa.m$^{1/2}$)	Young's Modulus (GPa)	Tensile Strength (MPa)
PLA	45–55	0.70–1.10	3.45–3.83	48.0–60.0
ABS	62–77	1.90–2.10	2.00–2.90	30.0–50.0
PA12	90–130	3.32–3.66	1.33–1.65	58.5–71.5

The mechanical behavior of the spring support was evaluated by three-point bending and tensile tests (room temperature). The PLA and ABS specimens submitted to three-point bending tests were produced by MEX and the Nylon specimen was manufactured by Selective Laser Sintering (SLS), both with dimensions of 60 × 10 × 2 mm^3. Three-point bending tests were carried out according to ASTM D790-10, using a Shimadzu Autograph AG-X Universal Tester Machine equipped with a 5 kN load cell, and the TRAPEZIUM X software for data processing, at a displacement rate of 2 mm/min and gauge length of 32 mm.

The specimens submitted to the tensile tests had the geometry shown in Figure 5 with dimensions of 100 × 10 × 2 mm^3. The tensile tests were performed according to ASTM D 3039, using a Shimadzu Autograph AG-X Universal Tester Machine equipped with a 5 kN load cell, and the TRAPEZIUM X software for data processing, at a load rate of 5 mm/min and a gauge length of 62 mm.

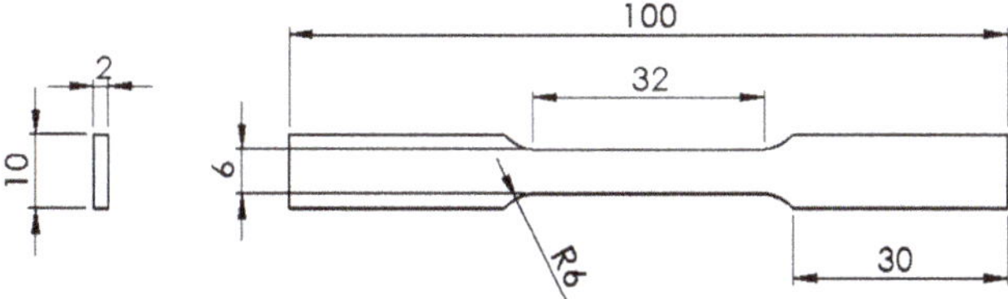

Figure 5. Geometry and dimensions of the specimens for tensile tests.

A prototype was developed using one spring (Figure 6a,b) for the thermal resistance test. In this test, a 9 V DC serial was applied at the springs to evaluate the operating temperature in contact with different materials during ten cycles. The integrity of the polymeric support of the prototype was visually evaluated, and showed no softening or damaging.

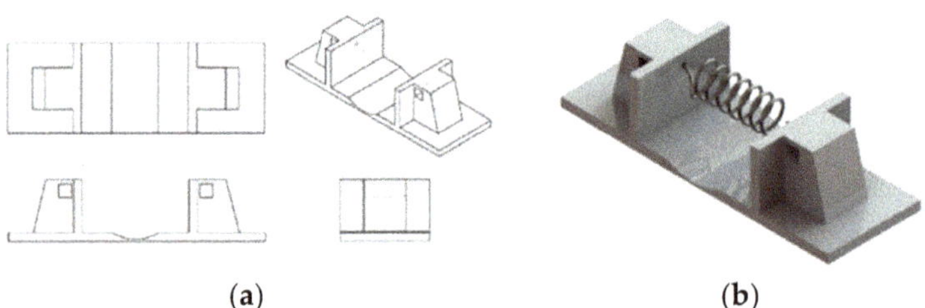

Figure 6. Prototype (one spring): (**a**) schema; (**b**) support with SMA springs.

2.2. Prototypes Fabrication

Two additive manufacturing techniques were selected, to produce the support for the prototype (Figure 2a, MEX and SLS). The dimensions of the support were constant and independent of the number of springs. The MEX parameters were infill density of 40%, gyroid fill patterns for layers, 2 perimeters layers, fill scan pattern angle of 45°, printing speed rate of 50 mm/s, 0.2 mm of layer height, and a printing surface temperature of 215 °C. The equipment used was a Prusa MK3S 3D Printing. The second additive manufacturing technology was supported by a 3D Systems ProX® SLS 6100 3D Printer, where the parameters were as follows: chamber temperature 169 °C, warm-up rate from room temperature 1 °C min^{-1}, laser power contour 18W, laser power infilling 60W, laser scan linear speed 237.5 mms^{-1}, scan spacing 0.25 mm, layer height 0.1 mm.

The functional characteristics of NiTi springs were studied as actuators in the prototypes in this particular system (Figure 2d). The phase transition temperature of the NiTi spring was attained by a serial connection circuit and a 9 V DC power supply, which can induce the spring to achieve a temperature of between 50 °C and 90 °C. The extended and retracted spring times, with and without induced cooling, and the final angle of inclination reached by the prototype, were measured. The number of springs was changed to evaluate the angle of inclination and the number of springs needed; these variations was tested using two, four, and six springs. In order to guarantee a uniform current, a superelastic NiTi wire makes the electric contact between the springs.

3. Results and Discussion

3.1. Prototype Details

3.1.1. Actuator

The results of the NiTi spring tensile tests (Figure 7) highlight the maximum force achieved in the S10, S20 and S30 cycles (heating and cooling cycle times), which corresponds to 10 s, 20 s and 30 s, respectively. This represents the transformation from martensite to austenite through heating. When the spring was subjected to cycles of 10 s (S10), the average force reached was 2.1 N. In longer cycles, the average force reached increased proportionally with time, and was 2.2 N for cycles of 20 s (S20), and 2.3 N in cycles of 30 s (S30) (Table 2).

Table 2. Maximum force function of the thermal cycle (NiTi springs).

Cycle Heating/Cooling	Maximum Force (N)
S10	2.1
S20	2.2
S30	2.3

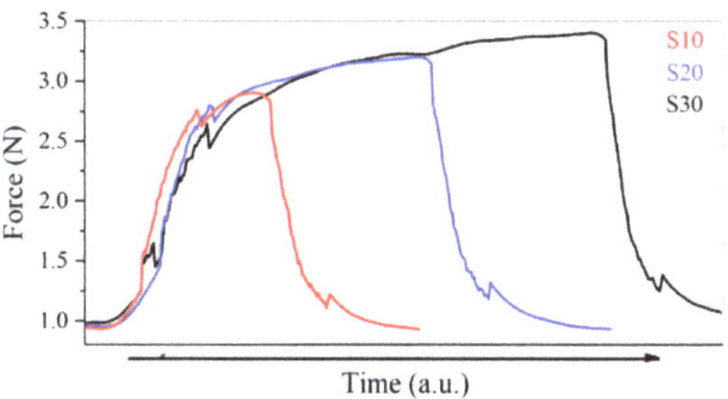

Figure 7. Force vs. time of the thermal cycle of the NiTi spring (tensile test) (10 s (S10), 20 s (S20), and 30 s (S30)).

3.1.2. Support

The visual analyses of the effects of the thermal resistance test using one NiTi spring reveal a significant problem with the PLA after one cycle (Figure 8a) and with the ABS after a few cycles (Figure 8b). When in direct contact with the NiTi spring during heating, imperfections, damage, and some deformation occur at the contact points. These make it impossible to use these materials for supports that work at thermal cycles above the phase transformation temperature of NiTi springs (A_f = 69 °C). The PLA shows a service temperature from 45 to 55 °C and ABS between 62–77 °C, which supports the observation of damage. For the degradation temperature when comparing PLA and ABS, the latter one support higher temperatures than the PLA. Using dynamic mechanical analysis (DMA), K. Arunprasath et.al. (2021) observed that a decrease in the degree of crystallinity of both materials (PLA and ABS) increases their strength.

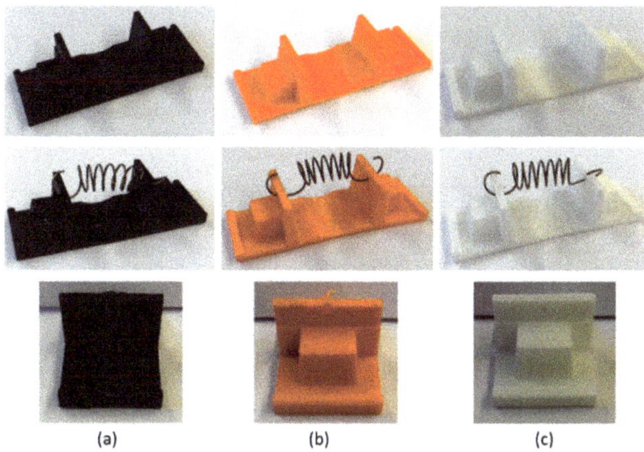

Figure 8. Visual aspect after thermal resistance tests on the supports (one spring): (**a**) PLA; (**b**) ABS; (**c**) PA12.

The maximum effective displacement of the actuator support, and consequently of the system, leaded to modify its design creating a zone where it could easily bend without breaking.

3.2. Prototype Tests

After the selection of the prototype materials, the supports for the prototypes in PA12 were manufactured by Selective Laser Sintering (SLS), and the NiTi springs were assembled to the supports (Figure 9).

The characterization test values for the prototype are summarized in Table 3. The highest angle of inclination of the prototype (Figure 10) was attained for the highest number of springs tested (6). It is possible to vary the angle reached and the extended spring according to the number of springs in the prototype. The retracted time with cooling induction was considerably shorter than without cooling. Figure 11 highlights the linear relationships between the angle of inclination value of the prototype and opening time versus the number of SMA springs.

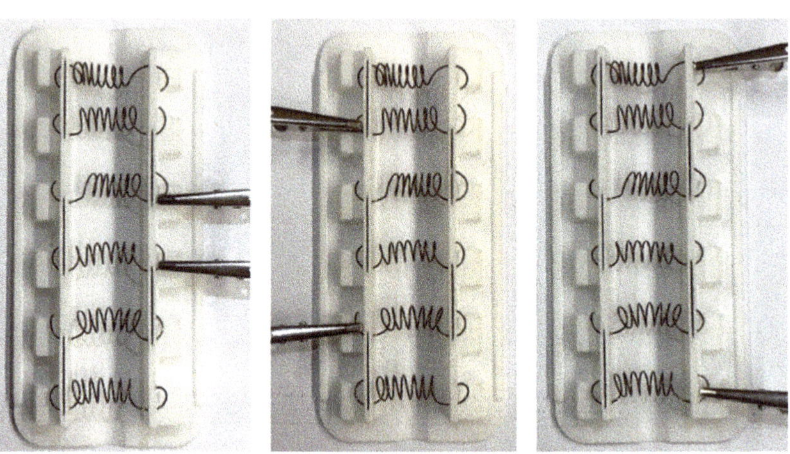

Figure 9. Prototype with six springs and heating system (across different points).

Table 3. Extended and retracted time with and without induced cooling and maximum angle of inclination.

Number of Springs	Extended Time (s)	Retracted Time with Induced Cooling-Air (s)	Retracted Time without Induced Cooling-Air (s)	Maximum Angle (°)
2	~19	20–35	75–90	10–12
4	~23	20–35	75–90	18–23
6	~25	20–35	75–90	25–30

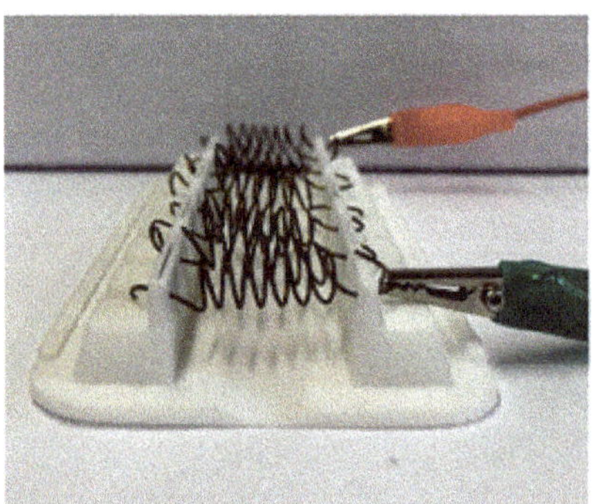

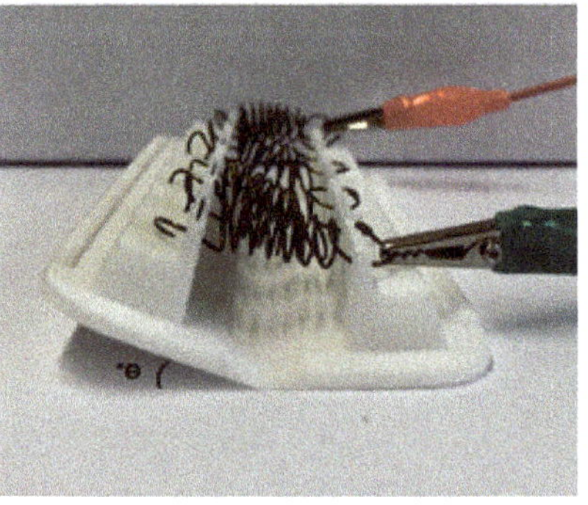

Figure 10. Evaluation of angle of inclination of the prototype.

Nevertheless, it would be interesting to evaluate the prototype's behavior for different distances between the springs and a more significant number of springs.

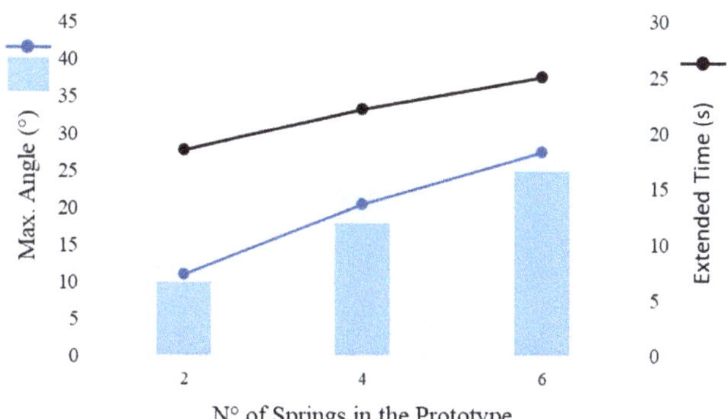

Figure 11. Maximum angle of inclination of the prototype (blue) and extended time as a function of the number of NiTi springs (black) in PA12 support.

4. Conclusions

The linear actuation stroke of NiTi (SMA) wire actuators is improved significantly by converting them into coil spring structures (spring). The present study contributes to establishing the main characteristics of a prototype constituted by spring actuators based on NiTi (SMA) with unique thermo-mechanical performances for wing aircraft system. A PA (nylon) support produced by additive manufacturing was selected because it is strong enough for the forces applied and resistant to the temperatures developed during the application.

The main characteristics for the industrial application of NiTi actuators are as follows:

- The maximum inclination angle of the support is a linear function of the number of springs;
- Cooling of the prototype has significant importance regarding prototype recovery time;
- The maximum inclination angle of the prototype is a linear function of the temperature imposed on the spring.

Author Contributions: R.B. conceptualization, formal analysis, investigation, methodology, visualization, writing—original draft preparation. P.F.R., H.C., P.C. and M.T.V. writing—review and editing, supervision, methodology, validation. All authors have read and agreed to the published version of the manuscript.

Funding: This research was funded by: Project PTDC/CTM-CTM/29101/2017-POCI-01-0145-FEDER-029101 funded by FEDER funds through COMPETE2020-Programa Operacional Competitividade e Internacionalização (POCI) and by national funds (PIDDAC) through FCT/MCTES, and Project POCI-01-0247-FEDER-033758 funded by Agência Nacional de Inovação, S.A. PT2020-SII&DT-Copromoção. This research was also supported by FEDER funds through the program COMPETE-Programa Operacional Factores de Competitividade, and by national funds through FCT-Fundação para a Ciência e a Tecnologia, under the project UIDB/EMS/00285/2020.

Institutional Review Board Statement: Not applicable.

Informed Consent Statement: Not applicable.

Acknowledgments: This research was funded by: Project PTDC/CTM-CTM/29101/2017-POCI-01-0145-FEDER-029101 funded by FEDER funds through COMPETE2020-Programa Operacional Competitividade e Internacionalização (POCI) and by national funds (PIDDAC) through FCT/MCTES, and Project POCI-01-0247-FEDER-033758 funded by Agência Nacional de Inovação, S.A. PT2020- SII&DT-Copromoção. This research was also supported by FEDER funds through the program COMPETE-Programa Operacional Factores de Competitividade, and by national funds through FCT-Fundação para a Ciência e a Tecnologia, under the project UIDB/EMS/00285/2020, UIDB/04044/2020, UIDP/04044/2020. This study was also supported by PAMI-ROTEIRO/0328/2013 (N° 022158), MATIS (CENTRO-01-0145-FEDER-000014-3362) and 4DComposites (POCI-01-0247-FEDER-033758)

Conflicts of Interest: The authors declare no conflict of interest.

References

1. Abdullah, E.J.; Gaikwad, P.S.; Azid, N.; Abdul Majid, D.L.; Mohd Rafie, A.S. Temperature and strain feedback control for shape memory alloy actuated composite plate. *Sens. Actuators A Phys.* **2018**, *283*, 134–140. [CrossRef]
2. Bertacchini, O.W.; Lagoudas, D.C.; Calkins, F.T.; Mabe, J.H. Thermomechanical cyclic loading and fatigue life characterization of nickel rich NiTi shape-memory alloy actuators. Behavior and Mechanics of Multifunctional and Composite Materials. *Proc. SPIE* **2008**, *6929*, 692916. [CrossRef]
3. Mabe, J.H.; Calkins, F.T.; Bushnell, G.S.; Bieniawski, S.R. Aircraft Systems with Shape Memory Alloy (SMA) Actuators, and Associated Methods. U.S. Patent 7878459B2, 1 February 2011.
4. Antonucci, V.; Martone, A. Phenomenology of Shape Memory Alloys. In *Shape Memory Alloy Engineering: For Aerospace, Structural and Biomedical Applications*; Elsevier Inc.: Amsterdam, The Netherlands, 2015; pp. 33–56. [CrossRef]
5. Saburi, T. Ti-Ni Shape memory Alloys. In *Shape Memory Materials*, 1st ed.; Wayman, C.M., Otsuka, K., Eds.; Cambridge University Press: New York, NY, USA, 1998; pp. 49–96.
6. Schmahl, W.W.; Khalil-Allafi, J.; Hasse, B.; Wagner, M.; Heckmann, A.; Somsen, C. Investigation of the phase evolution in a super-elastic NiTi shape memory alloy (50.7 at.% Ni) under extensional load with synchrotron radiation. *Mater. Sci. Eng. A* **2004**, *378*, 81–85. [CrossRef]
7. Miyazaki, S.; Imai, T.; Igo, Y.; Otsuka, K. Effect of Cyclic Deformation on the Pseudoelasticity Characteristics of Ti-Ni Alloys. *Metall. Trans. A Phys. Metall. Mater. Sci.* **1986**, *17*, 115–120. [CrossRef]
8. Frenzel, J.; George, E.P.; Dlouhy, A.; Somsen, C.; Wagner, M.F.X.; Eggeler, G. Influence of Ni on martensitic phase transformations in NiTi shape memory alloys. *Acta Mater.* **2010**, *58*, 3444–3458. [CrossRef]
9. Otsuka, K.; Ren, X. Physical metallurgy of Ti–Ni-based shape memory alloys. *Prog. Mater. Sci.* **2005**, *50*, 511–678. [CrossRef]
10. Abeykoon, C.; Sri-Amphorn, P.; Fernando, A. Optimization of fused deposition modeling parameters for improved PLA and ABS 3D printed structures. *Int. J. Light. Mater. Manuf.* **2020**, *3*, 284–297. [CrossRef]
11. Shimoga, G.; Kim, T.H.; Kim, S.Y. An Intermetallic NiTi-Based Shape Memory Coil Spring for Actuator Technologies. *Metals* **2021**, *11*, 1212. [CrossRef]
12. Yuan, H.; Fauroux, J.C.; Chapelle, F.; Balandraud, X. A review of rotary actuators based on shape memory alloys. *J. Intell. Mater. Syst. Struct.* **2017**, *28*, 1863–1885. [CrossRef]
13. Jani, J.M.; Leary, M.; Subic, A. Designing shape memory alloy linear actuators: A review. *J. Intell. Mater. Syst. Struct.* **2016**, *28*, 1699–1718. [CrossRef]
14. Koh, J.S. Design of Shape Memory Alloy Coil Spring Actuator for Improving Performance in Cyclic Actuation. *Materials* **2018**, *11*, 2324. [CrossRef] [PubMed]
15. Ko, W.-S.; Maisel, S.B.; Grabowski, B.; Jeon, J.B.; Neugebauer, J. Atomic scale processes of phase transformations in nanocrystalline NiTi shape-memory alloys. *Acta Mater.* **2017**, *123*, 90–101. [CrossRef]
16. Seok, S.; Onal, C.D.; Cho, K.J.; Wood, R.J.; Rus, D.; Kim, S. Meshworm: A peristaltic soft robot with antagonistic nickel titanium coil actuators. *IEEE/ASME Trans. Mechatron.* **2013**, *18*, 1485–1497. [CrossRef]
17. Costanza, G.; Radwan, N.; Tata, M.E.; Varone, E. Design and characterization of linear shape memory alloy actuator with modular stroke. *Procedia Struct. Integr.* **2019**, *18*, 223–230. [CrossRef]
18. Costanza, G.; Tata, M.E. Shape memory alloys for aerospace, recent developments, and new applications: A short review. *Materials* **2020**, *13*, 1856. [CrossRef] [PubMed]
19. Lima, P.C.; Rodrigues, P.F.; Ramos, A.S.; da Costa, J.D.M.; Braz Fernandes, F.M.; Vieira, M.T. Experimental Analysis of NiTi Alloy during Strain-Controlled Low-Cycle Fatigue. *Materials* **2021**, *14*, 4455. [CrossRef] [PubMed]
20. Arunprasath, K.; Vijayakumar, M.; Ramarao, M.; Arul, T.G.; Pauldoss, S.P.; Selwin, M.; Radhakrishnan, B.; Manikandan, V. Dynamic mechanical analysis performance of pure 3D printed polylactic acid (PLA) and acrylonitrile butadiene styrene (ABS). *Mater. Today Proc.* **2021**, *50*, 1559–1562. [CrossRef]
21. Yaagoubi, H.; Abouchadi, H.; Janan, M.T. Numerical simulation of heat transfer in the selective laser sintering process of Polyamide12. *Energy Rep.* **2021**, *7*, 189–199. [CrossRef]
22. Mokrane, A.; Boutaous, M.; Xin, S. Process of selective laser sintering of polymer powders: Modeling, simulation, and validation. *Comptes Rendus Mécanique* **2018**, *346*, 1087–1103. [CrossRef]

Article

In Situ Observation of Thermoelastic Martensitic Transformation of Cu-Al-Mn Cryogenic Shape Memory Alloy with Compressive Stress

Zhenyu Bian [1,2], Jian Song [1], Pingping Liu [1,*], Farong Wan [1,*], Yu Lei [3], Qicong Wang [1], Shanwu Yang [1], Qian Zhan [1], Liubiao Chen [4] and Junjie Wang [4]

1 School of Materials Science and Engineering, University of Science and Technology Beijing, Beijing 100083, China; bianzhenyu@c.ccs.org.cn (Z.B.); zybian@ccs.org.cn (J.S.); qwea142311@163.com (Q.W.); yangsw@mater.ustb.edu.cn (S.Y.); qzhan@mater.ustb.edu.cn (Q.Z.)
2 China Classification Society Certification Co., Ltd., Beijing 100006, China
3 Faculty of Engineering, Hokkaido University, Sapporo 060-8628, Japan; 18610132651@163.com
4 Key Laboratory of Cryogenics, Technical Institute of Physics and Chemistry, Chinese Academy of Sciences, Beijing 100190, China; chenliubiao@mail.ipc.ac.cn (L.C.); wangjunjie@mail.ipc.ac.cn (J.W.)
* Correspondence: ppliu@ustb.edu.cn (P.L.); wanfr@mater.ustb.edu.cn (F.W.)

Abstract: The thermoelastic martensitic transformation and its reverse transformation of the Cu-Al-Mn cryogenic shape memory alloy, both with and without compressive stress, has been dynamically in situ observed. During the process of thermoelastic martensitic transformation, martensite nucleates and gradually grow up as they cool, and shrink to disappearance as they heat. The order of martensite disappearance is just opposite to that of their formation. Observations of the self-accommodation of martensite variants, which were carried out by using a low temperature metallographic in situ observation apparatus, showed that the variants could interact with each other. The results of in situ synchrotron radiation X-ray and metallographic observation also suggested there were some residual austenites, even if the temperature was below M_f, which means the martensitic transformation could not be 100% accomplished. The external compressive stress would promote the preferential formation of martensite with some orientation, and also hinder the formation of martensite with other nonequivalent directions. The possible mechanism of the martensitic reverse transformation is discussed.

Keywords: cryogenic shape memory alloys; compressive stress; cryogenic metallography; in situ observation; residual austenite; synchrotron radiation

Citation: Bian, Z.; Song, J.; Liu, P.; Wan, F.; Lei, Y.; Wang, Q.; Yang, S.; Zhan, Q.; Chen, L.; Wang, J. In Situ Observation of Thermoelastic Martensitic Transformation of Cu-Al-Mn Cryogenic Shape Memory Alloy with Compressive Stress. *Materials* **2022**, *15*, 3794. https://doi.org/10.3390/ma15113794

Academic Editor: Salvatore Saputo

Received: 27 February 2022
Accepted: 16 May 2022
Published: 26 May 2022

1. Introduction

Thermoelastic martensitic transformation and its reverse transformation are the root of the shape memory effect (SME) and a unique part of martensitic transformation research. Shape memory alloy (SMA) based on SME has been widely used in many areas. Cu-based SMA, for example, has been widely used in intelligent valves, connectors, dampers and seismic attenuation due to its SME and high damping properties [1–3]. It also has the great potential of being used in cryogenic sealing because of its pseudoelasticity, which is also based on the thermoelastic martensite [4–6].

SME is now generally well understood and generally considered to be associated with the reversibility of the martensitic transformation, that is, the thermoelastic martensitic transformation. As early as 1938, Greninger and Mooradian first discovered the phenomenon of martensite growth and disappearance with temperature fall and rise in the Cu-Zn alloy during their investigation of copper alloys (copper–zinc and copper–tin) [7]. Then, in about 1949, Kurdjumov and Khandros named this reversible transformation as "thermoelastic martensitic transformation" after a detailed study of thermoelasticity in Cu-Zn and Cu-Al-Ni alloys [1,8].

Many aspects of thermoelastic martensitic transformation have been studied in detail [9], including the associated crystallographic behavior [10], the thermodynamics [11,12], the effects of chemical composition [13–16] and post-heat treatment [3,17–22]. A notable phenomenon in the thermoelastic martensitic transformation of SMA is that an austenite phase grain can transform to different martensitic variants during the martensitic transformation [23], but all the martensite variants have to transform to the same original austenite phase grain instead of forming different austenite phase grains during the revise transformation. This phenomenon is very important because it is the basis of SME. Otsuka and Shimizu [24] discussed the effects of ordering on the crystallographic reversibility of the martensitic transformation and concluded that the complete reversibility of the martensitic transformation is characteristic of ordered alloys. However, they also note that the fcc-to-fct (face-centered-tetragonal) transformation is an "exception". Bhattacharya et al. [25] provide an explanation for the reversibility on the basis of the symmetry change during the transformation. They show, through rigorous mathematical theory and numerical simulation, that irreversibility is inevitable in a "reconstructive" phase transformation, but not in a "weak" martensitic transformation, in which the symmetry group of both the parent and product phases are included in a common finite symmetry group (which includes symmetry breaking). Our previous work shows that there are always some residual austenites, even if the temperature is lower than M_f and the last formed martensite on cooling firstly disappeared on heating, through a low temperature metallographic in situ observation in a Cu-Al-Mn alloy [26]. However, so far, there is still much work remaining to understanding the effects of external compressive stress on the process of martensitic transformation and its reverse transformation. The direct observation and detailed knowledge of the reverse thermoelastic martensitic transformation of SMA are essential for the mechanism of thermoelastic martensitic transformation and the proper understanding of SME.

In this paper, a Cu-Al-Mn cryogenic SMA (martensitic transformation start temperature, Ms, is as low as about 100 K) was prepared. The thermoelastic martensitic transformation and its reverse transformation of the alloy have been observed in situ with and without compressive stress. The possible mechanism of martensitic reverse transformation is also discussed.

2. Experimental Procedure

The material analyzed in this study was a Cu-Al-Mn cryogenic SMA. The chemical composition and Ms point are listed in Table 1. The Cu-Al-Mn alloy was melted in a vacuum induction furnace with high purity Cu, Al and Mn [26]. The bulk specimen with a size of 20 mm × 15 mm × 5 mm was cut from the sample by wire cutting, and then heat treated by water quenching after holding at 900 °C for 10 min. The Ms measurement was performed using a PPMS-9 (Quantum Design, China). The Ms and martensitic transformation finish temperature (M_f) are shown in Figure 1. Ms is about 108 K and M_f is about 80 K.

Table 1. Chemical composition and Ms point of the sample.

Composition (wt.%)			Ms (K)
Cu	Al	Mn	
76.8	12.5	10.7	108

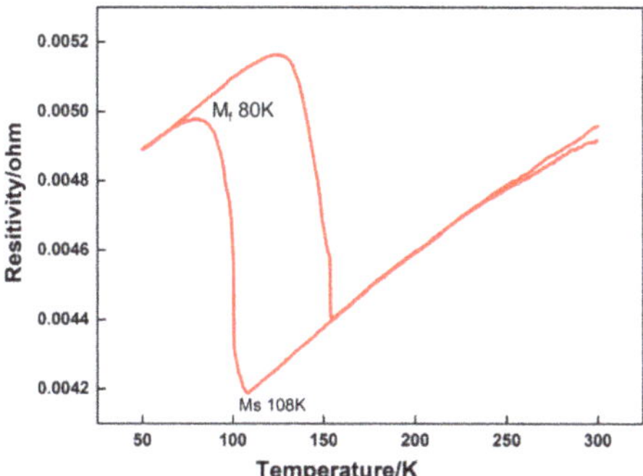

Figure 1. The Ms temperature of Cu-Al-Mn shape memory alloys.

The working schematic diagram of the low temperature metallographic in situ observation instrument with deformation excitation unit is shown in Figure 2. The deformation-adjusting device on one side of the instrument can apply compressive stress to the sample, as shown in Figure 3. The schematic diagram is shown in Figure 3a–c. The sample was an ordinary metallographic one with maximum size of 20 mm × 20 mm × 8 mm. Cooling medium was liquid nitrogen. The temperature control accuracy was ±2 K, the minimum temperature could be reduced to 77 K, and the cooling rate was 45 K/min. The phase transformation process of the Cu-Al-Mn cryogenic SMA with and without compress stress could be observed through the quartz glass window of this instrument. In situ XRD experiments were carried out at the beamline 4B9A in Beijing Synchrotron Radiation Facilities (BSRF). Two XRD patterns were, respectively, collected at 293 K and 77 K under low-vacuum conditions (0.1 Pa) with an incident X-ray wavelength of 1.54 Å. The data collection time for the low temperature XRD pattern was about 2 h.

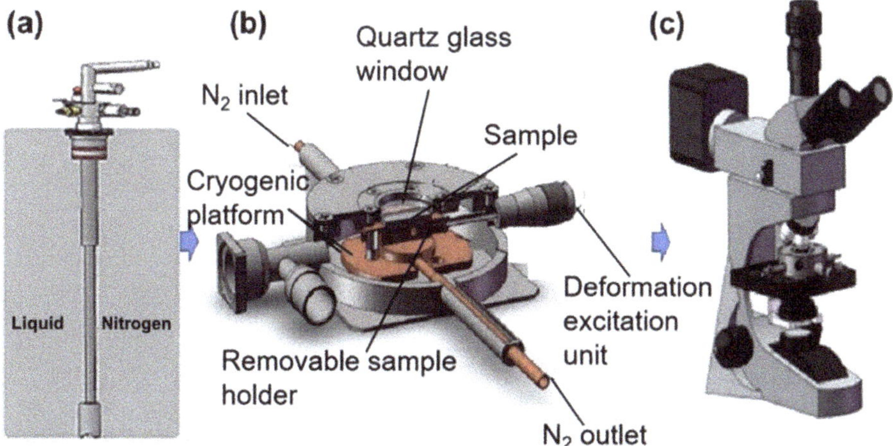

Figure 2. In situ observation apparatus of cryogenic metallographic with deformation excitation unit. (**a**) the pipe connecting the sample table and the liquid nitrogen tank, through which liquid nitrogen is introduced into the sample stage (**b**) sample stage with deformation excitation unit and (**c**) the metallographic microscope is equipped with micro image processing system (MIPs).

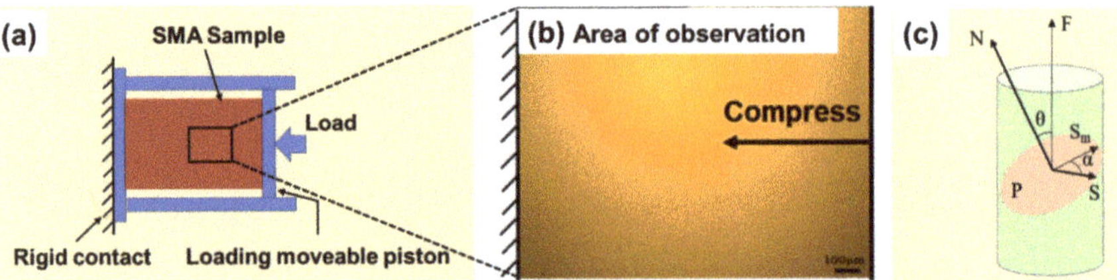

Figure 3. A schematic diagram of the area of observation with external stress. (**a**) Schematic diagram of observation window of copper sample table; (**b**) the picture of area of observation and (**c**) schematic diagram of compressive stress.

3. Experimental Results

3.1. Thermoelastic Martensitic Transformation and Reverse Transformation without Compressive Stress

The cryogenic metallographic images of the surface of the Cu-Al-Mn alloy with the change of the temperature are shown in Figure 4. The martensitic transformation process during cooling is shown in images 1–5 in Figure 4, and the reverse martensitic transformation process during heating is shown in images 6–10 in Figure 4; no loading pressure was applied to the sample when cooling and heating.

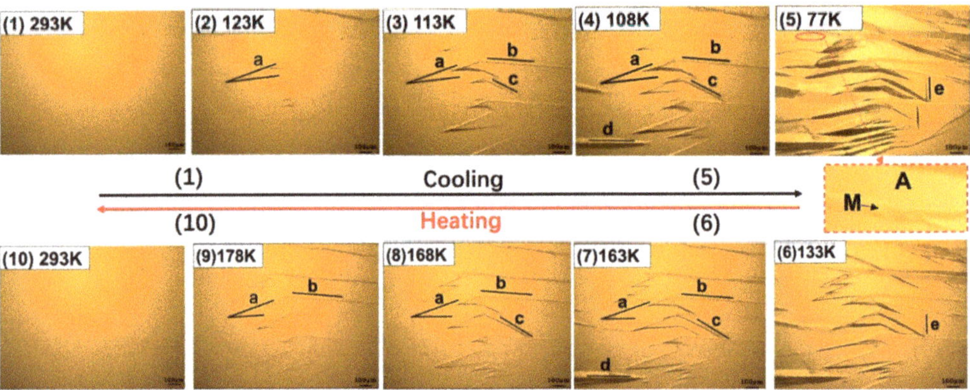

Figure 4. In situ observation of the thermoelastic martensitic transformation (1–5) and reverse transformation (6–10) without compressive stress. 'A' is austenite and 'M' is martensite.

The original sample surface is smooth, as shown in image Figure 4(1), which is at the fully complete β1 phase at 293 K. As the temperature decreases, the martensite variants with mark of "a", "b", "c", "d" and "e" gradually appear on the sample surface, in the order of "a" → "b" → "c" → "d" → "e", as shown in the images 2–5 of Figure 4. The martensites with different orientations grow continuously in their length and width directions. These preferentially grown martensite strips can interact with other oriented martensites subsequently formed. When the temperature was reduced to 77 K, and was held for a while, it was shown that there were still a few regions which remained in their original austenite phases and would not translate into martensites, as indicated by a ellipse mark in image 5 of Figure 4, and in the larger version of this figure.

As shown in images 6–10 of Figure 4, the martensitic variants of five orientations faded away and the sample surface returned to its original smoothness once again. The order of martensite disappearance is "e" → "d" → "c" → "b" → "a", which is just opposite to the order of their formation.

3.2. Thermoelastic Martensitic Transformation and Reverse Transformation with Compressive Stress

The thermoelastic martensitic transformation during cooling and reverse transformation process during heating with compressive stress are shown in images 1–5 of Figure 5 and images 6–10 of Figure 5, respectively. The compressive stress was applied to the same sample as shown in Figure 4, and the schematic diagram is shown in Figure 3b,c. The martensitic transformation process under compressive stress was observed in the same field of view and with same experimental parameters as shown in Figure 4, and the other experimental parameters were also same as those in Figure 4.

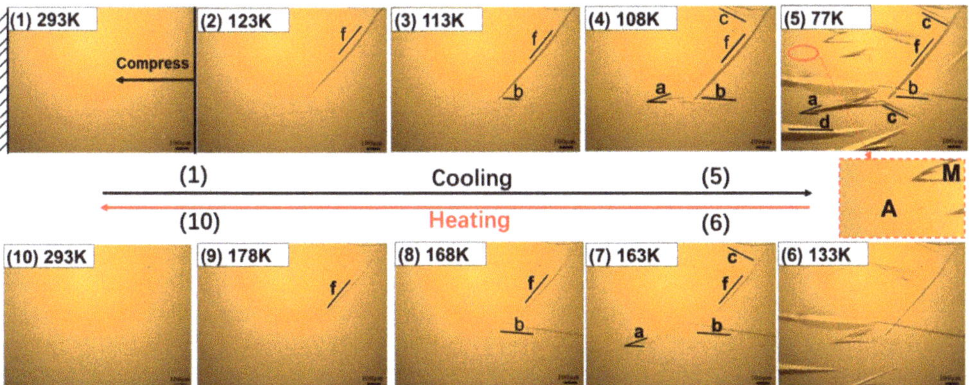

Figure 5. In situ observation of the thermoelastic martensitic transformation (1–5) and reverse transformation (6–10) with compressive stress. 'A' is austenite and 'M' is martensite.

As the temperature decreases, a martensite variant with a new orientation marked with "f" formed firstly, while the martensite with the orientation of "e" did not appear under the external compressive stress. The number of martensitic variants with orientations "a", "b", "c" and "d" decreased. The temperature corresponding to the initial appearance of the martensitic variants with each orientation also decreased. The order of appearance also changed to be "f" → "b" → "a" → "c" → "d".

As shown in images 6–10 of Figure 5, the martensitic variants of five orientations ("f","b", "a", "c", "d") faded away, and the sample surface returned to its original smoothness again. The order of disappearance was "d" → "c" → "a" → "b" → "f", which is also just opposite to the order of their formation. It is noted that the number of martensites with compressive stress was smaller than that without compressive stress.

3.3. Synchrotron Radiation X-ray Diffraction

Figure 6 shows the synchrotron radiation X-ray diffraction spectrum of the same sample area of the Cu-Al-Mn alloy at 293 K and 77 K. As shown in Figure 6a, the parent phase of the Cu-Al-Mn cryogenic SMA at room temperature is $AlCu_2Mn$ phase, which is the DO_3 structure. Figure 6b shows that there are $(10\overline{8})$, (020), $(12\overline{2})$, $(12\overline{8})$, (1216), (2020) and (3218) diffraction peaks in the XRD pattern, which are unique to the M18R structure of martensite, and (200), (220) and (422) diffraction peaks of parent phase with DO_3 structures.

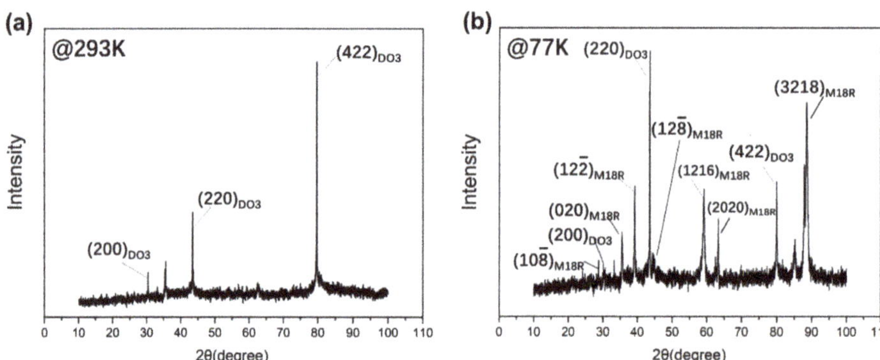

Figure 6. Synchrotron radiation X-ray diffraction spectrum of Cu-Al-Mn alloy at 293 K (**a**) and 77 K (**b**).

According to the martensitic transformation temperature measurement results shown in Figure 1, the martensitic transformation should be completed at 77 K. However, the diffraction peaks of the parent/austenite phase are still there, as shown in Figure 6b, when the temperature is lower than Mf, which indicates that some parent/austenite phase remains in the sample after the martensitic transformation. The results of Figure 6b are consistent with the metallographic observation results of Figures 4 and 5.

4. Discussion

During the process of thermoelastic martensitic transformation in SMA, it is assumed that there is a crystallographic relationship, "F", between the martensite and the austenite: $(h_1k_1l_1)_M//(h_2k_2l_2)_P$, $[u_1v_1w_1]_M//[u_2v_2w_2]_P$. Based on this relationship, some martensites with crystallographically equivalent directions can be formed, which are named martensite variants. An austenite "A" with a certain crystallographic orientation would transform into many martensites, M_1, M_2, M_3, ... M_n, during cooling. Here, we assumed that M_n is the last martensite during this transformation process. Then, all these martensites, M_1, M_2, M_3, ... M_n, have to reverse to the same austenite "A" following the inverse relationship of relationship F, with the shape restoration of the SMA sample.

A question naturally arises. The martensite M_n may transform into any austenite A_i which has the crystallographically equivalent direction according to the relationship, "F", during the reverse transformation process. However, the other austenite A_i may not keep the equivalence relation with the other martensite variants M_i. This means that not all of the martensite variants transform into the same A_i following the relationship "F". Another A_j must be formed, and the original shape of the materials cannot be restored. However, the restoration of the original shape of the materials (the parent phase) can always be observed in the experiment. It is suggested that the austenite can transform into different martensite variants during thermoelastic martensitic transformation, but all the martensite variants have to transform to the same original austenite during the reverse transformation. This is a kind of "asymmetry". It is further speculated that the transformation from austenite to martensite cannot be completed to 100%, as a little parental austenite phase will remain. This is consistent with the current in situ experimental observations.

From the perspective of thermodynamics, the asymmetry between the transformation from austenite to martensite and its reverse transformation is possible. A Gibbs free energy change of a system upon the martensite transformation may be written as [27]:

$$\Delta G^{P\to M} = \Delta G_c^{P\to M} + \Delta G_i^{P\to M} + \Delta G_e^{P\to M} \quad (1)$$

where $\Delta G_c^{P\to M}$ is a chemical energy term (with a negative value, the driving force for the transformation), $\Delta G_i^{P\to M}$ is an interface energy term (with a positive value, the resistance for the transformation) and $\Delta G_e^{P\to M}$ is an elastic strain energy term around the martensite (with a positive value, the resistance for the transformation). As a shear transformation, the martensitic transformation requires the adaptive deformation of austenite adjacent to martensite, which will produce an elastic strain energy $\Delta G_e^{P\rightarrow M}$. This elastic strain energy is the main resistance of the martensitic transformation as the low interface energy of the coherent interface between the martensite and the parent phase. During the martensitic transformation, the formation of multiple martensite variants can reduce the elastic strain energy by a self-coordinating effect.

During the reverse transformation, a Gibbs free energy change may be defined as Formula (2). If the reverse transformation is the growth process of residual austenites, this transformation process is the reverse shear process of the phase interface reverse migration (the disappearance order of variants in reverse transformation is opposite to that in martensitic transformation, which is the result of the reverse shear process). Martensitic transformation is a process of accumulating strain in the austenite phase, so the reverse transformation is a process of eliminating strain. This leads to an essential difference between Formula (2) and Formula (1); that is, the elastic strain energy term in Formula (2) will become negative and become the driving force of the phase transition. However, if the austenite phase is formed by nucleation in the martensite phase, this driving force is missing. Therefore, the growth of residual austenites has a driving force advantage over the nucleation of the austenite phase in the reverse transformation, which can also explain why it always changes back to the original austenite phase grain through the reverse transformation.

$$\Delta G^{M\rightarrow P} = \Delta G_c^{M\rightarrow P} + \Delta G_i^{M\rightarrow P} + \Delta G_e^{M\rightarrow P} \quad (2)$$

Therefore, it is speculated that the growth of residual austenites will dominate the reverse transformation. Different martensitic variants transform into the original austenite grain, and in turn, the shape of the alloys is restored.

The direction of compressive stress applied to the Cu-Al-Mn SMA in this experiment is shown in Figure 3b. The relationship of the external stress (F) and the shear stress (S) in the martensite habit plane can be expressed as [28]:

$$S = F \cdot \sin(\theta) \cdot \cos(\alpha) \quad (3)$$

The schematic diagram of force analysis is also shown in Figure 3c. The stress S will provide part of the driving force for the martensite transformation, which promotes the preferential formation of martensite with this orientation (such as f orientation here). While in the other nonequivalent direction, the compressive stress could hinder the formation of martensite, which will make the martensite variants with orientations of "a", "b", "c" and "d" appear later or even not appear (such as "e" orientation). With constant applied stress and heating, the order of disappearance of the thermoelastic martensitic variants in the inverse phase transition is just opposite to the order of their formation during in the martensitic phase transition. It is suggested that constant external stresses do not affect the release of the strain energy stored in the martensitic phase transition.

5. Conclusions

The thermoelastic martensitic transformation and its reverse transformation of the Cu-Al-Mn cryogenic SMA with and without compressive stress were dynamically in situ observed by using a self-designed cryogenic metallographic device. Some thermoelastic martensitic transformation rules were summarized. The size and number of martensites gradually increase up with the decrease of temperature, while they decrease with the increase of temperature, with and without the compressive stress. Different martensite variants grow continuously in the length and width directions. These preferentially grown martensite strips can interact with other oriented martensites subsequently formed. The order of martensite disappearance is just opposite to that of their formation. In the process of thermoelastic martensite transformation, even when the temperature is lower than M_f, there are still some austenites which are not completely transformed into martensites. It means that the transformation from austenite to martensite was not completed to 100%. It is speculated that the martensite transformation of SMA would follow two stages, nucleation and growth, but its reverse transformation can occur only by the growth of residual austenites. If the reverse transformation can only occur by the growth of residual austenites, this can also explain why the deformed SMA remembers its original shape. The external compressive stress would change the appearance order of martensite variants, could induce new martensite variant and reduce some original martensite variants. Compared with the case without compressive stress, the growth of the original martensite variants could be restrained by the new stress field.

Author Contributions: Conceptualization, P.L. and F.W.; Data curation, J.S., Y.L. and Q.W.; Formal analysis, Z.B., S.Y. and Q.Z.; Funding acquisition, F.W.; Methodology, L.C. and J.W.; Resources, P.L. and F.W.; Supervision, P.L. and F.W.; Validation, Z.B.; Writing—original draft, Z.B.; Writing—review & editing, P.L., F.W., S.Y. and Q.Z. All authors have read and agreed to the published version of the manuscript.

Funding: This work was funded by the National Magnetic Confinement Fusion Program (grant no. 2022YFE03110003 and 2019YFE03130002), the Fundamental Research Funds for the Central Universities, with grant nos. FRF-IDRY-20-002 and FRF-GF-20-02A, and the National Natural Science Foundation of China (grant nos. 51971019 and 51971030).

Institutional Review Board Statement: Not applicable.

Informed Consent Statement: Not applicable.

Data Availability Statement: The data can be made available from the corresponding author upon reasonable request.

Acknowledgments: This work is supported by the National Magnetic Confinement Fusion Program (grant no. 2022YFE03110003 and 2019YFE03130002), the Fundamental Research Funds for the Central Universities, with grant nos. FRF-IDRY-20-002 and FRF-GF-20-02A, and the National Natural Science Foundation of China (grant nos. 51971019 and 51971030). The authors also would like to acknowledge the beamline 4B9A in Beijing Synchrotron Radiation Facilities (BSRF).

Conflicts of Interest: The authors declare no conflict of interest.

References

1. Lecce, L.; Concilio, A. *Shape Memory Alloy Engineering*; Butterworth-Heinemann Elsevier: Oxford, UK, 2015.
2. Arun, D.I.; Chakravarthy, P.; Arockia, K.R.; Santhosh, B. *Shape Memory Materials*; CRC Press: Boca Raton, FL, USA, 2018.
3. Li, M.; Liu, J.; Yan, S.; Yan, W.; Shi, B. Effect of aging treatment on damping capacity in Cu–Al–Mn shape memory alloy. *J. Alloys Compd.* **2020**, *821*, 153213. [CrossRef]
4. Araki, Y.; Endo, T.; Omori, T.; Sutou, Y.; Koetaka, Y.; Kainuma, R.; Ishida, K. Potential of superelastic Cu-Al-Mn alloy bars for seismic applications. *Earthq. Eng. Struct. Dyn.* **2011**, *40*, 107–115. [CrossRef]
5. Omori, T.; Koeda, N.; Sutou, Y.; Kainuma, R.; Ishida, K. Superplasticity of Cu-Al-Mn-Ni shape memory alloy. *Mater. Trans.* **2007**, *48*, 2914–2918. [CrossRef]
6. Omori, T.; Kawata, S.; Kainuma, R. Orientation Dependence of Superelasticity and Stress Hysteresis in Cu-Al-Mn Alloy. *Mater. Trans.* **2020**, *61*, 55–60. [CrossRef]
7. Greninger, A.B.; Mooradian, V.G. Strain transformation in metastable beta copper-zinc and beta copper-tin alloys. *Trans. AIME* **1938**, *138*, 337–368.
8. Kurdjumov, G.; Khandros, L. On the thermoelastic equilibrium on martensitic transformations. *Dokl. Akad. Nauk SSSR* **1949**, *2*, 211–214.
9. Kainuma, R. Recent Progress in Shape Memory Alloys. *Mater. Trans.* **2018**, *59*, 327–331. [CrossRef]
10. Jani, J.M.; Leary, M.; Subic, A.; Gibson, M.A. A review of shape memory alloy research, applications and opportunities. *Mater. Des.* **2014**, *56*, 1078–1113. [CrossRef]
11. Salzbrenner, R.J.; Cohen, M. On the thermodynamics of thermoelastic martensitic transformations. *Acta Metall.* **1979**, *27*, 739–748. [CrossRef]
12. Tong, H.C.; Wayman, C.M. Thermodynamics of thermoelastic martensitic transformations. *Acta Metall.* **1975**, *23*, 209–215. [CrossRef]
13. Mallik, U.S.; Sampath, V. Effect of composition and ageing on damping characteristics of Cu-Al-Mn shape memory alloys. *Mater. Sci. Eng. A Struct.* **2008**, *478*, 48–55. [CrossRef]

14. Mallik, U.S.; Sampath, V. Influence of aluminum and manganese concentration on the shape memory characteristics of Cu-Al-Mn shape memory alloys. *J. Alloys Compd.* **2008**, *459*, 142–147. [CrossRef]
15. Sutou, Y.; Kainuma, R.; Ishida, K. Effect of alloying elements on the shape memory properties of ductile Cu-Al-Mn alloys. *Mater. Sci. Eng. A Struct.* **1999**, *273*, 375–379. [CrossRef]
16. Zak, G.; Kneissl, A.C.; Zatulskij, G. Shape memory effect in cryogenic Cu-Al-Mn alloys. *Scr. Mater.* **1996**, *34*, 363–367. [CrossRef]
17. Shivaramu, L.; Shivasiddaramaiah, A.G.; Mallik, U.S.; Prashantha, S. Effect of Ageing on Damping Characteristics of Cu-Al-Be-Mn Quaternary Shape Memory Alloys. *Mater. Today Proc.* **2017**, *4*, 11314–11317. [CrossRef]
18. Jiao, Y.Q.; Wen, Y.H.; Li, N.; He, J.Q.; Teng, J. Effect of solution treatment on damping capacity and shape memory effect of a CuAlMn alloy. *J. Alloys Compd.* **2010**, *491*, 627–630. [CrossRef]
19. Mielczarek, A.; Kopp, N.; Riehemann, W. Ageing effects after heat treatment in Cu-Al-Mn shape memory alloys. *Mater. Sci. Eng. A Struct.* **2009**, *521*, 182–185. [CrossRef]
20. Sutou, Y.; Koeda, N.; Omori, T.; Kainuma, R.; Ishida, K. Effects of aging on stress-induced martensitic transformation in ductile Cu-Al-Mn-based shape memory alloys. *Acta Mater.* **2009**, *57*, 5759–5770. [CrossRef]
21. Sutou, Y.; Omori, T.; Wang, J.J.; Kainuma, R.; Ishida, K. Effect of grain size and texture on superelasticity of Cu-Al-Mn-based shape memory alloys. *J. Phys. IV* **2003**, *112*, 511–514.
22. Sutou, Y.; Omori, T.; Okamoto, T.; Kainuma, R.; Ishida, K. Effect of grain refinement on the mechanical and shape memory properties of Cu-Al-Mn base alloys. *J. Phys. IV* **2001**, *11*, 185–190. [CrossRef]
23. Saburi, T.; Wayman, C.M.; Takata, K.; Nenno, S. The shape memory mechanism in 18R martensitic alloys. *Acta Metall.* **1980**, *28*, 15–32. [CrossRef]
24. Otsuka, K.; Shimizu, K. On the crystallographic reversibility of martensitic transformations. *Scr. Metall.* **1977**, *11*, 757–760. [CrossRef]
25. Bhattacharya, K.; Conti, S.; Zanzotto, G.; Zimmer, J. Crystal symmetry and the reversibility of martensitic transformations. *Nature* **2004**, *428*, 55–59. [CrossRef] [PubMed]
26. Lei, Y.; Qin, X.; Wan, F.; Liu, P.; Chen, L.; Wang, J. In-situ observation of martensitic transformation in Cu-Al-Mn cryogenic shape memory alloy. *Fusion Eng. Des.* **2017**, *125*, 603–607. [CrossRef]
27. Otsuka, K.; Wayman, C.M. *Shape Memory Materials*; Cambridge University Press: Cambridge, UK, 1998.
28. Zhao, L.C.; Cai, W.; Zheng, Y.F. *Shape Memory Effect and Superelasticity in Alloys*; National Defense Industry Press: Beijing, China, 2002.

 materials

MDPI

Article

The Effect of Heat Treatment on Damping Capacity and Mechanical Properties of CuAlNi Shape Memory Alloy

Ivana Ivanić [1,*], Stjepan Kožuh [1], Tamara Holjevac Grgurić [2], Ladislav Vrsalović [3] and Mirko Gojić [1]

1 Department of Physical Metallurgy, Faculty of Metallurgy, University of Zagreb, Aleja Narodnih Heroja 3, 44000 Sisak, Croatia; kozuh@simet.unizg.hr (S.K.); gojic@simet.unizg.hr (M.G.)
2 Faculty of Medicine, Catholic University of Croatia, Ilica 242, 10000 Zagreb, Croatia; tamara.grguric@unicath.hr
3 Department of Electrochemistry and Materials Protection, Faculty of Chemistry and Technology, University of Split, Ruđera Boškovića 35, 21000 Split, Croatia; ladislav@ktf-split.hr
* Correspondence: iivanic@simet.unizg.hr; Tel.: +385-44533379 (ext. 203)

Abstract: This paper discusses the effect of different heat treatment procedures on the microstructural characteristics, damping capacities, and mechanical properties of CuAlNi shape memory alloys (SMA). The investigation was performed on samples in the as-cast state and heat treated states (solution annealing at 885 °C/60′/H_2O and after tempering at 300 °C/60′/H_2O). The microstructure of the samples was examined by light microscopy (LM) and scanning electron microscopy (SEM) equipped with a device for energy dispersive spectrometry (EDS) analysis. Light and scanning electron microscopy showed martensitic microstructure in all investigated samples. However, the changes in microstructure due to heat treatment by the presence of two types of martensite phases ($\beta_{1'}$ and $\gamma_{1'}$) influenced alloy damping and mechanical properties by enhancing alloy damping characteristics. Heat treatment procedure reduced the alloys' mechanical properties and increased hardness of the alloy. Fractographic analysis of the alloy showed a transgranular type of fracture in samples after casting. After solution annealing, two types of fracture mechanisms can be noticed, transgranular and intergranular, while in tempered samples, mostly an intergranular type of fracture exists.

Keywords: shape memory alloys; heat treatment; microstructure; damping capacity; tensile strength; hardness

Citation: Ivanić, I.; Kožuh, S.; Grgurić, T.H.; Vrsalović, L.; Gojić, M. The Effect of Heat Treatment on Damping Capacity and Mechanical Properties of CuAlNi Shape Memory Alloy. *Materials* **2022**, *15*, 1825. https://doi.org/10.3390/ma15051825

Academic Editors: Salvatore Saputo and Carmine Maletta

Received: 31 December 2021
Accepted: 25 February 2022
Published: 28 February 2022

1. Introduction

The main goal of the application of materials with high damping capacity is a reduction in mechanical vibrations by energy dissipation. For this type of application, materials must also have good mechanical strength, and good electrical and thermal conductivities [1]. However, the damping capacity of shape memory alloys (SMAs) is far greater than that of standard materials [2].

The functional properties of shape memory alloys are influenced by thermoelastic martensitic transformation, which operates in a certain temperature range, depending on the alloy's chemical composition [3]. These functional properties are affected by the mobile nature of the interfaces (twin boundaries, austenite/martensite (A/M) phase boundaries, different martensite variants) as well as the type of material, grain size, and defects in the structure [4,5].

Copper based SMAs are of interest for investigation due to their high thermal stability. Moreover, they exhibit good damping due to their unique martensitic transition characteristics [3,6].

Depending on alloy composition, applied stress, or temperature, the CuAlNi alloy can transform into different types of martensitic phase, $\gamma_{1'}$ (2H), $\beta_{1'}$ (18R1), and $\alpha_{1'}$ (6R). There is a change in martensitic phase due to the variation in compositions (Cu-(11-14)Al-(3-4.5)Ni, wt.%) from the β-$\beta_{1'}$ transformation to the β-$\gamma_{1'}$ by aging in the austenite phase [2]. Due to susceptibility to microstructural changes by aging, damping properties are also likely to substantially change [3,7].

The damping capacity, which consists of transferring one form of energy into another (the dissipation of mechanical energy into heat) is a valuable property that characterizes almost all materials including shape memory materials. In shape memory alloys, damping capacity is related to changes during martensitic transformation. The interfaces between A/M phases or between the different variants of martensite influence alloy damping capacity as well as twin boundaries inside the martensitic phase. Although the martensitic transformation has the most significant impact on the functional properties of alloys, other microstructural defects such as dislocations, vacancies, etc. can cause significant changes [8,9].

Hysteresis observed in pseudoelasticity is one of the energy dissipation manifestations [8]. According to the results of Wu et al. [10], in Cu-xZn-11Al (x = 7.0, 7.5, 8.0, 8.5, and 9.0 wt.%), a higher amount of martensite is transformed during martensite transformation, so the alloy exhibits a higher amount of energy dissipation due to higher hysteresis.

For shape memory alloys, it also depends on the difference between the operating and transformation temperatures. In general, three damping regimes can be distinguished in SMAs: (a) in the austenitic phase, the damping capacity is small; (b) an increase in damping capacity for operating below M_f temperature; and (c) the damping capacity reaches its maximum by stress induced martensite [2].

In recent research [11–13], the focus was primarily on phase field modeling and the simulation of martensite transformation and structural defect interaction under the thermal loading or martensite reorientation and de-twinning process. It can be seen that the evolution of microstructural constituents helps to understand microstructural mechanisms and their overall effect on the optimum performances of alloys. Li et al. [12] explained that not only does the twinning affect alloy damping, but also the reorientation in the martensite microstructure itself.

The aim of this study was to investigate how the changes in microstructure affected by heat treatment procedures influence the Cu-12.8 Al-4.1 Ni shape memory alloy damping and mechanical properties.

2. Materials and Methods

The sample of Cu-12.8 Al-4.1 Ni (wt.%) shape memory alloy was prepared in a vacuum induction furnace by the vertical continuous casting technique. The obtained 8 mm diameter rods were mechanically prepared for different stages of investigation. Dimensions of the investigated samples were ø 8 mm × 10 mm. One sample was left in the as-cast state and two samples were subjected to the heat treatment procedure. First, solution annealing was performed at 885 °C with a retention time of 60 min following water quenching. After solution annealing, one sample was subjected to tempering at 300 °C for 60 min following water quenching. The samples for microstructural observation were cut off in the shape of a cylinder, placed in conductive mass, ground, polished, and etched by the procedure explained in our previous work [14]. In order to reveal microstructural constituents, microstructural analysis after etching was performed using a light microscope (LM) OLYMPUS GX 51 with a digital camera and scanning electron microscope TESCAN VEGA TS 5136 MM (SEM) equipped with energy dispersive spectrometry (EDS). Fracture surface morphology was investigated using a scanning electron microscope JEOL JSM 5610 at several different magnifications.

For grain size measurements, the linear intercept procedure was used according to ASTM-E112-13 on optical micrographs at a magnification of 100×. The method involves an actual count of the number of grain boundary intersections by a test line, per unit length of test line. The average value of grain size in the as-cast state sample was calculated according to 512 counted intersections.

Phase transformation temperatures and the alloy's ability for damping capacity were tested using the dynamic mechanical analysis technique on a TA Instruments device,

DMA 983, at a constant frequency of 1 Hz with amplitude of 0.5 mm and at a heating rate of 2 °C min^{-1}.

Investigations of mechanical properties were carried out on universal tensile testing machine (Zwick/Roell Z050). Samples were mechanically prepared by turning at the dimensions shown in Figure 1. Tensile testing was performed on three samples for each state at room temperature with a tensile testing rate of 0.001 mm s^{-1}.

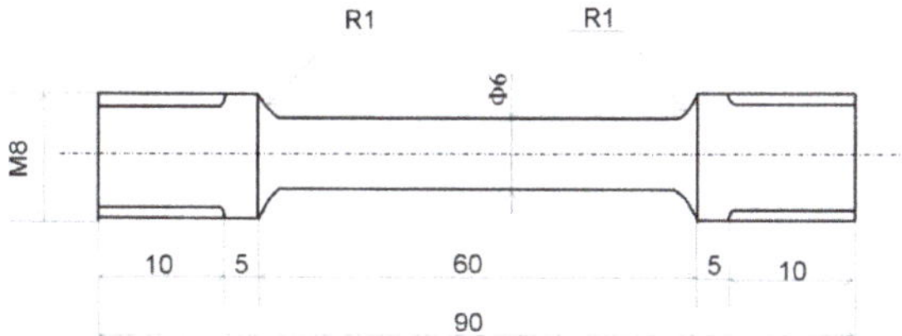

Figure 1. Schematic illustration of the tensile test sample.

3. Results and Discussion

In order to explain the damping capacity and mechanical properties of the investigated alloy, it is extremely important to examine how process parameters (casting, heat treatment) affect the microstructural characteristics of the alloy.

3.1. Microstructural Analysis of CuAlNi Shape Memory Alloy

The results of the light and scanning electron microscopy are presented in Figures 2 and 3, respectively. The martensitic microstructure was clearly visible in all of the samples, even in the as-cast state sample (Figures 2a and 3a). Often, during solidification of Cu-based SMAs, a residual austenitic phase or brittle γ_2 phase (Cu_9Al_4 phase) can be found, which strongly influences an alloy's functional properties in the as-cast state [4]. Hence, the heat treatment procedure cannot be avoided to improve the alloy's properties. The heat treatment process by solution annealing the CuAlNi SMA was carried out to achieve a fully martensitic microstructure. In addition to the formation of the martensitic phase from the initial austenitic (β) phase, there was a change in grain size that depends on the conditions of the heat treatment process (heat treatment temperature, retention time at certain temperature, and the choice of cooling medium).

Changes in the CuAlNi alloy microstructure vary depending on the heat treatment process [15]. In Figure 2b,c, we see the martensitic microstructure after solution annealing and tempering. In the micrographs obtained by light microscopy, it can be observed that during solution annealing and tempering, the grain size increased within the microstructure compared to the as-cast state sample. Grain size is an extremely important characteristic of the microstructure because other mechanical and functional properties of the alloy depend on it. The disadvantage of shape memory alloys is the coarse-grained microstructure, which negatively affects the behavior of the alloy. Grain size in the as-cast state was 158.76 μm (average value). Average value of the grain size in the as-cast state was obtained by the linear intercept procedure through 512 measurements of line intercepting points. ASTM grain size (G) was 2.02 and the number of grains per unit area (N_A) was 31.62 grains per square millimeter.

In addition, the grain sizes changed from the edge toward the middle of the rod because of solidification (i.e., directed heat dissipation where the rod is in contact with the crystallizer, small grains appear). After solution annealing and tempering, larger uniform grains in the microstructure could be observed, even in the order of 1 mm (Figure 2b,c). Kök et al. [16] investigated the thermal stability of the quaternary CuAlNiTa alloy and concluded that the heat treatment procedure enhanced grain size. Furthermore, Xi et al. [13] performed a phase field study of the grain size effect on the thermomechanical behavior of a NiTi shape memory alloy thin film and concluded that the grain size had an inhibiting effect on the temperature induced martensitic transformation.

The martensitic microstructure was confirmed on all samples and by scanning electron microscopy (Figure 3). Different orientations of martensite needles within a single grain can be explained by the nucleation of groups of martensitic needles at numerous sites within the grain and the creation of a local stress within the grain that allows for the formation of multiple groups of differently oriented needles. Martensite originated primarily as needle martensite. In some samples, after solution annealing and tempering (Figure 3b,c), the V-shape of martensite could be observed. The morphology of the resulting martensitic microstructure was a typical self-accommodating zig-zag morphology, which is primarily characteristic of $\beta_{1'}$ martensite in CuAlNi shape memory alloys [17,18].

It has also been confirmed in the literature [18,19] that different variants are characteristic of self-accommodating martensite in CuAlNi alloys and that these are most often the two types of heat-induced martensite ($\beta_{1'}$ and $\gamma_{1'}$ (18R and 2H)) that coexist in the microstructure. The $\beta_{1'}$ martensite appears as lath type martensite with thin plates (needle-like morphology) and $\gamma_{1'}$ is thicker with plate-like morphology (coarse variants) [5]. It has been reported that the $\gamma_{1'}$ martensite with a 2H structure in the CuZnAl alloy has abundant movable twins [10].

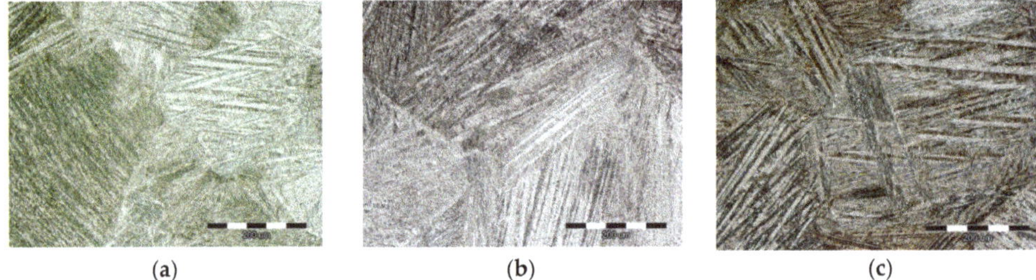

(a) (b) (c)

Figure 2. Light micrographs of the CuAlNi shape memory alloys in the as-cast state (**a**), after solution annealing at 885 °C/60′/H_2O (**b**), and after solution annealing at 885 °C/60′/H_2O and tempering at 300 °C/60′/H_2O (**c**), magnification 200×.

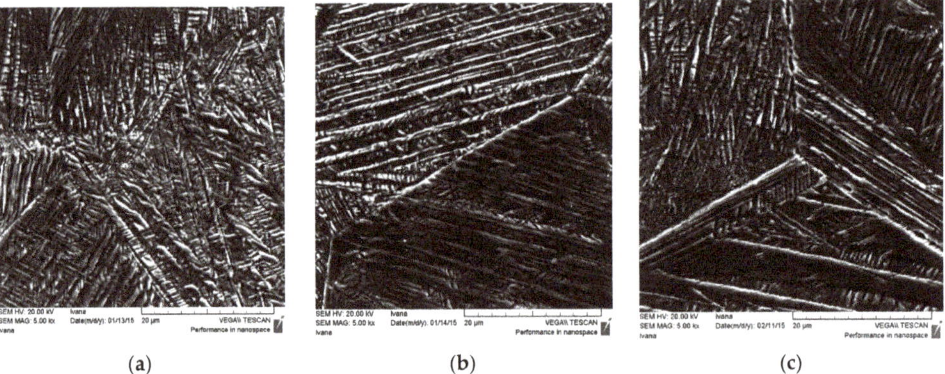

(a) (b) (c)

Figure 3. SEM micrographs of the CuAlNi shape memory alloys in the as-cast state (**a**), after solution annealing at 885 °C/60′/H_2O (**b**), and after solution annealing at 885 °C/60′/H_2O and tempering at 300 °C/60′/H_2O (**c**).

3.2. Results of Dynamic-Mechanical Analysis

Dynamic-mechanical analysis (DMA) is a technique of the thermal analysis of materials by which we can monitor the response of materials to cyclic loading during the controlled heating of materials at different temperature, time, frequency, stress, atmosphere, or a combination of these parameters. The sinusoidal cyclic stress of the material results in deformations that change sinusoidally with time at the same frequency [20]. The storage modulus (E′) is related to the properties of the elastic component and is proportional to the stored energy, which is returned as mechanical energy during periodic deformation. The stress component bound to the viscous component is determined by the size of the loss modulus (E″) proportional to the lost mechanical energy in the form of heat. The phase shift angle is given by the ratio of the loss module (E″) and the storage module (E′) and is a measure of the energy loss in the material due to viscous friction [21].

The results of the dynamic-mechanical analysis measurements are shown in Figures 4–6. The storage modulus (E′), loss modulus (E″), and mechanical damping parameter (tan δ) are shown as a function of temperature during heating.

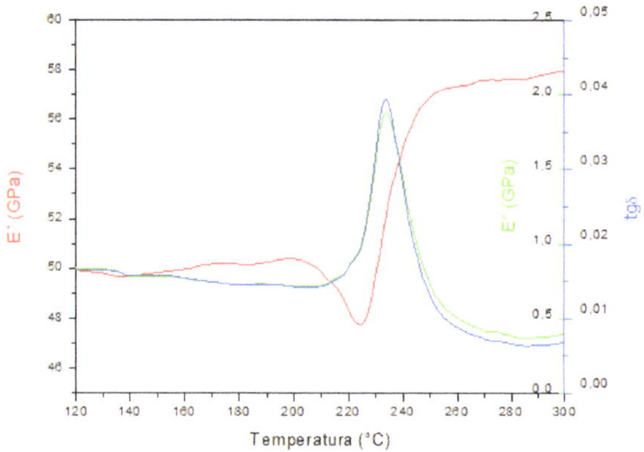

Figure 4. DMA spectrum of the CuAlNi shape memory alloy in the as-cast state.

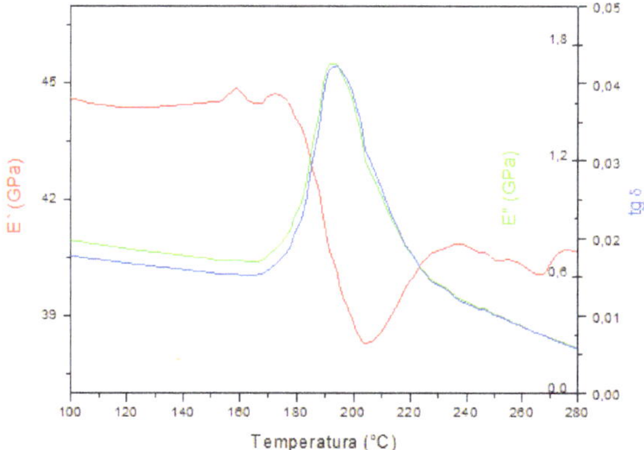

Figure 5. DMA spectrum of the CuAlNi shape memory alloy after solution annealing at 885 °C/60′/H_2O.

Dynamic-mechanical analysis showed the dependence of the tangent of the phase shift angle, which is a measure of the energy loss in the material (tan δ), the storage modulus (E′), and the loss modulus (E″) with temperature. The DMA spectra of the investigated samples indicated that the highest temperature of austenitic transformation (A_s = 220 °C, A_f = 250 °C) was in the samples in the as-cast state, which is in accordance with the obtained results of the differential scanning calorimetry reported in our previous work [22]. Temperatures of A_s and A_f were higher than the DSC results (A_s = 179 °C, A_f = 212 °C), which is a common measurement difference between the two techniques (DSC and DMA), reported in the research by Graczykowski et al. [23]. The maximum intensity of the loss modulus (E″) and the change in the storage modulus (E′) indicated a slightly lower austenitic transformation temperature in the solution annealed and tempered alloy and the lowest temperature in the solution annealed alloy at 885 °C/60′/H_2O. The values of the storage modulus were highest in the martensitic structure of the as-cast alloy and decreased with the obtained heat treatment procedure, which correlated with the tests of the mechanical properties and tensile strength (Figure 7, Table 1). The lowest values of tan δ and E″ could be observed in the sample in the as-cast state. The phase shift angle tan δ increased for the solution annealed and tempered state, and it can be concluded that the energy loss ability in the material was higher for the samples after solution annealing and tempering. In addition, the loss modulus (E″) increased for the solution annealed state sample, and was slightly lower for the tempered state sample. Since heat treatment leads to changes in the microstructure, it can be concluded that the newly formed intermediate boundaries of martensite ($\beta_{1'}$ and $\gamma_{1'}$ martensite) favorably affect the ability for damping of the CuAlNi shape memory alloy. Ursanu et al. investigated CuAlMn shape memory alloys by DMA and discovered that after heat treatment process by aging above 300 °C, the alloy's damping capacity decreased due to formations of brittle γ_2 precipitates, which restricted the mobility of the martensite interfaces [24].

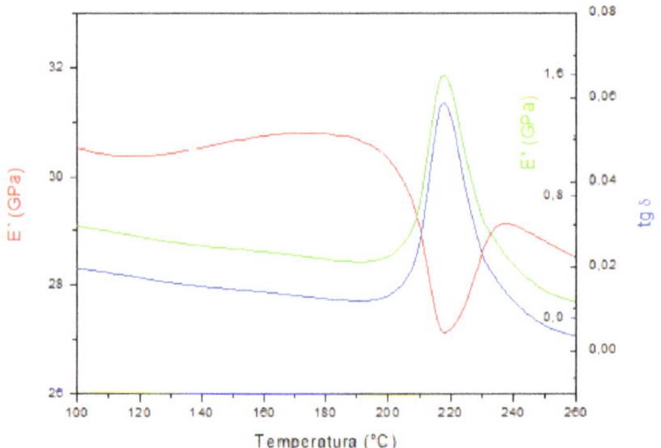

Figure 6. DMA spectrum of the CuAlNi shape memory alloy after solution annealing at 885 °C/60′/H_2O and tempering at 300 °C/60′/H_2O.

Table 1. Mechanical properties of the CuAlNi alloy after casting and heat treatment.

Sample	Tensile Strength, MPa	Elongation, %	Hardness, HV1
L (as-cast state)	475.5 ± 9.3	4.78 ± 0.28	344.0 ± 18.4
K-2 (885 °C/60′/H_2O)	367.5 ± 48.8	2.72 ± 0.46	480.0 ± 25.0
K-2-4 (885 °C/60′/H_2O + 300 °C/60′/H_2O)	241.7 ± 11.9	1.36 ± 0.18	483.0 ± 14.4

The difference in tan δ values between the solution annealed and tempered state was also noticed by Suresh and Ramamurty [3] (80% higher tan δ in the solution annealed sample). Damping capacity of the shape memory alloys can be significantly altered by an artificial aging process. Shape memory alloys have a high damping capacity due to the ability to move the austenite/martensite interface during the phase transformation. However, it is difficult to move the interfacial boundary due to the precipitation of the low-temperature γ_2 phase and unfavorable influence on alloy damping capacity [3].

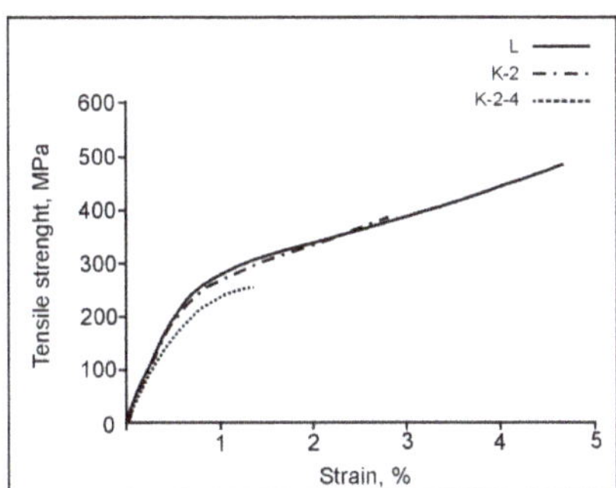

Figure 7. Stress vs. strain in the CuAlNi SMA.

Chang [25] investigated the damping capacity by dynamic mechanical analysis on a Cu-X% Al-4% Ni (X = 13.0–14.1 wt.%) SMA. He found that an alloy with 14% Al, thanks to the high concentration of moving twins interfaces in $\gamma_{1'}$ (2H) martensite, satisfied the application in which the vibration damping property is required under isothermal conditions.

3.3. *Results of Mechanical Properties of the CuAlNi Shape Memory Alloy*

The investigated CuAlNi shape memory alloy showed satisfactory results of stress, strain, and hardness in the as-cast state (Table 1, Figure 7). With heat treatment procedure, the values of tensile strength and strain decreased, while the hardness increased significantly. The reason for such behavior can be explained by changes in the obtained microstructure and grain size. Although the mechanical properties notably decreased after heat treatment, the development of favorable martensite phases produced better damping properties. Two types of martensite can be found in this alloy, $\beta_{1'}$ and $\gamma_{1'}$, respectively [26]. Appearance of $\gamma_{1'}$ martensite improves the damping capacity of the alloy due to movement of twin boundaries in the martensite phase and higher hysteresis in comparison to $\beta_{1'}$ martensite [3]. Improvement in the mechanical properties can also be achieved by grain refinement procedures (microalloying or rapid solidification techniques) in order to affect or limit the movement of internal dislocations in the alloy [27].

Significant changes in fracture mechanism could be noted by SEM fractography analysis, Figure 8. In the as-cast state, a mainly transgranular type of fracture could be seen including areas with small and shallow dimples, assuming that certain plasticity in the sample had appeared. After solution annealing, the appearance of the intergranular type of fracture was visible, and after tempering at 300 °C/60′/H_2O, an almost completely intergranular type of fracture appeared. The degradation of fracture mechanisms, or shift from a transgranular type of fracture to intergranular type after heat treatment, can be explained by the alloy's high elastic anisotropy [28,29] and large grain size [30]. Stress level concentration is high at large grain boundaries and nucleation of the cracks appears specifically in these places due to low cohesion. It was shown that because of the low ductility of the CuAlNi SMA, cracking of the samples occurred even during elastic deformation. This investigation is in accordance with the results that Pushin et al. [31] obtained in the studies of different NiTi-based and Cu-based shape memory alloys by confirming that large grains in the NiTi alloy revealed brittle fracture as well as classic brittle fracture along the grain boundaries in the investigated CuAlNi SMA.

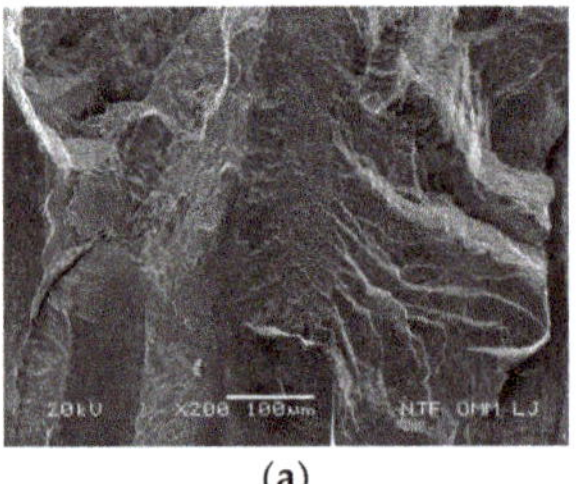

(**a**)

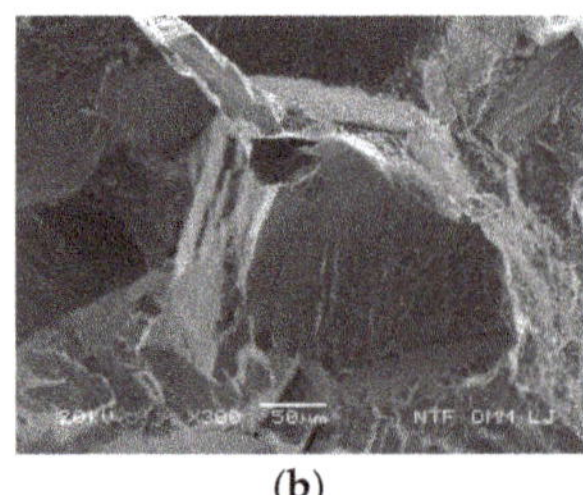

(**b**)

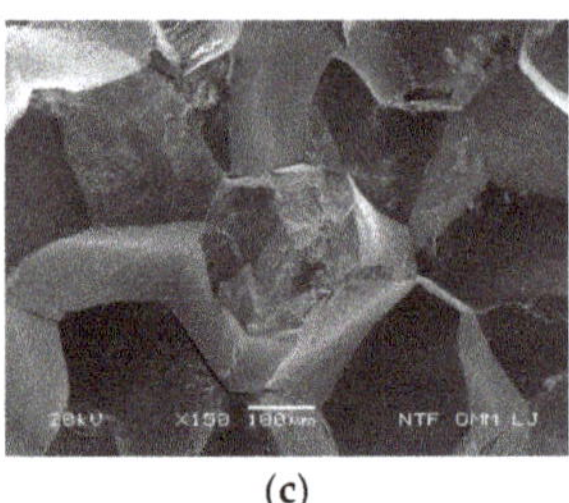

(**c**)

Figure 8. Fracture surface morphology after tensile testing in the as-cast state (**a**), after solution annealing at 885 °C/60′/H_2O (**b**), and after solution annealing at 885 °C/60′/H_2O and tempering at 300 °C/60′/H_2O (**c**).

4. Conclusions

An investigation of the microstructural and dynamic-mechanical properties of the CuAlNi shape memory alloy was carried out. From the detailed analysis, the following conclusions can be drawn:

- Microstructural analysis by light microscopy and scanning electron microscopy revealed the martensitic phase in the as-cast, solution annealed, and tempered state samples. The appearance of two types of martensite phases ($\beta_{1'}$ and $\gamma_{1'}$) was noted.
- Dynamic-mechanical analysis showed a higher damping capacity for samples after solution annealing and tempering. The newly formed interfacial boundaries of martensite ($\beta_{1'}$ and $\gamma_{1'}$ martensite) favorably affected the damping properties of the CuAlNi shape memory alloy.
- Results of the mechanical properties showed degradation of stress, strain, and hardness in the heat treated states due to large grain size, which appeared after the heat treatment process.
- Fractographic analysis showed a transgranular type of fracture in the as-cast state sample, transgranular and intergranular type of fracture in the solution annealing state, and mostly intergranular type of fracture in the tempered state sample. It can be concluded that the brittleness of the alloy arose from this heat treatment process.

Author Contributions: Conceptualization, I.I. and S.K.; methodology, I.I. and T.H.G.; formal analysis, I.I., and T.H.G.; investigation, I.I. and T.H.G.; resources, M.G.; data curation, I.I.; writing—original draft preparation, I.I.; writing—review and editing, L.V. and M.G.; visualization, I.I.; supervision, I.I., S.K., L.V.; project administration, M.G.; funding acquisition, M.G. All authors have read and agreed to the published version of the manuscript.

Funding: This work was fully supported by the Croatian Science Foundation under project IP-2014-09-3405.

Institutional Review Board Statement: Not applicable.

Informed Consent Statement: Not applicable.

Data Availability Statement: The data presented in this study are available on request from the corresponding author.

Conflicts of Interest: The authors declare no conflict of interest.

References

1. Mielczarek, A.; Kopp, N.; Riehemann, W. Ageing effects after heat treatment in Cu-Al-Mn shape memory alloys. *Mater. Sci. Eng. A* **2009**, *521–522*, 182–185. [CrossRef]
2. Patoor, E.; Lagoudas, D.C.; Entchev, P.B.; Brinson, L.C.; Gao, X. Shape memory alloys, Part I: General properties and modeling of single crystals. *Mech. Mater.* **2006**, *38*, 391–429. [CrossRef]
3. Suresh, N.; Ramamurty, U. Aging response and its effect on the functional properties of Cu-Al-Ni shape memory alloys. *J. Alloys Compd.* **2008**, *449*, 113–118. [CrossRef]

4. Van Humbeeck, J. Damping capacity of thermoelastic martensite in shape memory alloys. *J. Alloys Compd.* **2003**, *355*, 58–64. [CrossRef]
5. Sampath, V. Studies on the effect of grain refinement and thermal processing on shape memory characteristics of Cu-Al-Ni alloys. *Smart Mater. Struct.* **2005**, *14*, 253–260. [CrossRef]
6. Sutou, Y.; Omori, T.; Koeda, N.; Kainuma, R.; Ishida, K. Effects of grain size and texture on damping properties of Cu–Al–Mn-based shape memory alloys. *Mater. Sci. Eng. A* **2006**, *438–440*, 743–746. [CrossRef]
7. Otsuka, K.; Wayman, C.M. *Shape Memory Materials*; University of Cambridge: Cambridge, UK, 1998; pp. 27–36.
8. Strittmatter, J.; Clipa, V.; Gheorghita, V.; Gümpel, P. Characterization of NiTi Shape Memory Damping Elements designed for Automotive Safety Systems. *J. Mater. Eng. Perform.* **2014**, *23*, 2696–2703. [CrossRef]
9. Alaneme, K.K.; Umar, S. Mechanical behaviour and damping properties of Ni modified Cu-Zn-Al shape memory alloys. *J. Sci. Adv. Mater. Dev.* **2018**, *3*, 371–379. [CrossRef]
10. Wu, S.K.; Chan, W.J.; Chang, S.H. Damping Characteristics of Inherent and Intrinsic Internal Friction of Cu-Zn-Al Shape Memory Alloys. *Metals* **2017**, *7*, 397. [CrossRef]
11. Javanbakht, M.; Ghaedi, M.S. Interaction of martensitic transformations and vacancy diffusion at the nanoscale under thermal loading: A phase field model and simulations. *Acta Mech.* **2021**, *232*, 4567–4582. [CrossRef]
12. Li, X.; Su, Y. A phase-field study of the martensitic detwinning in NiTi shape memory alloys under tension or compression. *Acta Mech.* **2020**, *231*, 1539–1557. [CrossRef]
13. Xi, S.; Su, Y. A phase field study of the grain-size effect on the thermomechanical behavior of polycrystalline NiTi thin films. *Acta Mech.* **2021**, *232*, 4545–4566. [CrossRef]
14. Ivanić, I.; Gojić, M.; Kožuh, S.; Kosec, B. Microstructural and Fractographic Analysis of CuAlNi Shape Memory Alloy before and after Heat Treatment. *Defect Diffus. Forum* **2020**, *405*, 100–106.
15. Lexcellent, C. *Shape-Memory Alloys Handbook*; John Wiley & Sons: New York, NY, USA, 2013; pp. 16–25.
16. Kök, M.; Qader, I.N.; Mohammed, S.S.; Öner, E.; Dağdelen, F.; Aydogdu, Y. Thermal stability and some thermodynamics analysis of heat treated quaternary CuAlNiTa shape memory alloy. *Mater. Res. Express* **2020**, *7*, 015702. [CrossRef]
17. Gojić, M.; Kožuh, S.; Anžel, I.; Lojen, G.; Ivanić, I.; Kosec, B. Microstructural and phase analysis of CuAlNi shape-memory alloy after continuous casting. *Mater. Technol.* **2013**, *47*, 149–152.
18. Sari, U.; Aksoy, I. Micro-structural analysis of self-accommodating martensites in Cu–11.92 wt%Al–3.78 wt%Ni shape memory alloy. *J. Mater. Process. Technol.* **2008**, *195*, 72–76. [CrossRef]
19. Sari, U.; Aksoy, I. Electron microscopy study of 2H and 18R martensites in Cu-11.92 wt% Al-3.78 wt% Ni shape memory alloy. *J. Alloys Compd.* **2006**, *417*, 138–142. [CrossRef]
20. Bashir, M.A. Use of Dynamic Mechanical Analysis (DMA) for Characterizing Interfacial Interactions in Filled Polymers. *Solids* **2021**, *2*, 108–120. [CrossRef]
21. Holjevac Grgurić, T. Modificiranje i Stabilnost Plastomernih Mješavina sa Stiren-Etilen/Butilen-Stiren Blok Kopolimerom. Ph.D. Thesis, University of Zagreb Faculty of Chemical Engineering and Technology, Zagreb, Croatia, 19 September 2006.
22. Ivanić, I.; Kožuh, S.; Holjevac Grgurić, T.; Kosec, B.; Gojić, M. The Influence of Heat Treatment on Microstructure and Phase Transformation Temperatures of Cu-Al-Ni Shape Memory Alloy. *Chem. Ind.* **2019**, *68*, 111–118. [CrossRef]
23. Graczykowski, B.; Biskupski, P.; Mroz, B.; Mielcarek, S.; Nò, M.L.; San Juan, J. Elastic properties of Cu–Al–Ni shape memory alloys studied by dynamic mechanical analysis. *Smart Mater. Struct.* **2010**, *19*, 015010. [CrossRef]
24. Ursanu, A.I.; Stanciu, S.; Pricop, B.; Săndulache, F.; Cimpoeșu, N. Dynamic mechanical analyze of superelastic CuMnAl shape memory alloy. *Mater. Sci. Eng.* **2016**, *147*, 012032.
25. Chang, S.H. Influence of chemical composition on the damping characteristics of Cu-Al-Ni shape memory alloys. *Mater. Chem. Phys.* **2011**, *125*, 358–363. [CrossRef]
26. Ivanić, I.; Kožuh, S.; Kurajica, S.; Kosec, B.; Anžel, I.; Gojić, M. XRD analysis of CuAlNi shape memory alloy before and after heat treatment. In Proceedings of the Mechanical Technologies and Structural Materials 2016, Split, Croatia, 22–23 September 2016; pp. 55–60.
27. Zhang, Q.; Cui, B.; Sun, B.; Zhang, X.; Dong, Z.; Liu, Q.; Cui, T. Effect of Sm Doping on the Microstructure, Mechanical Properties and Shape Memory Effect of Cu-13.0Al-4.0Ni Alloy. *Materials* **2021**, *14*, 4007. [CrossRef]
28. Ivanić, I.; Gojić, M.; Kožuh, S.; Kosec, B. Microstructural analysis of CuAlNiMn shape-memory alloy before and after the tensile testing. *Mater. Tehnol.* **2014**, *48*, 713–718.
29. Sutou, Y.; Omori, T.; Kainuma, R.; Ono, N.; Ishida, K. Enhancement of superelasticity in Cu-Al-Mn-Ni shape-memory alloys by texture control. *Metall. Mater. Trans. A* **2002**, *33*, 2817–2824. [CrossRef]
30. Sari, U. Influences of 2.5 wt% Mn addition on the microstructure and mechanical properties of Cu-Al-Ni shape memory alloys. *Metall. Mater.* **2010**, *17*, 192–198.
31. Pushin, V.; Kuranova, N.; Marchenkova, E.; Pushin, A. Design and Development of Ti–Ni, Ni–Mn–Ga and Cu–Al–Ni-Based Alloys with High and Low Temperature Shape Memory Effects. *Materials* **2019**, *12*, 2616. [CrossRef]

 materials

MDPI

Article

Optimized Neural Network Prediction Model of Shape Memory Alloy and Its Application for Structural Vibration Control

Meng Zhan [1,*], Junsheng Liu [2], Deli Wang [3], Xiuyun Chen [1], Lizhen Zhang [1] and Sheliang Wang [4]

1 College of Construction Engineering, HuangHuai University, No. 76 Kaiyuan RD., Zhumadian 463000, China; Cxy0396@126.com (X.C.); zhenzhen_5255@sina.com (L.Z.)
2 Shaanxi Institute of Building Science, No. 272 Huancheng RD., Xi'an 710082, China; jsliu1973@163.com
3 Taizhou Urban and Rural Planning and Design Institute, No. 465 Shifu RD., Taizhou 318000, China; wangdeli2012@sina.com
4 College of Civil Engineering, Xi'an University of Architecture and Technology, No. 13 Yanta RD., Xi'an 710055, China; sheliangw@163.com
* Correspondence: zhanyi313@163.com

Abstract: The traditional mathematical model of shape memory alloy (SMA) is complicated and difficult to program in numerical analysis. The artificial neural network is a nonlinear modeling method which does not depend on the mathematical model and avoids the inevitable error in the traditional modeling method. In this paper, an optimized neural network prediction model of shape memory alloy and its application for structural vibration control are discussed. The superelastic properties of austenitic SMA wires were tested by experiments. The material property test data were taken as the training samples of the BP neural network, and a prediction model optimized by the genetic algorithm was established. By using the improved genetic algorithm, the position and quantity of the SMA wires were optimized in a three-storey spatial structure, and the dynamic response analysis of the optimal arrangement was carried out. The results show that, compared with the unoptimized neural network prediction model of SMA, the optimized prediction model is in better agreement with the test curve and has higher stability, it can well reflect the effect of loading rate on the superelastic properties of SMA, and is a high precision rate-dependent dynamic prediction model. Moreover, the BP network constitutive model is simple to use and convenient for dynamic simulation analysis of an SMA passive control structure. The controlled structure with optimized SMA wires can inhibit the structural seismic responses more effectively. However, it is not the case that the more SMA wires, the better the shock absorption effect. When SMA wires exceed a certain number, the vibration reduction effect gradually decreases. Therefore, the seismic effect can be reduced economically and effectively only when the number and location of SMA wires are properly configured. When four SMA wires are arranged, the acceptable shock absorption effect is obtained, and the sum of the structural storey drift can be reduced by 44.51%.

Keywords: SMA; prediction model; BP neural network; genetic algorithm; seismic response

Citation: Zhan, M.; Liu, J.; Wang, D.; Chen, X.; Zhang, L.; Wang, S. Optimized Neural Network Prediction Model of Shape Memory Alloy and Its Application for Structural Vibration Control. *Materials* **2021**, *14*, 6593. https://doi.org/10.3390/ma14216593

Academic Editors: Salvatore Saputo and Radim Kocich

Received: 13 September 2021
Accepted: 27 October 2021
Published: 2 November 2021

1. Introduction

Shape memory alloy (SMA) is a novel functional material, which not only has two unique properties of shape memory and super elasticity, but also has the advantages of high damping, fatigue, and corrosion resistance [1–3]. Therefore, SMA has been widely used in the field of civil engineering for reinforcement and vibration control [4–6]. Dehghani and Aslani developed hooked-end pseudoelastic shape memory alloy fibres (PSMAF) which can provide re-centring and crack-closing behaviour in cementitious composites [7]. Rezapour et al. investigated the effects of SMA post-tensioning reinforcements on originally unreinforced masonry walls using Abaqus soft [8]. Siddiquee et al. evaluated the seismic performance of concrete frame buildings reinforced with superelastic SMA rebar in terms of the collapse margin ratio, and the results are compared with that of a reinforced concrete structure with regular steel rebar only [9]. Schranz et al. studied the effectiveness of two different prestressed strengthening methods for flexural behavior of concrete members using ribbed Fe-SMA bars [10]. Peng et al. proposed a shape memory alloy cable-double friction pendulum bearing which combines a double friction pendulum bearing with superelastic SMA cables [11]. Based on the pseudoelasticity of SMA and the electrodeformation of piezoelectric transition ceramic, Zhan et al. designed a novel SMA/PZT composite control device and investigated its energy dissipation performance and neural network constitutive model [12]. In addition, some researchers discussed the control of shape memory alloy actuator using PID controller and pulse width modulation [13–16].

The constitutive model of SMA material is the basis of theoretical analysis and experimental study for properties of SMA [17,18]. A series of constitutive models of SMA have been established based on material experiments. Falk established a theoretical constitutive model of a single crystal based on Landau's theory, but it only applies to single crystal and only considers the shearing motion of crystal [19]. Abeyartane and Knowle proposed a one-dimensional constitutive model of SMA material by using the thermodynamic bar theory and combining the Helmholzt free energy principle and the expression of thermodynamic relation, but this model did not consider the redirection of stress-induced martensite variation under non-proportional load [20]. Boyd and Lagoudas presented a meso-mechanical constitutive model based on meso-mechanics and thermodynamics, but this model is too complex to be applied in engineering [21]. Based on the constitutive relationship between thermodynamics and kinetics, Brinson proposed a phenomenological constitutive model with the characteristics of plasticity theory, which is the most widely used constitutive model of SMA [22]. Liu et al. explored a macroscopic phenomenological constitutive model involving strain amplitude and loading rate by introducing an internal variable evolution equation, considering the difference of characteristic parameters during the positive/reverse transformation of martensite and the hardening properties of martensite under large strain amplitudes [23]. Based on the Landau theory of phase transitions, Du et al. proposed a differential model to describe the hysteresis of magnetic shape memory alloys caused by magnetic field induced martensite reorientation [24]. However, for shape memory alloy, there are many factors affecting its constitutive model, so it is impossible to establish an accurate mathematical model to express the influence of all various factors on it, and it can only be simplified and approximated artificially. The artificial neural network does not need to achieve the precise constitutive relation of the material, but only needs to consider the influencing factors and expected objectives, and it avoids the inevitable error in the traditional modeling method and provides a new way to establish the accurate SMA constitutive model. Lee and Lee used a multi-layer perceptron neural network to forecast the resilience of SMA [25]. Ren et al. used a radial basis function network to forecast the hysteresis behavior of SMA at the same strain amplitude with the number of cycles, loading information, and strain value as the input of neurons [26]. However, the change of initial weight/threshold of neurons has a great influence on the prediction results of the artificial neural network, so the initial weight/threshold should be optimized in the neural network to improve the accuracy and stability.

The damping effect of the SMA damper on an engineering structure is mainly determined by its own performance, installation position, and layout number [27,28]. Even if its damping force is large and the shock absorption effect is good, the ideal damping effect may not be achieved if the installation position and number are inappropriate [29–31]. Therefore, in order to achieve the efficiency and economy of the SMA control system, it is necessary to optimize the installation position and quantity of SMA wires in the engineering application. By idealizing the hysteretic curve of SMA damper and the transient tangent stiffness matrix of structure, Mulay and Shmerling presented a simplified nonlinear equation model to design SMA dampers in three-dimensional asymmetric plane structures [32]. Pang et al. proposed a novel seismic risk-based optimization design method of SMA restrained sliding bearings in highway bridges suffered from near-fault ground motions via the particle swarm optimization [33]. Zhan et al. discussed the optimal arrangement of SMA piezoelectric composite dampers in a space truss structure using the adaptive immune memory cloning algorithm with the performance indicator based on the modal controllable degree as antigen-antibody affinity function [34].

This paper explored a novel SMA prediction model considering loading rate and loading history via a back-propagation (BP) neural network optimized by the genetic algorithm (GA). SMA wires were selected to control the dynamics response of a three-storey spatial structure, and the application of artificial neural network to SMA constitutive model and the damping effect of SMA wires on the seismic response of spatial structure are discussed. Firstly, the sum of lateral inter-story displacement is taken as the control objective, the optimization of the number and position of SMA wires is carried out by applying the improved genetic algorithm. Then, the Newmark-β method is employed to accomplish the seismic response control analysis of the spatial structure with an optimal allocation of SMA wires via MATLAB soft.

2. SMA Wire Mechanical Performance Test

2.1. Test Loading Scheme

The chemical composition of the austenitic SMA wire used in the test is Ti-51%atNi [35,36] with diameters 0.5, 0.8, 1.0, and 1.2 mm. Phase transition temperature: Martensite finish temperature M_f is −42 °C; Martensite start temperature M_s is −38 °C; Austenite start temperature A_s is −6 °C; Austenite finish temperature A_f is −2 °C. Taiwan Hongda HT-2402 computer servo-controlled material testing machine is used in Material Science Laboratory of Xi'an University of Technology. The axial force is measured by the force sensor that comes with the testing machine, and the axial deformation is measured by the displacement extensometer. The gauge length is 33.5 mm, and all data are automatically collected by computer.

This test mainly considers the influence of the number of loading cycles, strain amplitude, loading rate, and material diameter on the stress–strain curve, energy dissipation capacity, equivalent damping ratio, and equivalent secant modulus of austenitic SMA wire. In this experiment, the loading rate was 10 mm/min, 30 mm/min, 60 m/min, 90 mm/min, and the strain amplitude was 3%, 6%, and 8%. To ensure the accuracy of the test, a pre-tension of 10~30 MPa is applied to the SMA wire before each working condition starts to work so that the SMA wire can be straightened and tightened. To eliminate the influence of the test piece length on SMA, the length of the test piece used is 300 mm with an effective length of 100 mm. In each cycle, only when the SMA wire strain reaches the amplitude strain, loading stops, and when the axial force of the SMA wire was less than 5N, unloading stops. Loading cycle was 30 cycles in each working condition.

2.2. Test Results and Analysis

The stress of four points (σ_a, σ_b, σ_c, σ_d) is used to replace the critical stress of SMA transformation, as shown in Figure 1. The starting point of the stress–strain curve platform of the loading section is taken as point a. The point where the slope of the loading section curve is significantly increased after the loading platform is taken as point b. The point where the stress–strain drops start to deviate from the linear relation is taken as point c. The point where the stress and strain begin to decrease at the end is taken as point d. Substituting the above-mentioned points for the phase transition critical point of austenite SMA, although there is a certain error between the characteristic point stress and the true phase transition stress of SMA, the phase transition process of SMA can still be reflected.

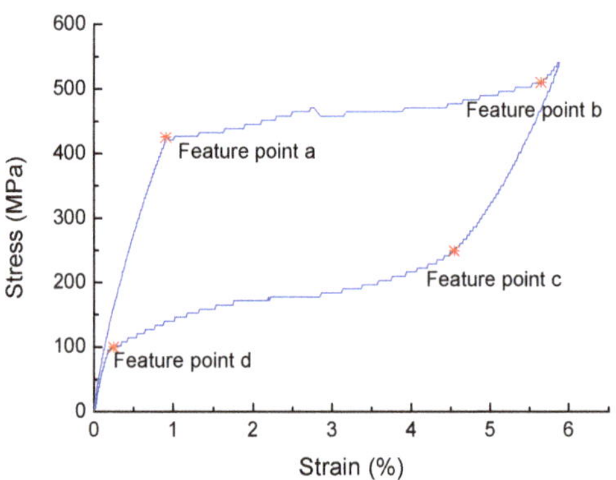

Figure 1. Characteristic points of austenite SMA constitutive curve.

The mechanical performance indicators are defined as follows: ΔW represents the energy consumption value of the SMA wire in a single cycle, that is, the graphic area enclosed by the stress–strain curve; K_s represents the equivalent secant stiffness of a single cycle; δ represents the equivalent damping ratio of a single cycle.

$$K_s = \frac{\sigma_{max} - \sigma_{min}}{\varepsilon_{max} - \varepsilon_{min}} \tag{1}$$

$$\delta = \frac{\Delta W}{2\pi K_s \varepsilon^2} \tag{2}$$

where σ_{max} (ε_{max}) and σ_{min} (ε_{min}) respectively represent the maximum stress (strain) and minimum stress (strain) of each cycle, and ε is the amplitude strain.

According to the experimental results, the effects of cyclic loading, strain amplitude, loading rate and material diameter on the mechanical properties of austenitic SMA wire are analyzed.

(1) The effect of cyclic loading on the mechanical properties of superelasticity. The stress–strain curve with a diameter of 1.0 mm, a loading rate of 10 mm/min, and a strain amplitude of 3% is shown in Figure 2, and the parameters are shown in Table 1. It can be seen that, with the increase of the number of cycles, the cumulative residual deformation of the austenitic SMA wire gradually increases, but the residual deformation of the single cycle gradually becomes smaller, and stabilizes after 15 cycles, and the residual strain is zero. With the increase in the number of cycles of loading, the performance of austenitic SMA wire gradually stabilizes, the stress–strain curve gradually becomes smooth, the energy dissipation capacity and equivalent damping ratio of the SMA wire gradually decrease, and the equivalent secant stiffness slightly decreased, but stabilized after 15 loading/unloading cycles. The number of cycles has a great influence on the mechanical properties of austenitic SMA wire. In actual engineering applications, in order to obtain stable superelastic properties, the SMA wire must be cyclically loaded in advance, and it usually takes about 20 cycles.

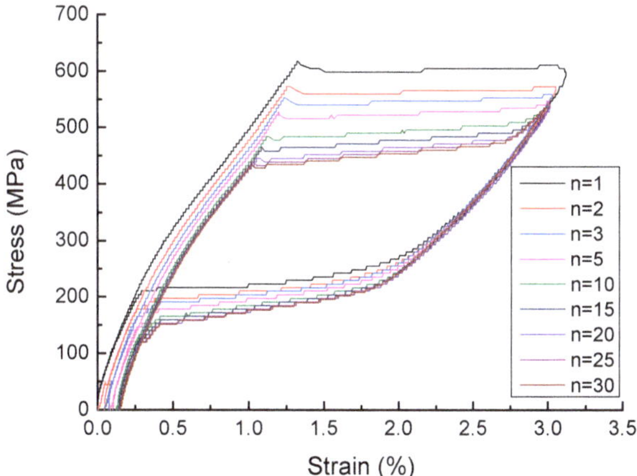

Figure 2. Stress-strain curve of austenite SMA wire with different cycles.

Table 1. Mechanical properties of SMA wires with different cycles.

Cycles	σ_a (MPa)	σ_b (MPa)	σ_c (MPa)	σ_d (MPa)	ΔW (MJ.m^{-3})	ζ_a (%)	K_s (GPa)
1	604.79	604.79	273.75	178.25	6.84	6.11	19.78
2	560.23	572.96	254.65	171.89	6.19	5.81	18.84
3	541.13	560.23	241.92	171.89	5.80	5.44	18.82
5	515.66	541.13	241.92	165.52	5.48	5.18	18.72
10	483.83	509.30	222.82	159.15	5.04	4.76	18.71
15	464.73	496.56	222.82	159.15	4.77	4.48	18.81
20	439.27	483.83	216.45	152.79	4.60	4.37	18.62
25	432.90	477.46	216.45	152.79	4.46	4.18	18.88
30	432.90	477.46	216.45	152.79	4.44	4.16	18.85

(2) The effect of strain amplitude on the mechanical properties of superelasticity. A stress–strain curve with a diameter of 1.0 mm, a loading rate of 10 mm/min, and a strain amplitude of 3% is shown in Figure 3, and the parameters are shown in Table 2. As the strain amplitude of SMA wire increases, the cumulative residual deformation gradually increases. When the strain amplitude is small, the austenitic SMA wire is basically in the elastic stage, and the elastic modulus is approximately 450 MPa after stabilization. When the strain amplitude exceeds 1%, the SMA wire will undergo martensitic transformation and austenite transformation, showing super-elastic performance, and the greater the strain amplitude, the better its super-elastic performance, and the greater the energy dissipation capacity. The strain amplitude is the most significant factor affecting the energy dissipation capacity of SMA wires. When the strain amplitude increases from 3% to 8%, the single-turn energy consumption of the SMA wire increases from 4.46 MJ·m^{-3} to 20.76 MJ·m^{-3}, which increases the energy consumption by 3.65 times. As the strain amplitude increases, the damping ratio gradually increases, the equivalent secant stiffness gradually decreases, and the energy consumption capacity continues to increase.

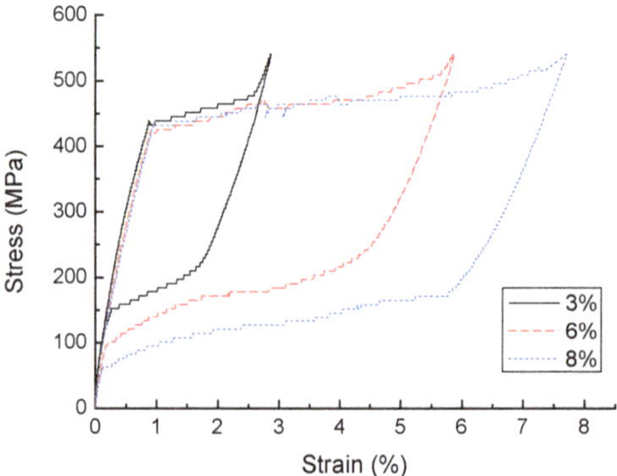

Figure 3. Stress-strain curve of austenite SMA wire with different strain amplitudes.

Table 2. Mechanical properties of SMA wires with different strain amplitudes.

Strain Amplitudes	σ_a (MPa)	σ_b (MPa)	σ_c (MPa)	σ_d (MPa)	ΔW (MJ.m^{-3})	ζ_a (%)	K_s (GPa)
3%	432.90	496.56	260.65	120.96	4.46	4.18	18.88
6%	420.17	509.30	254.65	101.86	12.70	6.09	9.21
8%	432.90	515.66	254.65	70.03	20.76	6.60	7.81

(3) The effect of loading rate on the mechanical properties of superelasticity. The stress–strain curve of the 30th cycle with a diameter of 1.0 mm, and a strain amplitude of 6% is shown in Figure 4, and the parameters are shown in Table 3. As the loading rate increases, the single-cycle energy consumption of the austenitic SMA wire gradually decreases, and the shape of the stress–strain curve changes significantly. In the phase of unloading, the initial stress of the change increases significantly. The stress–strain shape gradually transitions from a rectangle and a diamond to a trapezoid and a narrower triangle. The area enclosed by the hysteresis curve gradually decreases. The equivalent damping ratio and equivalent stiffness generally show a decreasing trend and energy consumption gradually decreases. This is mainly because the heat generated during the loading process of the SMA wire causes the temperature rise of the SMA specimen, which reduces its own energy consumption.

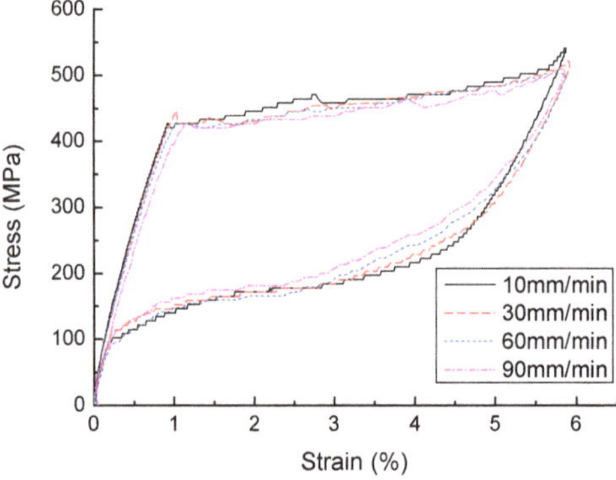

Figure 4. Stress-strain curve of austenite SMA wire with different loading rates.

Table 3. Mechanical properties of SMA wires with different loading rates.

Loading Rates	σ_a (MPa)	σ_b (MPa)	σ_c (MPa)	σ_d (MPa)	ΔW (MJ.m^{-3})	ζ_a (%)	K_s (GPa)
10 mm/min	420.17	509.30	254.65	101.86	12.70	6.09	9.21
30 mm/min	426.54	515.36	280.11	107.59	12.31	6.25	8.70
60 mm/min	420.17	502.93	326.04	109.86	11.93	6.15	8.58
90 mm/min	420.17	502.93	331.94	118.23	10.52	5.34	8.71

(4) The effect of diameters on the mechanical properties of superelasticity. The stress–strain curve of the 30th cycle with a loading ratio of 90 mm, and a strain amplitude of 6% is shown in Figure 5, and the parameters are shown in Table 4. As the diameter of the material increases, the stress–strain curve of the SMA wire tends to be smooth, but the number of cycles required to reach stability increases, and the cumulative residual deformation presents a gradually increasing trend. The stress of each characteristic point of SMA wire decreases with the increase of the material diameter. As the diameter of the material increases, the energy dissipation capacity and equivalent damping ratio show a significant decrease. This is mainly due to the increase in the diameter of the material, and the heat generated during the loading process cannot be dissipated in time, causing the specimen temperature to increase, which reduces the energy consumption of SMA wire. The equivalent stiffness is less affected by the diameter, and the change is not obvious. Therefore, in engineering applications, SMA wires with appropriate diameters should be selected for the passive control of seismic response.

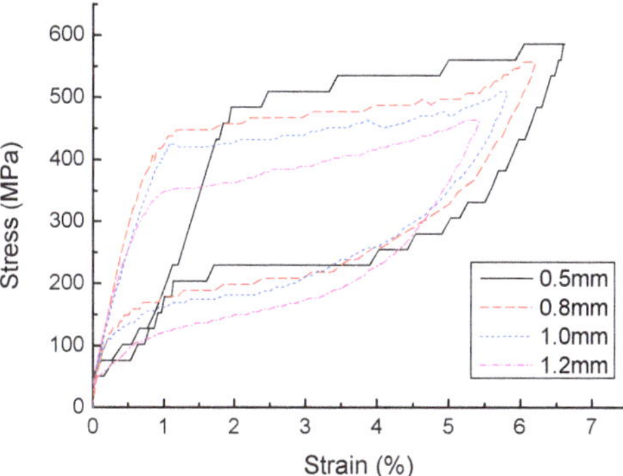

Figure 5. Stress-strain curve of austenite SMA wire with different diameters.

Table 4. Mechanical properties of SMA wires with different diameters.

Diameters	σ_a (MPa)	σ_b (MPa)	σ_c (MPa)	σ_d (MPa)	ΔW (MJ.m^{-3})	ζ_a (%)	K_s (GPa)
0.5 mm	483.83	585.69	331.04	203.72	12.43	6.49	8.47
0.8 mm	447.62	527.20	358.10	139.26	12.22	6.01	8.99
1.0 mm	420.17	502.93	331.94	118.23	10.52	5.34	8.71
1.2 mm	349.26	464.20	247.57	70.74	9.63	5.00	8.52

3. BP Neural Network Model Optimized by GA

The BP neural network is a multi-layer feedforward network composed of input layer, hidden layer, and output layer [37,38]. The initial weights/thresholds are randomly selected by system that are different between different each training. Therefore, the final weights/thresholds and models of the neural network obtained after training are different, especially when the training data are small. It is possible that the two neural network models are completely different, so the generalization ability of neural network is poor. When the training data are sufficient and extensive, although the difference between the neural network models after training is small, the training convergence speed will be too low. Using the genetic algorithm to search for the best initial weight/threshold value in the entire range of weight/threshold value, the error of BP neural network after training under the best initial weight/threshold value can be minimized. In this way, the difference of the BP neural network after training due to different initial weights/thresholds can be avoided, and the problem of network oscillation and non-convergence caused by improper initial weights/threshold values can also be prevented [39,40]. The process of optimizing the initial weight/threshold of the BP neural network by GA is shown in Figure 6.

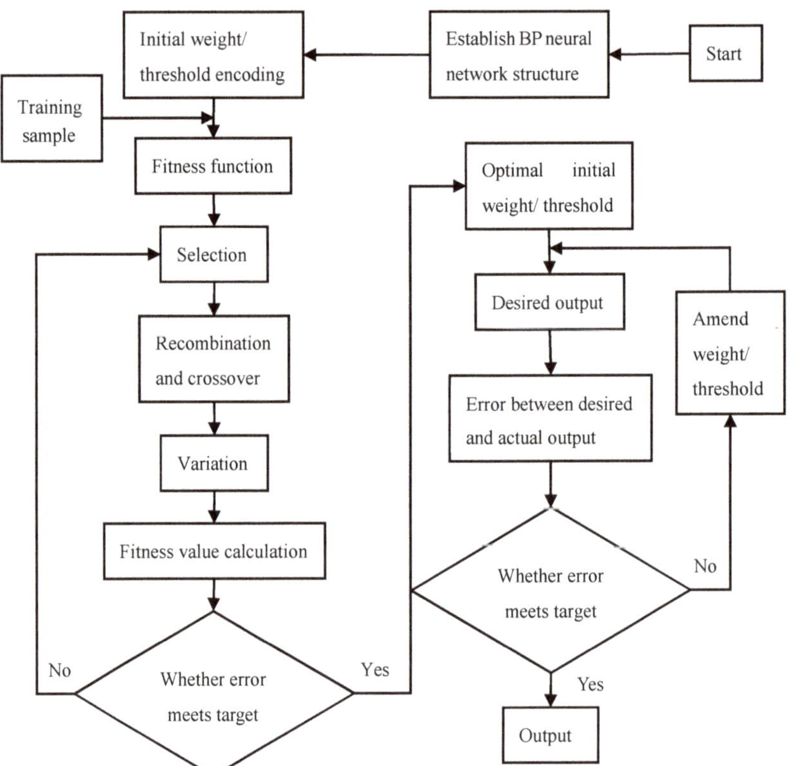

Figure 6. Flow chart of BP network optimized by genetic algorithm.

3.1. Structure of BP Neural Network

(1) Number of neurons in the input layer: When the diameter of the SMA wire is constant, the SMA constitutive relationship after stable performance is mainly affected by the loading rate and loading history. Therefore, the following variables can be determined as the input neurons of the BP neural network:

$$x_1 = v,\ x_2 = \sigma_{t-2},\ x_3 = \varepsilon_{t-2},\ x_4 = \sigma_{t-1},\ x_5 = \varepsilon_{t-1},\ x_6 = \varepsilon_t \quad (3)$$

where v is the loading rate of the SMA wire, σ_i and ε_i represent the stress and strain at time *I*, respectively.

(2) Number of neurons in the output layer: The variable required by the SMA constitutive model is the stress σ_t at time t, so $y = \sigma_t$ is determined as the output neuron of the BP neural network.

(3) Number of neurons in the hidden layer: The number of neurons in the hidden layer is a complex problem to be solved in the BP neural network. Currently, estimation methods [12] are usually used to determine the number of neurons in the hidden layer, and that is taken as 20.

(4) Neuron activation function: The activation function of the hidden layer neuron of the BP neural network is selected as logsig, and the activation function of the output layer neuron is selected as purelin.

3.2. Training Sample Collection and Processing

First, the training data is collected based on the test data of SMA wires. Since the SMA wire with a diameter of 1.0 mm is selected for structural vibration control, the neural network constitutive model for SMA with a diameter of 1.0 mm should be established. There are 48 working conditions in the SMA material property test, among which there are 12 working conditions with a diameter of 1.0 mm, and 4 of them are selected as the inspection data, respectively: ① The diameter is 1.0 mm, the loading rate is 10 mm/min, and the strain amplitude is 6%; ② The diameter is 1.0 mm, the loading rate is 30 mm/min, and the strain amplitude is 6%; ③ The diameter is 1.0 mm, the loading rate is 60 mm/min, and the strain amplitude is 6%; ④ The diameter is 1.0 mm, the loading rate is 90 mm/min, and the strain amplitude is 6%. The remaining 8 working conditions are for training data.

Then, the training data is normalized to obtain the samples required for training the BP neural network. Since the hidden layer neurons use the sigmoid activation function that is a saturation zone near the function value 0 and 1, and the function value changes very slowly, the normalization process can prevent the input data from entering the saturated region due to excessive absolute value.

3.3. Optimization Parameters of GA

The initial weight/threshold value of the unoptimized BP neural network is randomly assigned by the system, and that of the optimized BP neural network is determined by GA. According to the BP neural network structure, the weights to be determined in the BP neural network are $6 \times 20 + 20 \times 1 = 140$, and there are $20 + 1 = 21$ thresholds to be determined. The variables of GA are weights and thresholds, so the total number of variables is 161. Since the weight/threshold value can be any real number, in order to improve the accuracy and efficiency of the genetic algorithm, real-valued coding is used, and the chromosome length of the genetic algorithm is $140 + 21 = 161$. The objective function is the sum of squared errors between the expected output and the actual output obtained from the input of the training sample. Other parameters of GA: the initial population number is 40; the random traversal sampling selection function is used, and the generation gap is 0.9; the intermediate recombination crossover operator is selected; the real-valued mutation operator is used, and the mutation probability is 0.01; the maximum genetic algebra is 50.

3.4. Simulation Results and Analysis

The first step is to create a BP neural network model of austenite SMA, and a BP neural network model optimized by GA. MATLAB2013b neural network toolbox and gatbx toolbox are used to write simulation program code. The data of the last cycle of the test condition are used as the training sample of BP neural network. The training function is selected as trainlm, the maximum number of training times is 1000, the target error is 10^{-5}, and the learning rate is 0.1. The topological structure of BP neural network can be obtained by running the program as shown in Figure 7, and the validation performance chart is shown in Figure 8.

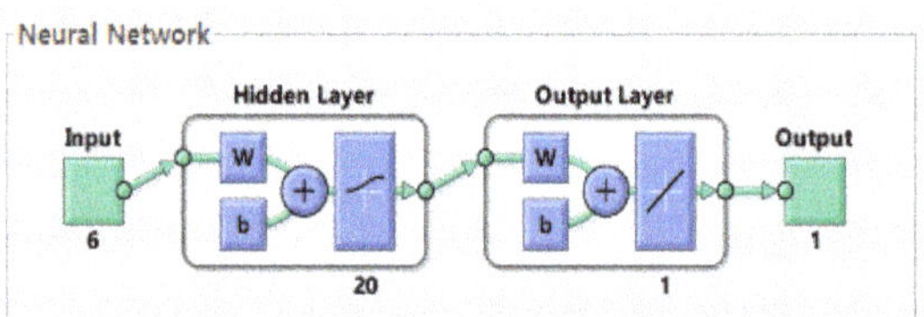

Figure 7. BP neural network topology.

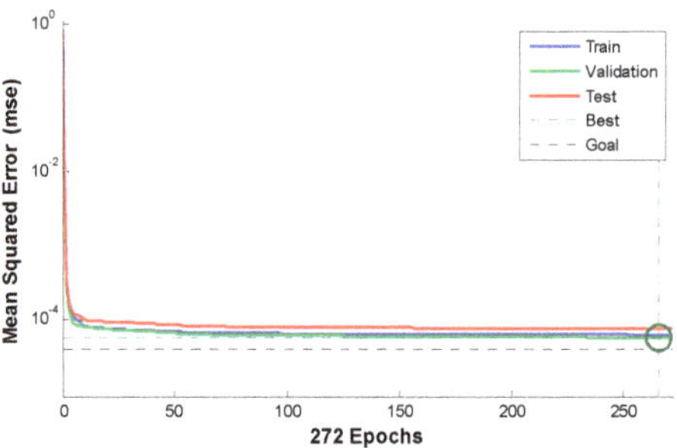

Figure 8. Validation Performance Chart.

The second step is to compare the accuracy of unoptimized and optimized BP neural network models. Four groups of normalized inspection data are input into the unoptimized BP neural network and the optimized BP neural network in sequence, and the expected output is obtained respectively. The accuracy of the two constitutive models is obtained by comparing expected output with the actual output and error analysis.

Figure 9 is a comparison diagram of the BP neural network prediction curve and the experimental curve at the diameter of 1.0 mm, loading rate of 90 mm/min, and strain amplitude of 6%. It can be seen that, after each training, the BP neural network prediction curve without GA optimization has a large fluctuation range due to the randomness of the initial weight/threshold, but that optimized by GA is in good agreement with the test curve, avoiding the difference of the model obtained from each run of the BP neural network, and it is a constitutive model of SMA with good stability and high accuracy.

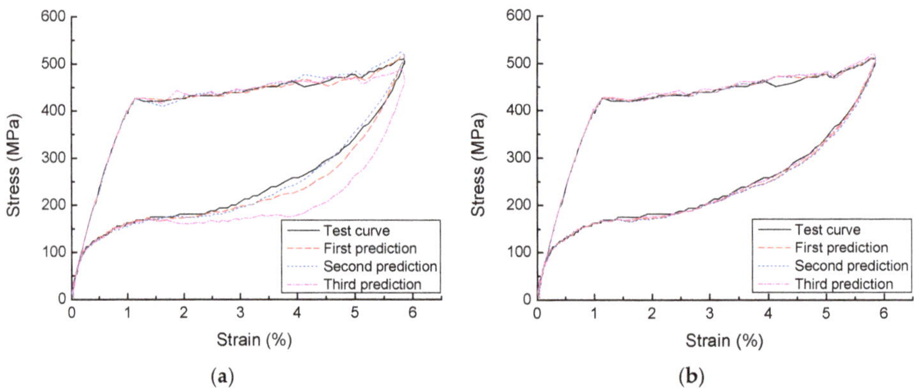

Figure 9. Comparison of BP network prediction curve and test curve (**a**) unoptimized BP network and (**b**) optimized BP network.

Since the network structure of the unoptimized BP neural network is different each time, the unoptimized BP neural network with better prediction results is selected for comparison with the optimized BP neural network. Figure 10 shows the comparison and error of the test curve with the unoptimized and optimized neural network prediction curve under different loading rates when the diameter is 1.0 mm and the loading amplitude is 6%. The average absolute percentage error E_{P} and E_{GP} of the prediction results of the two BP neural network models at the loading rate of 90 mm/min are:

$$E_{\mathrm{P}} = \frac{1}{n}\sum_{i=1}^{n}\frac{|Y_i - Y_{\mathrm{P}i}|}{Y_i} = 2.72\% \tag{4}$$

$$E_{\mathrm{GP}} = \frac{1}{n}\sum_{i=1}^{n}\frac{|Y_i - Y_{\mathrm{GP}i}|}{Y_i} = 2.13\% \tag{5}$$

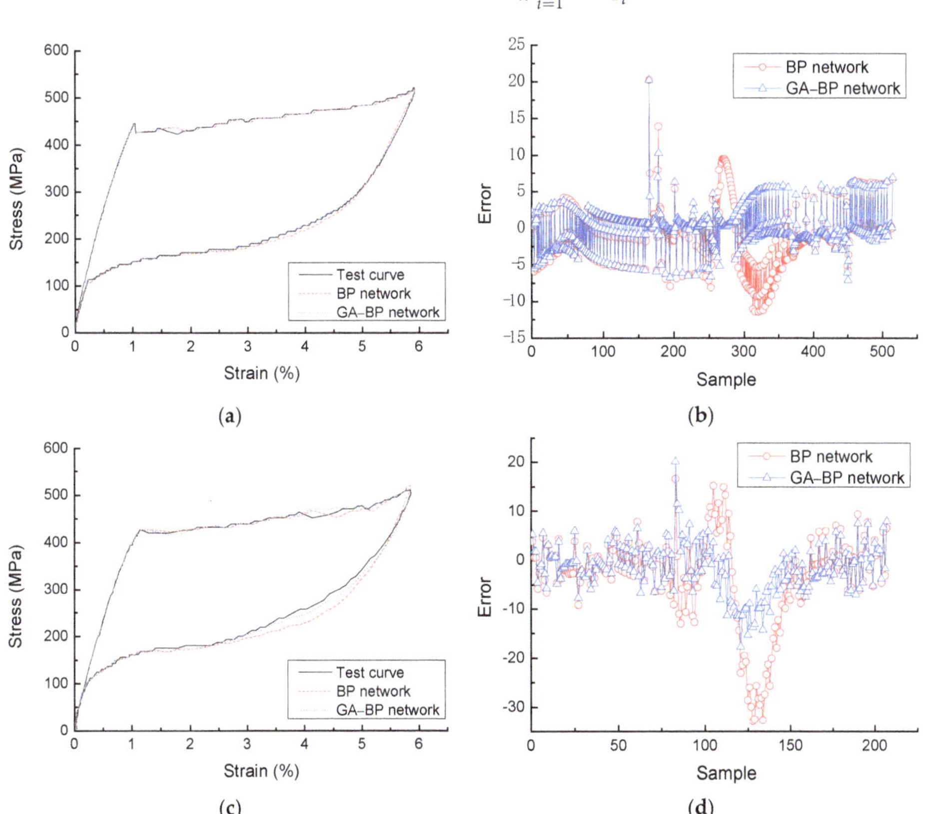

Figure 10. Comparison and corresponding error between test curve and prediction curves of un-optimized and optimized BP network by GA, (**a**) constitutive curves at 30 mm/min, (**b**) error at 30 mm/min, (**c**) constitutive curves at 90 mm/min, (**d**) error at 90 mm/min.

The linear coefficient of correlation γP and γGP of two prediction results are:

$$\gamma_{\mathrm{P}} = \frac{\sum_{i=1}^{n}(Y_i - \overline{Y})(Y_{\mathrm{P}i} - \overline{Y}_{\mathrm{P}})}{\sqrt{\sum_{i=1}^{n}(Y_i - \overline{Y})^2 \cdot \sum_{i=1}^{n}(Y_{\mathrm{P}i} - \overline{Y}_{\mathrm{P}})^2}} = 0.9983 \tag{6}$$

$$\gamma_{GP} = \frac{\sum_{i=1}^{n}(Y_i - \overline{Y})(Y_{GPi} - \overline{Y}_{GP})}{\sqrt{\sum_{i=1}^{n}(Y_i - \overline{Y})^2 \cdot \sum_{i=1}^{n}(Y_{GPi} - \overline{Y}_{GP})^2}} = 0.9995 \tag{7}$$

The root mean square error σ_P and σ_{GP} of two prediction results are:

$$\sigma_P = \sqrt{\frac{\sum_{i=1}^{n}(Y_i - Y_{Pi})^2}{n}} = 9.76 \tag{8}$$

$$\sigma_{GP} = \sqrt{\frac{\sum_{i=1}^{n}(Y_i - Y_{GPi})^2}{n}} = 5.43 \tag{9}$$

where Y_i and $\overline{Y}$ are the sample and mean stress measured in the experiment, Y_{Pi} and $\overline{Y}_P$ are the sample and mean stress predicted by the BP neural network model, and Y_{GPi} and $\overline{Y}_{GP}$ are the sample and mean stress predicted by the BP neural network model optimized by GA.

The analysis shows that: (a) With the change of SMA wire loading rate, the optimized BP neural network model can track the force behavior of SMA well, and the average absolute percentage error is 2.13%, the linear coefficient of correlation is 0.9995 and the root mean square error is 5.43. This shows that the neural network constitutive model based on GA optimization can well describe the change of SMA superelastic performance with loading rate, better predict the superelastic restoring force of SMA under repeated loading, and is a good rate-dependent dynamic constitutive model. (b) Although the unoptimized BP constitutive curve with better prediction effect is compared with the BP constitutive curve optimized by GA, the accuracy of the optimized model is still higher than that of the unoptimized model. The error distribution of BP constitutive curve optimized by GA and the experimental constitutive curve is relatively concentrated. Only the error of individual sample points is far from 0, and the absolute maximum error is smaller than that of the unoptimized BP network. The error distribution between the unoptimized neural network curve and the experimental curve is relatively scattered. More importantly, the optimal initial weight/threshold value obtained by optimization replaces the random value assigned by the system, so that the BP neural network has a fixed optimal initial weight/threshold value, avoiding the difference of the BP neural network after each training.

4. Optimization Control of Spatial Structure with SMA Wires

4.1. Dynamic Equation of SMA Passive Control System

According to the basic theory of structural dynamics [41,42], the motion equation of a structure equipped with an austenitic SMA passive control system under seismic excitation can be expressed as:

$$M\ddot{x} + C\dot{x} + Kx = -MI\ddot{x}_g + u \tag{10}$$

where $\ddot{x}_g$ is the acceleration of ground motion under seismic excitation; M, K, and C are the mass matrix, stiffness matrix and damping matrix of the structure, respectively; $\ddot{x}$, $\dot{x}$, and x are the acceleration, velocity and displacement column vector of the structure relative to the ground; I is the unit column vector; u is the passive control force column vector of the SMA wires acting on the structural nodes.

$$u = \left\{u_1{}^T, u_2{}^T \cdots u_n{}^T\right\}^T \tag{11}$$

$$\begin{cases} u_i = \sum_j \{F_{ij} \cdot \cos\theta_{ij}\} & \varepsilon_{ij} > 0 \\ u_i = 0 & \varepsilon_{ij} \leq 0 \end{cases} \tag{12}$$

$$F_{ij} = \sigma_{ij} \cdot A_{ij} \tag{13}$$

where u_i is the control force column vector of SMA wire acting on the structure node i; θ_{ij} represents the direction cosine column vector of the j-th SMA wire connected to the structural node i and the coordinate axes X, Y, and Z; F_{ij} represents the tension of the j-th SMA wire connected to the structural node i; σ_{ij}, ε_{ij}, and A_{ij} refer to the tensile stress, tensile strain and cross-sectional area of the j-th SMA wire connected to the structural node i.

MATLAB is used to compile the Newmark-β method calculation program to analyze the dynamic time history of the space model structure with the SMA passive control system. The simulation time step interval is very small, and the earthquake action is also very small in the first two-time steps, that is, the force of the SMA wire is very small in the initial stage. Therefore, it is approximately considered that in the first two-time steps of the earthquake, the σ_{ij} and ε_{ij} of all SMA wires are 0, and the structure is in an uncontrolled state. Stress σ_{ij} at any time later can be obtained via SMA BP network constitutive model optimized by GA. The vibration velocity at time t, the stress and strain at time $t-1$ and $t-2$, and the strain at time t are input into the optimized BP network constitutive model. Then, the stress of the SMA wire at time t can be obtained. Among them, the strain ε_{ij} (t) of SMA wire at any time can be obtained from the lateral displacement difference of the nodes at both ends of the SMA wire, and the strain of the oblique SMA wire arranged in the XY plane of the structure can be obtained by the following formula:

$$\begin{cases} \varepsilon_{ij}(t) = \varepsilon_{kj}(t) = \dfrac{\sqrt{h^2+(w+s_k(t)-s_i(t))^2}-\sqrt{h^2+w^2}}{\sqrt{h^2+w^2}} & s_k(t)-s_i(t)>0 \\ \varepsilon_{ij}(t) = \varepsilon_{kj}(t) = \dfrac{\sqrt{h^2+w^2}-\sqrt{h^2+(w+s_k(t)-s_i(t))^2}}{\sqrt{h^2+w^2}} & s_k(t)-s_i(t)\le 0 \end{cases} \tag{14}$$

where h and w are the layer height and single-span span of the structure respectively; when the angle between the j-th SMA wire and the positive X direction is less than 90°, S_k (t) refers to the lateral horizontal displacement of the k-th structural node connected to the upper end of the SMA wire, S_i (t) refers to the lateral horizontal displacement of the i-th structural node connected to the lower end of the SMA wire. When the angle between the j-th SMA wire and the positive X direction is greater than 90°, S_k (t) refers to the lateral horizontal displacement of the k-th structural node connected to the lower end of the SMA wire, and S_i (t) refers to the lateral horizontal displacement of the i-th structural node connected to the upper end of the SMA wire.

The structural dynamic time history analysis program is compiled to solve the seismic response of the structure. First, the genetic algorithm is employed to optimize the BP neural network constitutive model of SMA wire and the "save" command is used to save the optimized BP neural network model in the form of data structure. Then, the Newmark-β algorithm program is written to solve the dynamic Equation (10), in which the control force u can be obtained by calling the BP neural network model optimized by GA.

4.2. Optimization Criteria

The sum of the lateral displacements of the structure is taken as the optimized objective function, that is, the optimization criterion is

$$J = \sum_i |\Delta_i| = \sum_i |S_i - S_{i-1}| \tag{15}$$

$$S_i = \frac{\sum_j S_{ij}}{\sum_j j} \tag{16}$$

where J is the objective function value; $|\Delta_i|$ represents the absolute lateral displacement of the i-th layer; S_i represents the storey drift of the i-th layer, and it is the average of the lateral displacement of all nodes of the i-th layer; S_{ij} denotes the absolute lateral displacement of the node j of the i-th layer. The smaller the objective function value J, the smaller the dynamic response of the structure, and the better the control effect of the corresponding SMA wire arrangement on the structure. Based on the basic principles of GA, the following fitness function can be designed:

$$Fit = \frac{1}{J} = \frac{1}{\sum_i |S_i - S_{i-1}|} \tag{17}$$

this fitness function meets the design requirements of single value, continuous, non-negative, and maximization. The larger the fitness value corresponding to GA individual, the smaller the objective function value, and the better the corresponding SMA configuration scheme.

4.3. Optimization Control and Analysis of Spatial Structure

The spatial model structure is 2 spans along the X direction with each span of 500 mm in length, and along the Z direction is 1 span, the length is 600 mm, the height is 3 layers, and the height of each layer is 500 mm. In order to facilitate the installation and connection of the SMA wire, each layer of 6 nodes has an additional weight of 1 kg, and there are spiral holes for fixing the SMA wire. All rods are Q235 round steel pipes with an outer diameter of 10 mm and a wall thickness of 1 mm. The elastic modulus is 206 GPa, the Poisson's ratio is 0.3, and the density is 7.85×10^3 kg/m^3. The following assumptions are made in the analysis of structural seismic response: (a) All masses are concentrated at the nodes of each floor; (b) The initial working temperature of SMA wire is constant in its cross section and length direction.

As the vertical and horizontal SMA wires in the XY plane and the SMA wires in the YZ plane have little shock absorption effect on the structure under the action of the horizontal earthquake in the X direction. The SMA wires at these positions are not considered here. Only the diagonal SMA wires in the XY plane are considered, so the node numbers of the spatial model structure and the possible positions of the SMA wires are shown in Figure 11. The positions of the red lines in the figure represent all the possible positions of the SMA wires with a total number of 24.

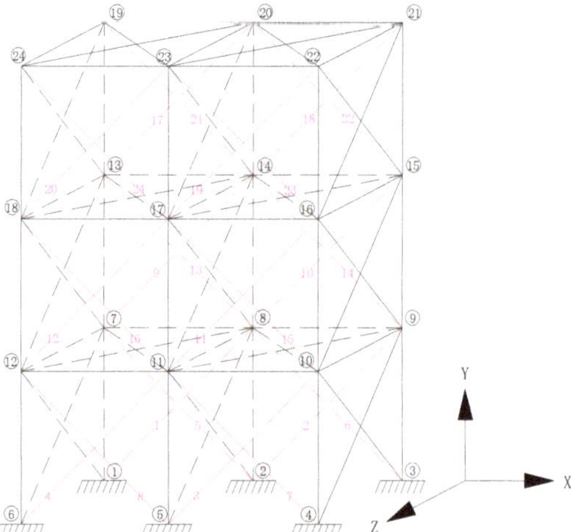

Figure 11. Possible locations of austenitic SMA wires and structural node numbers.

The operating parameters of GA are set as follows: real-valued coding is used, the initial population size is 40, the maximum number of generations is 50, and the crossover probability is 0.8. When the generation required to maintain the optimal individual unchanged does not exceeds 5, the mutation probability P_m is 0.05. When the generation required to maintain the optimal individual unchanged exceeds 20, the mutation probability GP_m is 0.2. When the optimal individual does not change and the required generation is between 5 and 20, the mutation probability is taken by linear interpolation between P_m and GP_m.

The El Centro wave with the duration of 20 s, the interval of 0.02 s, and the peak acceleration amplitude of 200 gal is selected as the seismic excitation of the spatial structure along the X direction. In the structure, 2, 4, 6, 8, 12, 16, 20, and 24 SMA wires are arranged for optimizing the position. The change of the objective function value with the evolutionary generation is shown in Figure 12. Table 5 presents the optimal positions of different numbers of SMA wires in the spatial structure, the value of the objective function, and the corresponding shock absorption effects.

Table 5. Optimal results of austenite SMA wires with different amounts.

Number of SMA Wires	Position Optimization Result	Objective Function Value (mm)	Suppression Ratio (%)
0 (non-control)	/	48.3	/
2	4, 15	33.7	30.23
4	4, 6, 12, 14	26.8	44.51
6	2, 6, 7, 13, 14, 20	23.5	51.35
8	4, 7, 8, 13, 14, 15, 21, 24	21.6	55.28
12	2, 8, 9, 10, 11, 13 14, 15, 19, 20, 22, 23	21.2	56.11
16	2, 3, 4, 5, 9, 10, 11, 12, 13 14, 15, 16, 17, 19, 21, 24	22.8	52.80
20	1, 3, 4, 5, 6, 9, 10, 11, 12, 13, 14 15, 16, 17, 19, 20, 21, 22, 23, 24	26.3	45.55
24	All	27.1	43.89

It can be seen from the optimization results that with the increase of genetic evolution generations, the optimal value of the objective function gradually decreases, and the average value of the objective function basically shows a gradually decreasing trend, which indicates that the optimal value and the average value are in a gradual convergence. During the process, the value of the objective function changes less and less. The optimized arrangement of the SMA wire has a better suppression effect on the seismic response of the space structure, but it is not that the more the number of SMA wires installed, the better the shock absorption effect. When the number of the SMA wire exceeds a certain number, the shock absorption effect decreases. When 4 austenitic SMA wires are arranged, the SMA wire layout optimized by GA can reduce the storey drift of the structure by 44.51%. Therefore, for the consideration of damping efficiency and economy, 4 SMA wires are selected to passively control the structure. The optimized layout is shown in Figure 13a, and the shaking table test model is shown in Figure 13b.

Using node 24, node 18, and node 12 to analyze the acceleration response and displacement response of the structure with and without control, the peak response is shown in Tables 6 and 7, and the time history curve of storey drift response and interlayer acceleration response is shown in Figure 14. It can be seen that the simulation results are in good agreement with the test results, which supports the rationality and feasibility of MATLAB simulation model for the seismic response analysis of space structure with SMA wires based on BP neural network. Meanwhile, the suppression effect of displacement response is more obvious than that of the acceleration response, the bottom layer has the best damping effect, the storey drift reduction rate is 52.98%, and the interlayer acceleration reduction rate is 25.89%.

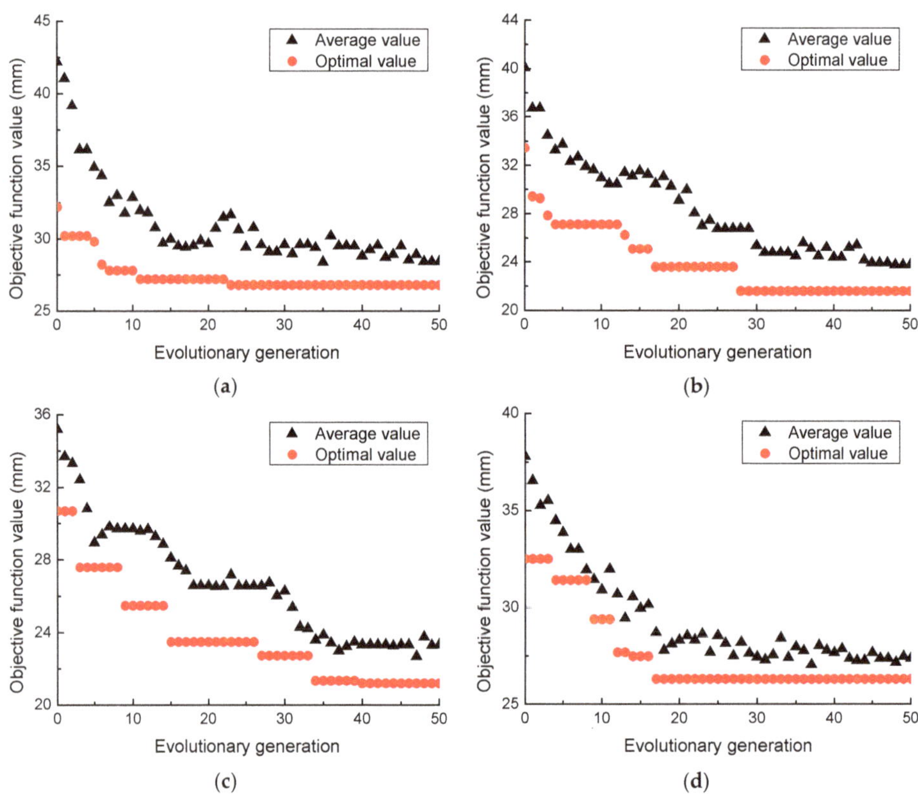

Figure 12. Objective function values with the change of evolutionary generations under different amounts of SMA wire, (**a**) 4 SMA wires, (**b**) 8 SMA wires, (**c**) 12 SMA wires, (**d**) 20 SMA wires.

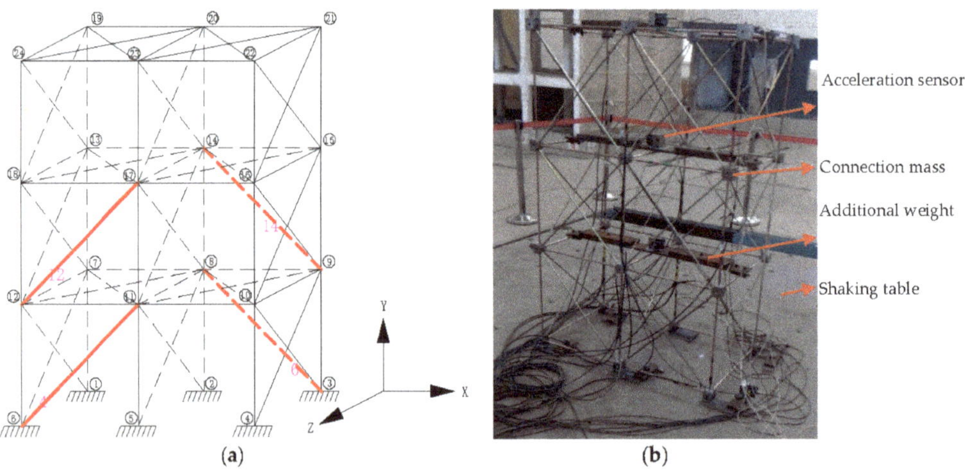

Figure 13. Spatial structure model with 4 austenite SMA wires (**a**) optimal position and (**b**) shaking table test.

Table 6. Peak storey drift and corresponding suppression rates under optimal and uncontrolled conditions.

Floor	Non-Control (mm)		Optimal Placement		Suppression Rate for Simulation Result (%)
	Simulation Result	Test Result	Simulation Result	Test Result	
1	23.052	25.182	10.838	13.431	52.98
2	16.352	15.258	8.702	9.287	46.78
3	8.910	7.641	7.236	6.484	18.79

Table 7. Peak interlayer acceleration and corresponding suppression rates under optimal and uncontrolled conditions.

Floor	Non-Control (m/s^2)		Optimal Placement (m/s^2)		Suppression Rate for Simulation Result (%)
	Simulation Result	Test Result	Simulation Result	Test Result	
1	4.734	4.407	3.508	3.116	25.89
2	4.028	3.696	3.153	2.519	21.73
3	2.517	1.719	2.068	1.423	17.83

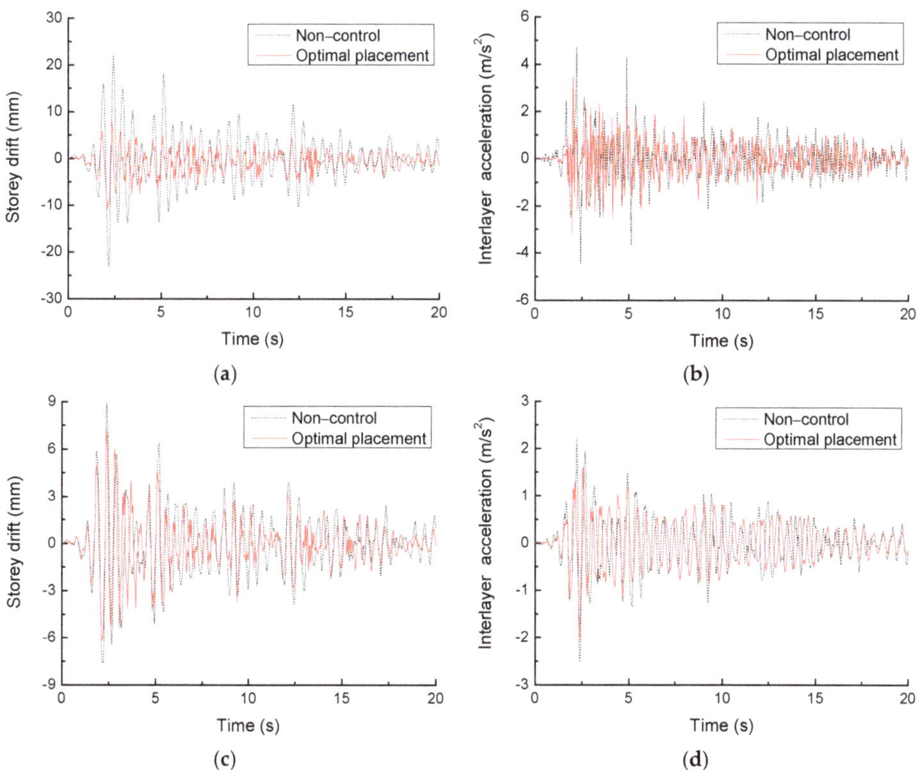

Figure 14. Time history curves of seismic response of spatial model structure with and without control, (**a**) storey drift of first floor, (**b**) interlayer acceleration of first floor, (**c**) storey drift of third floor, (**d**) interlayer acceleration of third floor.

5. Conclusions

(1) The mechanical tests of SMA wires show that with the increase of the number of cycles, the performance of SMA wires gradually stabilized, the stress–strain curve gradually becomes smooth, the accumulated residual deformation increases gradually, but the residual deformation of single cycle gradually decreases. After 15 cycles, the stress–strain curve tends to be stable, and the residual strain of single cycle is basically 0.

With the increase of strain amplitude, the energy dissipation capacity of SMA wires increases obviously. With the increase of loading rate and diameter, the energy dissipation capacity of SMA wires decreases, but not obviously. The strain amplitude is the most prominent factor affecting the energy dissipation capacity of SMA wires.

(2) Taking the material test data of SMA wires as the training sample and test sample of the BP neural network, the BP neural network prediction model optimized by genetic algorithm is established. The simulation results show that the prediction curve of optimized BP neural network is in good agreement with the test curve, and the average absolute percentage error is only 2.13%, the linear coefficient of correlation is 0.9995, and the root mean square error is 5.43. Thus, the model can well reflect the effect of loading velocity on the superelastic properties of SMA wires and is a velocity-dependent dynamic constitutive model with high precision for SMA.

(3) Since the initial weight/threshold is determined by the genetic algorithm, the optimized BP neural network avoids the difference of the prediction model in each run and reduces the phenomenon of network oscillation and non-convergence caused by the improper value of weight/threshold. Compared with the unoptimized BP neural network, it can predict the hysteretic behavior of SMA with better stability and higher accuracy.

(4) The BP neural network optimized by GA was employed to trace the stress–strain curve, and the optimization analysis of the SMA wires in a spatial structure model was carried out under the different seismic excitation. The simulation results are in good agreement with the test results, which supports the rationality and feasibility of MATLAB simulation model for the seismic response analysis of space structure with SMA wires based on BP neural network. Moreover, the results also show that the SMA wires after optimization can effectively reduce the seismic response of the structure, but it is not the case that the more SMA wires, the better the shock absorption effect. When the number of SMA wires exceeds a certain number, the vibration reduction effect gradually decreases. Therefore, the damping effect can be obtained economically and effectively only when the number and location of SMA wires are properly configured. When four SMA wires are arranged, a satisfactory control effect can be gained, the reduction rate of the sum of storey drift can reach 44.51%, and the reduction rate of storey drift and acceleration response at first storey are 52.98% and 25.89% respectively.

Author Contributions: The research was conceptualized by M.Z. and S.W., M.Z., D.W. and L.Z. carried out the mechanical property test of SMA. D.W. developed the preliminary numerical models. The numerical models were verified by J.L., D.W. and X.C. The draft manuscript was completed with the contribution of M.Z., X.C. and L.Z. The manuscript was revised and reviewed by all authors. All authors have read and agreed to the published version of the manuscript.

Funding: The research described in this paper was financially supported by National Natural Science Foundation of China (51678480), Science and Technology Project of ministry of Housing and Urban-rural Development (2020-K-127), Henan province key projects of science and technology (212102310976, 202102310248), Ningxia natural science foundation(2021AAC03189), Zhumadian major projects of science and technology (19005), Cultivating Project of National Natural Science Fund (XKPY-202009), Training program for young backbone teachers in Colleges and universities of Henan Province.

Institutional Review Board Statement: Not applicable.

Informed Consent Statement: Not applicable.

Data Availability Statement: Not applicable.

Conflicts of Interest: The authors declare no conflict of interest.

References

1. Wang, S.L. *Application of Shape Memory Alloy in the Structure Control*; Shaanxi Science & Technology Press: Xi'an, China, 2000. (In Chinese)
2. Zhou, B.; Yoon, S.H.; Leng, J.S. A three-dimensional constitutive model for shape memory alloy. *Smart Mater. Struct.* **2009**, *18*, 095016. [CrossRef]
3. Kosel, F.; Videnic, T. Generalized Plasticity and Uniaxial Constrained Recovery in Shape Memory Alloys. *Mech. Adv. Mater. Struct.* **2007**, *14*, 3–12. [CrossRef]
4. Ding, Y.L.; Chen, X.; Li, A.Q.; Zuo, X.B. A new isolation device using shape memory alloy and its application for long-span structures. *Earthq. Eng. Eng. Vib.* **2011**, *10*, 239–252. [CrossRef]
5. Li, T.; Wang, S.L.; Yang, T. Experiment and Simulation Study on Vibration Control of an Ancient Pagoda with Damping Devices. *Int. J. Struct. Stab. Dyn.* **2018**, *18*, 1850120. [CrossRef]
6. Habieb, A.B.; Valente, M.; Milani, G. Hybrid seismic base isolation of a historical masonry church using unbonded fiber reinforced elastomeric isolators and shape memory alloy wires. *Eng. Struct.* **2019**, *196*, 109281. [CrossRef]
7. Dehghani, A.; Aslani, F. Crack recovery and re-centring performance of cementitious composites with pseudoelastic shape memory alloy fibres. *Constr. Build. Mater.* **2021**, *298*, 123888. [CrossRef]
8. Rezapour, M.; Ghassemieh, M.; Motavalli, M.; Shahverdi, M. Numerical Modeling of Unreinforced Masonry Walls Strengthened with Fe-Based Shape Memory Alloy Strips. *Materials* **2021**, *14*, 2961. [CrossRef]
9. Siddiquee, K.N.; Billah, A.M.; Issa, A. Seismic collapse safety and response modification factor of concrete frame buildings reinforced with superelastic shape memory alloy (SMA) rebar. *J. Build. Eng.* **2021**, *42*, 102468. [CrossRef]
10. Schranz, B.; Michels, J.; Czaderski, C.; Motavalli, M.; Vogel, T.; Shahverdi, M. Strengthening and prestressing of bridge decks with ribbed iron-based shape memory alloy bars. *Eng. Struct.* **2021**, *241*, 112467. [CrossRef]
11. Peng, Z.; Wei, W.; Yibo, L.; Miao, H. Cyclic behavior of an adaptive seismic isolation system combining a double friction pendulum bearing and shape memory alloy cables. *Smart Mater. Struct.* **2021**, *30*, 075003. [CrossRef]
12. Zhan, M.; Wang, S.L.; Zhang, L.Z.; Chen, Z.F. Experimental evaluation of smart composite device with shape memory alloy and piezoelectric materials for energy dissipation. *J. Mater. Civ. Eng. ASCE* **2020**, *32*, 04020079. [CrossRef]
13. Villoslada, A.; Escudero, N.; Martín, F.; Flores, A.; Rivera, C.; Collado, M.; Moreno, L. Position control of a shape memory alloy 565 actuator using a four-term bilinear PID controller. *Sens. Actuators A Phys.* **2015**, *236*, 257–272. [CrossRef]
14. Kha, N.B.; Ahn, K.K. Position Control of Shape Memory Alloy Actuators by Using Self Tuning Fuzzy PID Controller. In Proceedings of the 1st IEEE Conference on Industrial Electronics and Applications, Singapore, 24–26 May 2006; pp. 1–5.
15. Samadi, S.; Koma, A.Y.; Zakerzadeh, M.R.; Heravi, F.N. Control an SMA-actuated rotary actuator by fractional order PID 569 controller. In Proceedings of the 5th RSI International Conference on Robotics and Mechatronics (ICRoM), Tehran, Iran, 25–27 October 2017; p. 570.
16. Ma, N.; Song, G. Control of shape memory alloy actuator using pulse width modulation. *Smart Mater. Struct.* **2003**, *12*, 12712. [CrossRef]
17. Fan, Y.J.; Sun, K.Q.; Zhao, Y.J.; Yu, B.S. A simplified constitutive model of Ti-NiSMA with loading rate. *J. Mater. Res. Technol.* **2019**, *8*, 5374–5383. [CrossRef]
18. Videnic, T.; Brojan, M.; Kunavar, J.; Kosel, F. A Simple One-Dimensional Model of Constrained Recovery in Shape Memory Alloys. *Mech. Adv. Mater. Struct.* **2014**, *21*, 376–383. [CrossRef]
19. Falk, F. One-dimensional model of shape memory alloys. *Arch Mech.* **1983**, *35*, 63–84.
20. Abeyaratne, R.; Knowles, J.K. A continuum model of a thermoelastic solid capable of undergoing phase transitions. *J. Mech. Phys. Solids* **1993**, *41*, 541–571. [CrossRef]
21. Boyd, J.G.; Lagoudas, D.C. A thermodynamical constitutive model for shape memory materials, Part II The SMA composite material. *Int. J. Plast.* **1996**, *12*, 843–874. [CrossRef]
22. Brinson, L.C. One-Dimensional Constitutive Behavior of Shape Memory Alloys: Thermomechanical Derivation with Non-Constant Material Functions and Redefined Martensite Internal Variable. *J. Intell. Mater. Syst. Struct.* **1993**, *4*, 229–242. [CrossRef]
23. Liu, B.; Wang, S.l.; Li, B.B.; Yang, T.; Li, H.; Liu, Y.; He, L. A Superelastic SMA Macroscopic Phenomenological Model Considering the Influence of Strain Amplitude and Strain Rate. *Mater. Rep.* **2020**, *34*, 14161–14167. (In Chinese)
24. Du, H.Y.; Han, Y.X.; Wang, L.X.; Melnik, R. A differential model for the hysteresis in magnetic shape memory alloys and its application of feedback linearization. *Appl. Phys. A* **2021**, *127*, 432. [CrossRef]
25. Lee, H.J.; Lee, J.J. Evaluation of the characteristics of a shape memory alloy spring actuator. *Smart Mater. Struct.* **2000**, *9*, 817–823. [CrossRef]
26. Ren, W.j.; Li, H.N.; Wang, L.Q. Superelastic shape memory alloy cyclic constitutive model based on neural network. *Rare Met. Mater. Eng.* **2012**, *9*, 243–246. (In Chinese)
27. Singh, M.P.; Moreschi, L.M. Optimal placement of dampers for passive response control. *Earthq. Eng. Struct. Dyn.* **2002**, *31*, 955–976. [CrossRef]
28. Amini, F.; Tavassoli, M.R. Optimal structural active control force, number and placement of controllers. *Eng. Struct.* **2005**, *27*, 1306–1316. [CrossRef]

29. Guo, K.M.; Jiang, J. Optimal Locations of Dampers/Actuators in Vibration Control of a Truss-Cored Sandwich Plate. In *Advances on Analysis and Control of Vibrations—Theory and Applications*; de la Hoz, M.Z., Pozo, F., Eds.; IntechOpen: London, UK, 2012. [CrossRef]
30. Chen, G.S.; Bruno, R.J.; Salama, M. Optimal placement of active/passive members in truss structures using simulated annealing. *AIAA J.* **1991**, *29*, 1327–1334. [CrossRef]
31. Rao, S.S.; Pan, T.S. Venkayya V B Optimal Placement of Actuators in Actively Controlled Structures Using Genetic Algorithms. *AIAA J.* **1991**, *29*, 942–943. [CrossRef]
32. Mulay, N.; Shmerling, A. Analytical approach for the design and optimal allocation of shape memory alloy dampers in three-dimensional nonlinear structures. *Comput. Struct.* **2021**, *249*, 106518. [CrossRef]
33. Pang, Y.T.; He, W.; Zhong, J. Risk-based design and optimization of shape memory alloy restrained sliding bearings for highway bridges under near-fault ground motions. *Eng. Struct.* **2021**, *241*, 112421. [CrossRef]
34. Zhan, M.; Wang, S.L.; Yang, T.; Liu, Y.; Yu, B.S. Optimum design and vibration control of a spatial structure with the hybrid semi-active control devices. *Smart Struct. Syst.* **2017**, *19*, 341–350. [CrossRef]
35. Fukuda1, T.; Takahata, M.; Kakeshita, T.; Saburi, T. Two-Way Shape Memory Properties of a Ti-51Ni Single Crystal Including Ti3Ni4 Precipitates of a Single Variant. *Mater. Trans.* **2001**, *42*, 323–328. [CrossRef]
36. Wang, W.; Fang, C.; Liu, J. Large size superelastic SMA bars: Heat treatment strategy, mechanical property and seismic application. *Smart Mater. Struct.* **2016**, *25*, 075001. [CrossRef]
37. Yun, C.B.; Yi, J.H.; Bahng, E.Y. Joint damage assessment of framed structures using a neural networks technique. *Eng. Struct.* **2001**, *23*, 425–435. [CrossRef]
38. Tsai, C.H.; Hsu, D.S. Diagnosis of Reinforced Concrete Structural Damage Base on Displacement Time History using the Back-Propagation Neural Network Technique. *J. Comput. Civ. Eng. ASCE* **2002**, *16*, 49–58. [CrossRef]
39. Liang, H.B.; Wei, Q.; Lu, D.Y.; Li, Z.L. Application of GA-BP neural network algorithm in killing well control system. *Neural Comput. Appl.* **2021**, *33*, 949–960. [CrossRef]
40. Yan, C.; Li, M.X.; Liu, W.; Qi, M. Improved adaptive genetic algorithm for the vehicle Insurance Fraud Identification Model based on a BP Neural Network. *Theor. Comput. Sci.* **2020**, *817*, 12–23. [CrossRef]
41. Li, J.; Chen, J.B. Dynamic response and reliability analysis of structures with uncertain parameters. *Int. J. Numer. Methods Eng.* **2005**, *62*, 289–315. [CrossRef]
42. Li, J.; Chen, J.B. The probability density evolution method for dynamic response analysis of non-linear stochastic structures. *Int. J. Numer. Methods Eng.* **2006**, *65*, 882–903. [CrossRef]

 *materials* 

Article

Modelling of SMA Vibration Systems in an AVA Example

Waldemar Rączka *, Jarosław Konieczny and Marek Sibielak

Department of Process Control, AGH University of Science and Technology, Al. Mickiewicza 30, 30-059 Krakow, Poland; koniejar@agh.edu.pl (J.K.); sibielak@agh.edu.pl (M.S.)
* Correspondence: waldemar.raczka@agh.edu.pl

Abstract: Vibration suppression, as well as its generation, is a common subject of scientific investigations. More and more often, but still rarely, shape memory alloys (SMAs) are used in vibrating systems, despite the fact that SMA springs have many advantages. This is due to the difficulty of the mathematical description and the considerable effortfulness of analysing and synthesising vibrating systems. The article shows the analysis of vibrating systems in which spring elements made of SMAs are used. The modelling and analysis method of vibrating systems is shown in the example of a vibrating system with a dynamic vibration absorber (DVA), which uses springs made of a shape memory alloy. The formulated mathematical model of a 2-DOF system with a controlled spring, mounted in DVA suspension, uses the viscoelastic model of the SMA spring. For the object, a control system was synthesised. Finally, model tests with and without a controller were carried out. The characteristics of the vibrations' transmissibility functions for both systems were determined. It was shown that the developed DVA can tune to frequency excitation changes of up to ±10%.

Keywords: modelling; vibration; shape memory alloys

Citation: Rączka, W.; Konieczny, J.; Sibielak, M. Modelling of SMA Vibration Systems in an AVA Example. *Materials* **2021**, *14*, 5905. https://doi.org/10.3390/ma14195905

Academic Editor: Salvatore Saputo

Received: 9 August 2021
Accepted: 1 October 2021
Published: 8 October 2021

1. Introduction

Due to its advantages, SMAs are materials that are increasingly used in many areas of our life. They are widely used in engineering, as various types of actuators, connecting elements, clamps and springs, and in medicine to create, e.g., stents, occluders, artificial heart valves, dental burs and many others. There are many applications using all effects that occur in SMAs (one-way, two-way shape memory or pseudoelasticity) in medicine, aerospace and general engineering. In this paper, we focus on SMA applications in vibrating systems and problems with their modelling. As these are materials with complex temperature- and stress-induced phase transformations depending on many factors, their usage is preceded by more or less labourious calculations using various types of mathematical models. Depending on the phenomenon we want to analyse, macroscopic, mesoscopic or microscopic models are used. Khandelwal and Buravalla in [1] made a valued review of various types of models developed, among others, by Birman [2], Bernardini and Pence in [3] and Paiva and Savi [4], Smith [5] and Lagoudas [6], Achenbach [7], Müller [8], Seelecke [9–11] and many others. They mainly focused on continuum models to describe phenomena that occur in SMAs. The second group of models describing SMAs is input-output models describing SMAs as a black box. These models usually describe the hysteresis phenomenon that occurs in SMAs. Such models are useful when the macroscopic effects of phenomena occurring in SMAs are the thing we are interested in the most. There are two main models of this type: Preisach [12–14] and Duhem–Madelung models [5]. These models describe the hysteresis phenomenon and are applicative in the description of one-dimensional SMA objects with lumped parameters. Both of them are usually used to obtain the time responses of the SMA object. However, if we want to get the object's response in the frequency domain, they are very labourious because each point of the chart should be determined separately. It is not easy to use these input-output models in frequency analyses, similar to the phenomenological models too. Thus, such analyses, due to their labour consumption,

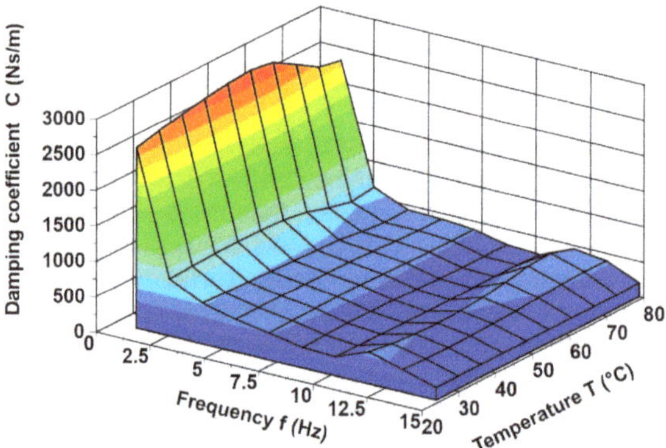

Figure 1. The damping of the spring as a function of frequency for selected temperatures.

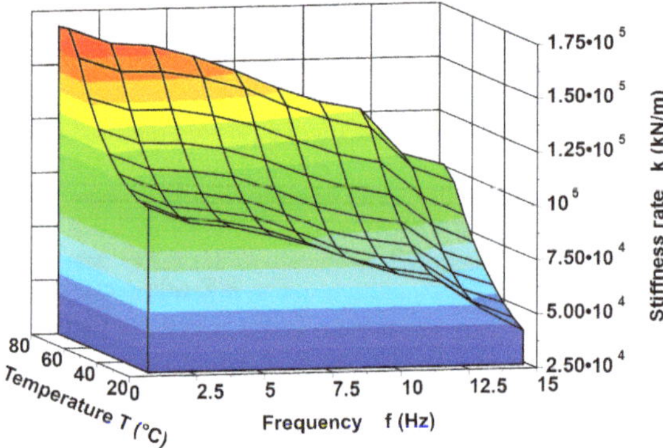

Figure 2. The spring rate k as a function frequency for selected temperatures.

2. Materials and Methods

The mathematical model (5), (6) of the AVA with a controlled dynamic damper was formulated based on its diagram shown in Figure 3. The absorber in the form of a mass $m_2 = 12$ kg is suspended by a SMA spring and protects the main mass $m_1 = 25$ kg. The SMA spring is represented by two elements, a controlled spring k_2 and a controlled damper c_2 connected in parallel. The protected mass m_1 is excited by the kinematic excitation $z_w = Asin\ (\omega t)$. For the sake of the notation simplification, the symbols $z_1 = z_1(t)$ and $Z_1 = Z_1(s)$ were adopted.

$$k_1(z_w - z_1) + c_1(\dot{z}_w - \dot{z}_1) = m_1\ddot{z}_1 + k_2(z_1 - z_2) + c_2(\dot{z}_1 - \dot{z}_2) \tag{5}$$

$$k_2(z_1 - z_2) + c_2(\dot{z}_1 - \dot{z}_2) = m_2\ddot{z}_2 \tag{6}$$

The SMA spring was described using a viscoelastic model with variable parameters (2), (3), (4) and is described above. After a Laplace transformation of the system of Equations (5) and (6), we obtained:

$$k_1(Z_w - Z_1) + c_1(Z_w - Z_1)s = m_1Z_1s^2 + k_2(Z_1 - Z_2) + c_2(Z_1 - Z_2)s \tag{7}$$

$$k_2(Z_1 - Z_2) + c_2(Z_1 - Z_2)s = m_2Z_2s^2 \tag{8}$$

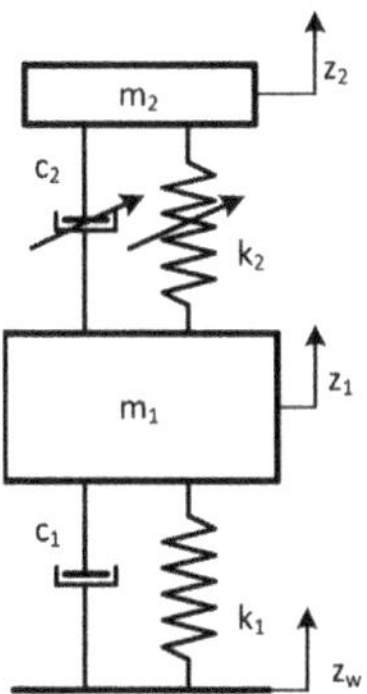

Figure 3. Calculation diagram of the vibration reduction system.

Equations (7) and (8) were written in matrix form:

$$Ax = Bu \tag{9}$$

where:

$$A = \begin{bmatrix} (-m_1s^2 - c_1s - k_1 - k_2 - c_2s) & (k_2 + c_2s) \\ (k_2 + c_2s) & -(m_2s^2 + c_2s + k_2) \end{bmatrix} \tag{10}$$

$$B = \begin{bmatrix} -(k_1 + sc_1) \\ 0 \end{bmatrix} \tag{11}$$

$$x = \begin{bmatrix} z_1 \\ z_2 \end{bmatrix} \tag{12}$$

$$u = z_w \tag{13}$$

To solve the system of Equation (9), we calculate determinants. The Formula (14) gives the principal determinant of A:

$$det(A) = \left(m_1s^2 + c_1s + k_1 + k_2 + c_2s\right)\left(m_2s^2 + c_2s + k_2\right) - (k_2 + c_2s)^2 \tag{14}$$

Determinant A_{z1} is written in the form:

$$det(A_{z1}) = z_w(k_1 + c_1s)\left(m_2s^2 + c_2s + k_2\right) \tag{15}$$

Determinant A_{z2} is written in the form:

$$det(A_{z2}) = z_w(k_1 + c_1s)(k_2 + c_2s) \tag{16}$$

The transfer function G_{z1zw} for input z_w and output z_1 is:

$$G_{z1zw}(s) = \frac{(k_1 + c_1s)(m_2s^2 + c_2s + k_2)}{(m_1s^2 + c_1s + k_1 + k_2 + c_2s)(m_2s^2 + c_2s + k_2) - (k_2 + c_2s)^2} \tag{17}$$

The transfer function G_{z2zw} for input z_w and output z_2 is:

$$G_{z2zw}(s) = \frac{(k_1 + c_1s)(k_2 + c_2s)}{(m_1s^2 + c_1s + k_1 + k_2 + c_2s)(m_2s^2 + c_2s + k_2) - (k_2 + c_2s)^2} \tag{18}$$

The transfer function G_{z2z1} for input z_1 and output z_2 is:

$$G_{z2z1}(s) = \frac{(k_2 + c_2s)}{(m_2s^2 + c_2s + k_2)} \tag{19}$$

Thus, the spectral transmittances of such an object for input displacement z_w and output displacements z_1 and z_2 are given by Equations (20) and (21), respectively.

$$G_{z1zw}(j\omega) = \frac{-m_2c_1j\omega^3 - P_1\omega^2 + P_2j\omega + k_1k_2}{m_1m_2\omega^4 - (m_2c_1 + Mc_2)j\omega^3 - (P_1 + Mk_2)\omega^2 + P_2j\omega + k_1k_2} \tag{20}$$

$$G_{z2zw}(j\omega) = \frac{-c_1c_2\omega^2 + P_2j\omega + k_1k_2}{m_1m_2\omega^4 - (m_2c_1 + Mc_2)j\omega^3 - (P_1 + Mk_2)\omega^2 + P_2j\omega + k_1k_2} \tag{21}$$

where:

$M = m_2 + m_1$,

$P_1 = m_2k_1 + c_1c_2$,

$P_2 = c_1k_2 + k_1c_2$.

The spectral transmittance of the absorber for the input z_1 protected the mass displacement, and the output z_2 damper displacement is given by the Equation (22).

$$G_{z2z1}(j\omega) = \frac{(c_2j\omega + k_2)}{(-m_2\omega^2 + c_2j\omega + k_2)} \tag{22}$$

3. Results

Figure 4 shows the vibration transmissibility function of the absorber with the mass m_2 described by the transfer function (22) as a function of the frequency of the displacement signal z_1 and the temperature of the SMA spring. The resonance frequency of the absorber increases with an increasing temperature from 11.1 Hz in the temperature 25 °C up to 14.5 Hz in the temperature 80 °C. It is a result of SMA spring features described by Equations (3) and (4). Thus, by controlling the spring temperature, we control the resonant frequency of the absorber in the range 11.1 Hz to 14.5 Hz. This means that by changing only the temperature of the SMA spring, we can tune the absorber to the frequency of disturbance z_w. Then, Figure 5 shows the phase shift between the displacements z_1 and z_2 for the absorber. Figures 6 and 7 show the vibration transmissibility functions and phase shifts of the absorber for selected temperatures (25 °C, 60 °C, 80 °C).

Figures 8–11 show similar graphs for the protected mass m_1. In turn, Figures 12–15 show the vibration transmissibility functions and the phase shifts between the displacements z_w and z_2 for the entire 2-DOF system described by the transmittance G_{z2zw}.

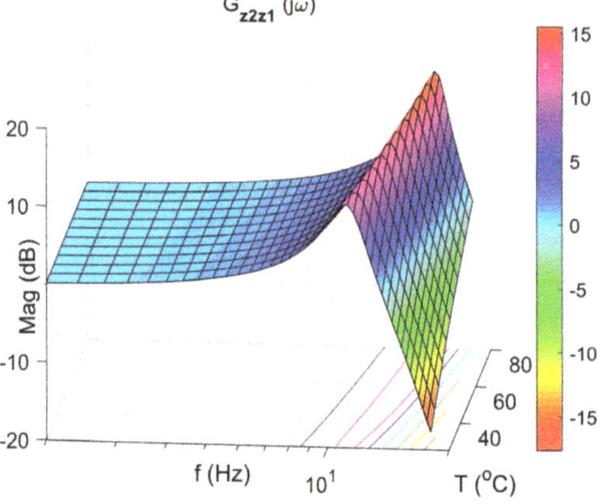

Figure 4. Vibration transmissibility function of the absorber as a function of frequency and temperature, the transfer function G_{z2z1}.

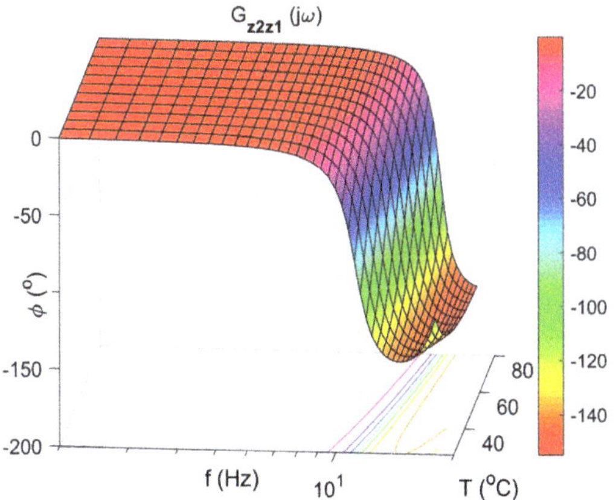

Figure 5. Absorber phase shift as a function of frequency and temperature, the transfer function G_{z2z1}.

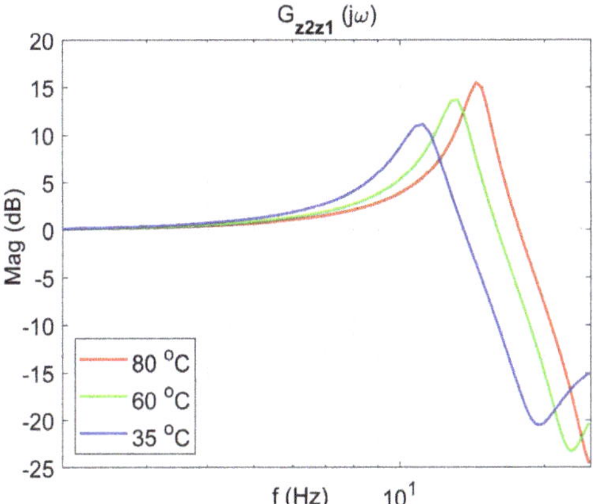

Figure 6. Vibration transmissibility functions of the absorber as a function of frequency for selected temperatures 25 °C, 60 °C, 80 °C, the transfer function G_{z2z1}.

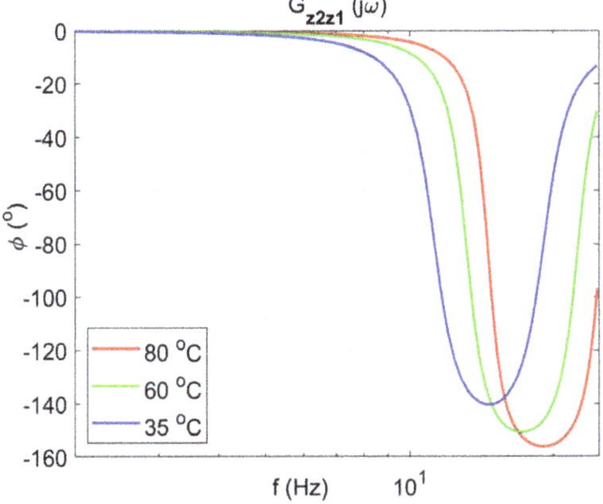

Figure 7. The absorber phase shifts as a function of frequency for selected temperatures 25 °C, 60 °C, 80 °C, the transfer function G_{z2z1}.

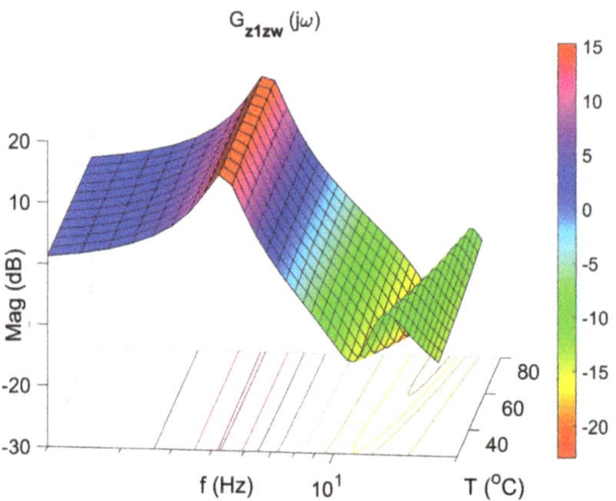

Figure 8. Vibration transmissibility function of disturbance z_w to the protected mass m_1 as a function of frequency and temperature, the transfer function G_{z1zw}.

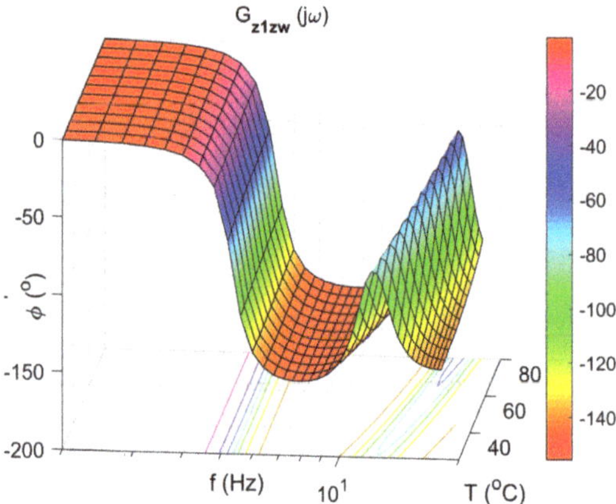

Figure 9. The phase shift of protected mass as a function of frequency and temperature, the transfer function G_{z1zw}.

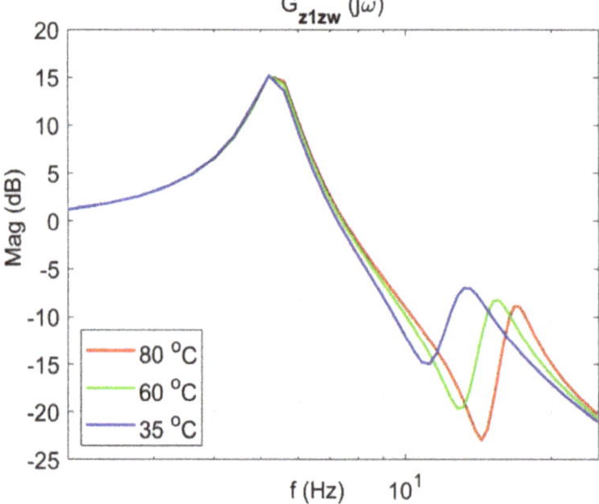

Figure 10. Vibration transmissibility functions of disturbance z_w to the protected mass m_1 as a function of frequency for selected temperatures 25 °C, 60 °C, 80 °C, the transfer function G_{z1zw}.

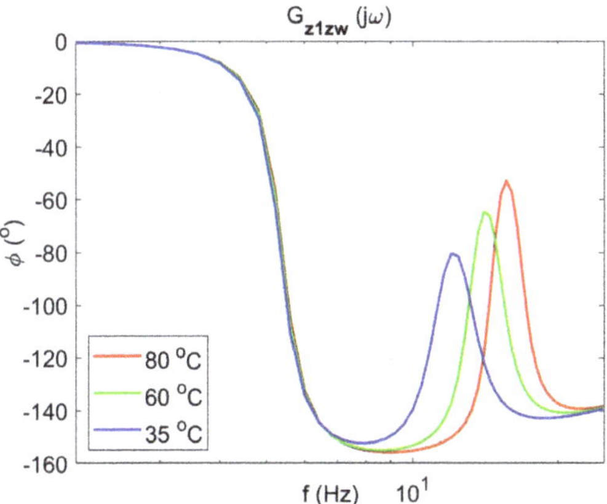

Figure 11. Absorber phase shift as a function of frequency for selected temperatures 25 °C, 60 °C, 80 °C, the transfer function G_{z2zw}.

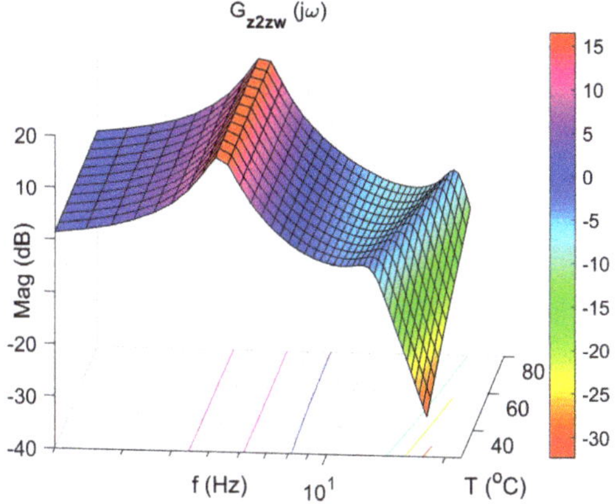

Figure 12. Vibration transmissibility function of disturbance z_w to the absorber mass m_2 as a function of frequency and temperature, the transfer function G_{z2zw}.

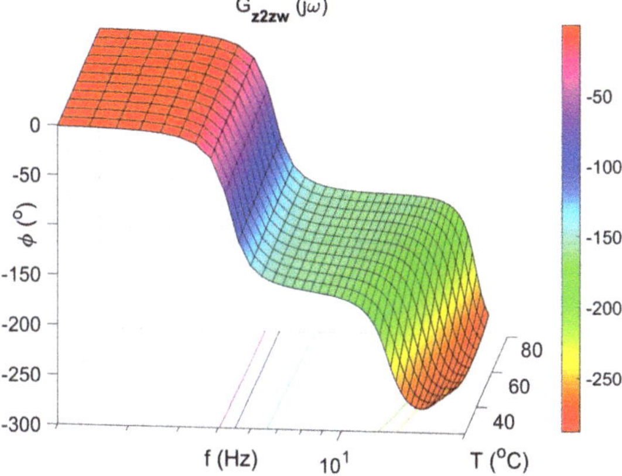

Figure 13. Protect mass phase shift between displacements z_w and z_2 as a function of frequency and temperature, the transfer function G_{z2zw}.

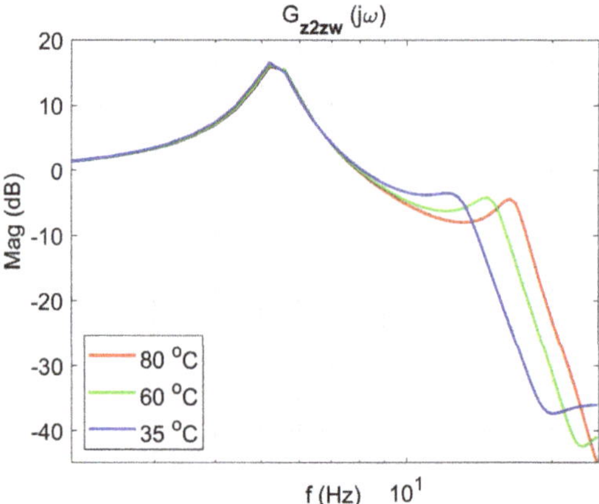

Figure 14. Vibration transmissibility functions of disturbance z_w to the absorber mass m_2 as a function of frequency for selected temperatures 25 °C, 60 °C, 80 °C, the transfer function G_{z2zw}.

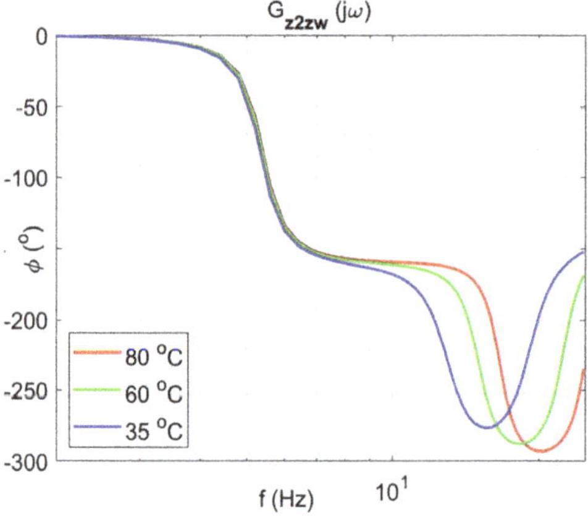

Figure 15. Absorber phase shift functions between displacements z_w and z_2 as a function of frequency for selected temperatures 25 °C, 60 °C, 80 °C, the transfer function G_{z2zw}.

Figure 12 shows the vibration transmissibility function of the system described by the transfer function (21), with the input z_w, the displacement of excitation and the output z_2 and the displacement of the mass of the absorber m_2. Figure 12 shows the significant change in the resonant frequency of the absorber due to the change in the stiffness and damping of the SMA spring caused by the change in its temperature. Then, Figure 16 shows the change in the natural frequency of the absorber (solid line) as a function of the spring temperature within the allowable range. The value of the natural frequency f_n of the dynamic damper can be calculated from the Formula (23). For comparison, in the figure, the change in the resonance frequency f_r of the absorber as a function of the temperature is shown too.

$$f_n = 2\pi \frac{a_2 + \sqrt{a_2{}^2 + 4(m - a_3)(a_1 + a_4 T)}}{2(m - a_3)} \tag{23}$$

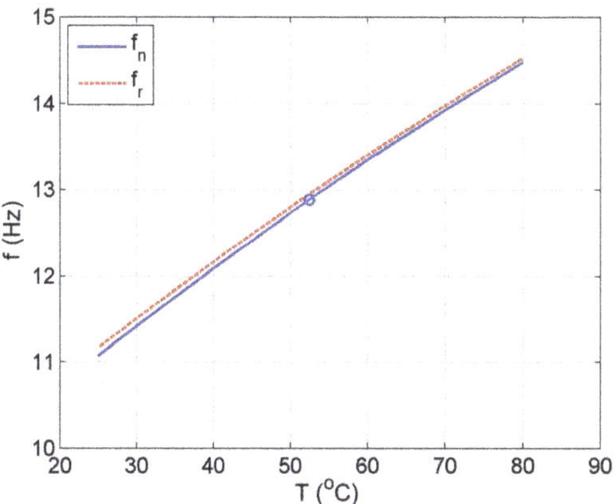

Figure 16. Chart of the natural frequency f_n (solid line) and the resonance frequency f_r (dashed line).

The ability to change the resonant frequency of the absorber enables it to be adjusted to the changing frequency of the disturbance z_w. Such tuning is rational if it is performed automatically. For this purpose, the control system shown in Figure 17 was proposed. The goal of the control system is to adjust the resonant frequency of the absorber to the disturbance frequency z_w. In this case, the control system uses the fact that the resonant frequency of the absorber can be estimated with the natural frequency, and the phase shift φ_{z2z1} between the displacement z_1 and the displacement z_2 is $-90°$. For this reason, it was decided that the feedback signal would be the cosine of the phase shift angle φ_{z2z1}. This signal is estimated in the "phase detector" block based on the Formula (25). In this case, the estimation error for the observation time T_o being multiple periods of forced oscillation is equal to zero. In general, the estimation error of the estimate is always inversely proportional to the observation time T_o.

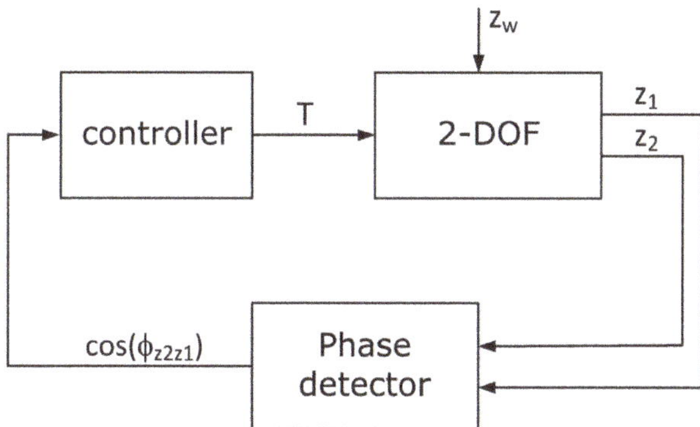

Figure 17. Control system block scheme.

In this example, it was assumed that the nominal natural frequency of the absorber is $f_{no} = 12.9$ Hz, and is equal to the nominal frequency of disturbance z_w. A circle in Figure 16 marks this value. This frequency is obtained in the SMA spring temperature $T_n = 52.5$ °C in the considered AVA. In this case, for an operating spring temperature range of 25 °C to 80 °C, the damper operating frequency range is 11.1 Hz to 14.5 Hz. The natural frequency can therefore be varied by more than ±10% from its nominal value f_{no}.

After the above consideration, the nonlinear controller was proposed in the form:

$$T = sat(K \cdot cos(\varphi_{z2z1})) \tag{24}$$

where:

$$sat(v) = \begin{cases} 80 \; for \; v > 80 \\ v \\ 25 \; for \; v < 25 \end{cases} \tag{25}$$

$$cos(\varphi_{z2z1}) = \frac{\frac{1}{T_o}\int_{t-T_o}^{t} z_w(\tau)z_2(\tau)d\tau}{\sqrt{\frac{1}{T_o}\int_{t-T_o}^{t} z_2^2(\tau)d\tau}\sqrt{\frac{1}{T_o}\int_{t-T_o}^{t} z_1^2(\tau)d\tau}} \tag{26}$$

Equation (26) follows directly from the definition of the dot product (27) between vectors in the functional space $L^2([0,T],R)$.

$$\langle v,w\rangle = \frac{1}{T}\int_0^T v(\tau)w(\tau)d\tau \tag{27}$$

The norm in the vector space $L^2([0,T],R)$ generated by this dot product is expressed by the following formula:

$$||v|| = \sqrt{\langle v,v\rangle} \tag{28}$$

In this case, the cosine of the angle φ between the two vectors v and w is expressed as follows:

$$cos(\varphi) = \frac{\langle v,w\rangle}{||v||\cdot||w||} \tag{29}$$

If vectors v and w are harmonic functions of the same frequency, and the time T is a multiple of the period, then the angle φ corresponds to the phase shift angle between these functions.

Spectral transmittances of the 2-DOF with the controller are given by Formulas (20)–(22) and (30), (31) formulas describing the SMA spring.

$$k_2(T,\omega) = a_1 + a_2\omega + a_3\omega^2 + a_4 sat(Kcos(\varphi_{z2z1})) \tag{30}$$

$$c_2(\omega) = b_1 + b_2\frac{1}{\omega} \tag{31}$$

The frequency response functions of the 2-DOF system with the controller for K = 50 are presented in Figures 18–20. The frequency characteristics of the closed system in the figures are marked in black. An analysis of these figures shows that the control system protects the mass m_1 the best. The reduction in mass m_1 vibrations (z_1) is better in a broader range than in controlled systems. Thanks to changing its parameters, the controlled system can tune to the disturbance and, therefore, reduce vibration better.

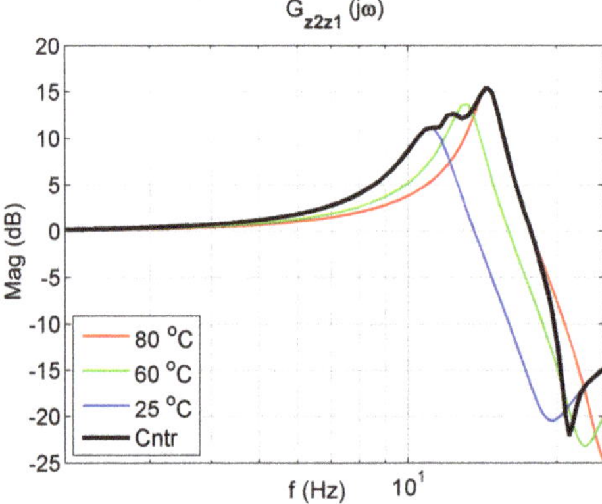

Figure 18. Vibration transmissibility functions of the passive absorber for selected temperatures of 25 °C, 60 °C, 80 °C and the controlled absorber (black). The transfer function G_{z2z1} describes the object.

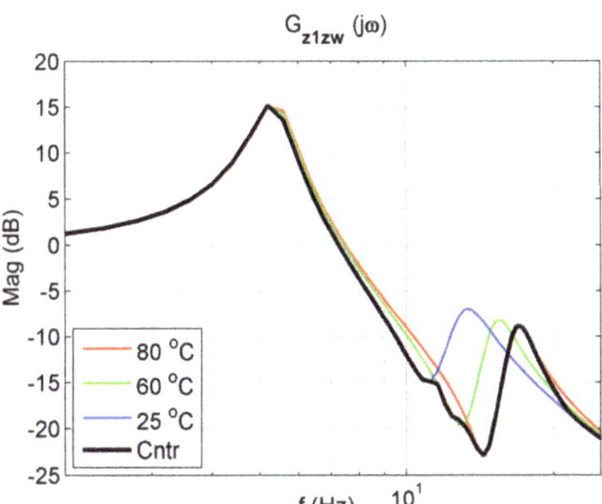

Figure 19. Vibration transmissibility functions of disturbance z_w to the protected mass m_1 of the passive absorber for selected temperatures of 25 °C, 60 °C, 80 °C and the controlled absorber (black). The transfer function G_{z1zw} describes the object.

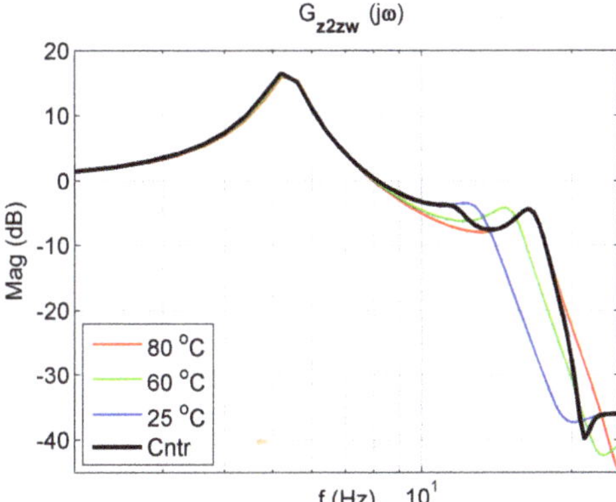

Figure 20. Vibration transmissibility functions of disturbance z_w to the mass m_2 of the passive absorber for selected temperatures of 25 °C, 60 °C, 80 °C and the controlled absorber (black). The transfer function G_{z2zw} describes the object.

4. Conclusions

The paper presents the developed mathematical model of the AVA system using a viscoelastic model of the SMA spring. The developed mathematical model enables numerical simulations in the frequency domains. The AVA was only an example of the application of this viscous model because, thanks to the use of the SMA spring, the AVA system can be tuned in real-time to the changing frequency of the disturbance in order to minimise the vibrations of the protected mass. As one can see, the viscous model enables the effective analysis of vibrating systems equipped with the SMA spring elements. Its characteristics can be easily determined using standard analytical methods. The results of the simulation tests of the model in the form of frequency characteristics are easy to obtain. Therefore, the work on the synthesis of the system is practical.

Additionally, it was shown that control systems can also be efficiently synthesised. For this purpose, a nonlinear control system was proposed and modelled. The tests of the AVA working in open and closed systems showed that thanks to the use of a controlled SMA spring, it is possible to perform an AVA which adjusts itself to the disturbance frequency on an ongoing basis, and it is better than passive systems because it operates in the broader frequency range.

Author Contributions: Conceptualization, W.R. and J.K.; methodology, M.S.; software, W.R.; validation, W.R., M.S. and J.K.; formal analysis, J.K. and M.S.; investigation, W.R.; resources, J.K.; data curation, M.S.; writing—original draft preparation, W.R.; writing—review and editing, W.R., J.K., M.S.; visualization, W.R.; supervision, J.K.; project administration, W.R.; funding acquisition, W.R. All authors have read and agreed to the published version of the manuscript.

Funding: This research was funded from national funds for science in the years 2009–2012 as the research projects No. N N502 266137 and from national funds for science in the years 2002–2005 as research project No. 4T07A 04129.

Institutional Review Board Statement: Not applicable.

Informed Consent Statement: Not applicable.

Data Availability Statement: The data presented in this study are available on request from the corresponding author.

Conflicts of Interest: The authors declare no conflict of interest.

References

1. Khandelwal, A.; Buravalla, V. Models for Shape Memory Alloy Behavior: An overview of modelling approaches. *Int. J. Struct. Chang. Solids* **2009**, *1*, 111–148.
2. Birman, V. Review of Mechanics of Shape Memory Alloy Structures. *Appl. Mech. Rev.* **1997**, *50*, 629–645. [CrossRef]
3. Schwartz, M.; Wiley, J. *Encyclopedia of Smart Materials Volume 1 and Volume 2*; John Wiley & Sons, Inc.: Hoboken, NJ, USA, 2002; Volume 1, ISBN 0471177806.
4. Paiva, A.; Savi, M.A. An overview of constitutive models for shape memory alloys. *Math. Probl. Eng.* **2006**, *2006*, 1–30. [CrossRef]
5. Smith, R.C. *Smart Material Systems*; Society for Industrial and Applied Mathematics: Philadelphia, PA, USA, 2005; ISBN 978-0-89871-583-5.
6. Lagoudas, D.C. *Shape Memory Alloys. Modeling and Engineering Applications*; Springer US: Boston, MA, USA, 2008; Volume 1, ISBN 978-0-387-47684-1.
7. Achenbach, M.; Muller, I. Simulation of material behaviour, of alloys with shape memory. *Arch. Mech. Mech.* **1985**, *37*, 573–585.
8. Achenbach, M.; Atanackovic, T.; Müller, I. A model for memory alloys in plane strain. *Int. J. Solids Struct.* **1986**, *22*, 171–193. [CrossRef]
9. Seelecke, S. Modeling the dynamic behavior of shape memory alloys. *Int. J. Non. Linear. Mech.* **2002**, *37*, 1363–1374. [CrossRef]
10. Heintze, O.; Seelecke, S. A coupled thermomechanical model for shape memory alloys—From single crystal to polycrystal. *Mater. Sci. Eng. A* **2008**, *481–482*, 389–394. [CrossRef]
11. Naso, D.; Seelecke, S. Passivity Analysis and Port-Hamiltonian Formulation of the Müller-Achenbach-Seelecke Model for Shape Memory Alloys: The Isothermal Case. *IFAC-PapersOnLine* **2018**, *51*, 713–718. [CrossRef]
12. Rączka, W.; Konieczny, J.; Sibielak, M.; Kowal, J. Discrete preisach model of a shape memory alloy actuator. *Solid State Phenom.* **2016**, *248*, 227–234. [CrossRef]
13. Ahn, K.K.; Kha, N.B. Modeling and control of shape memory alloy actuators using Preisach model, genetic algorithm and fuzzy logic. *Mechatronics* **2008**, *18*, 141–152. [CrossRef]
14. Mayergoyz, I.D. *Mathematical Models of Hysteresis and Their Applications*; Elsevier: Amsterdam, The Netherlands, 2003; ISBN 9780124808737.
15. Duval, L.; Noori, M.; Hou, Z.; Davoodi, H.; Seelecke, S. Random vibration studies of an SDOF system with shape memory restoring force. *Phys. B Condens. Matter* **2000**, *275*, 138–141. [CrossRef]
16. Ozga, A. Distribution of Random Pulses Forcing a Damped Oscillator Determined in a Finite Time Interval. *Acta Phys. Pol. A* **2014**, *125*, A159–A163. [CrossRef]
17. Jabłoński, M.; Ozga, A. Determining the distribution of values of stochastic impulses acting on a discrete system in relation to their intensity. *Acta Phys. Pol. A* **2012**, *121*, 174–178. [CrossRef]
18. Jabłoński, M.; Ozga, A. Statistical Characteristics of the Damped Vibrations of a String Excited by Stochastic Forces. *Arch. Acoust.* **2009**, *612*, 601–612.
19. Kwaśniewski, J.; Dominik, I.; Konieczny, J.; Lalik, K.; Sakeb, A. Application of self-excited acoustical system for stress changes measurement in sandstone bar. *J. Theor. Appl. Mech.* **2011**, *49*, 1049–1058.
20. Rzepecki, J.; Chraponska, A.; Mazur, K.; Wrona, S.; Pawelczyk, M. Semi-active reduction of device casing vibration using a set of piezoelectric elements. In Proceedings of the 20th International Carpathian Control Conference (ICCC), Kraków, Poland, 26–29 May 2019; pp. 648–652.
21. Sibielak, M.; Raczka, W.; Konieczny, J.; Kowal, J. Optimal control based on a modified quadratic performance index for systems disturbed by sinusoidal signals. *Mech. Syst. Signal Process.* **2015**, *64–65*, 498–519. [CrossRef]
22. Sibielak, M. Optimal controller for vibration isolation system with controlled hydraulic damper by piezoelectric stack. *Mech. Syst. Signal Process.* **2013**, *36*, 118–126. [CrossRef]

23. Rączka, W.; Sibielak, M.; Konieczny, J. Active vehicle suspension with a weighted multitone optimal controller: Considerations of energy consumption. In *Structural Health Monitoring, Photogrammetry & DIC*; Society for Experimental Mechanics Series; Springer: Cham, Switzerland, 2018; Volume 6.
24. Otsuka, K.; Wayman, C.M. *Shape Memory Materials*; Cambridge University Press: Cambridge, UK, 1998.
25. Bojarski, Z.; Morawiec, H. *Metale z Pamięcią Kształtu*; Państwowe Wydawnictwo Naukowe: Warszawa, Poland, 1988.
26. Kluszczyński, K.; Kciuk, M. Analytical Description of SMA Actuator Dynamics based on Fermi-Dirac Function. *Acta Phys. Pol. A* **2017**, *131*, 1274–1279. [CrossRef]
27. Rączka, W.; Konieczny, J.; Sibielak, M. Mathematical Model of a Shape Memory Alloy Spring Intended for Vibration Reduction Systems. *Solid State Phenom.* **2011**, *177*, 65–75. [CrossRef]
28. Tiseo, B.; Concilio, A.; Ameduri, S.; Gianvito, A. A Shape Memory Alloys Based Tuneable Dynamic Vibration Absorber For Vibration Tonal Control. *J. Theor. Appl. Mech.* **2010**, *48*, 135–153.
29. Klein, W.; Mezyk, A.; Switonski, E. Modelling and tuning the SMA absorber. In Proceedings of the ISMA2006: International Conference on Noise and Vibration Engineering, Heverlee, Belgium, 18–20 September 2006; pp. 413–420.
30. Rączka, W.; Sibielak, M.; Kowal, J.; Konieczny, J. Application of an SMA Spring for Vibration Screen Control. *J. Low Freq. Noise Vib. Act. Control* **2013**, *32*, 117–131. [CrossRef]
31. Ostachowicz, W.M.; Kaczmarczyk, S. Vibrations of composite plates with SMA fibres in a gas stream with defects of the type of delamination. *J. Compos. Struct.* **2001**, *54*, 305–311. [CrossRef]
32. Ni, Q.; Zhang, R.; Natsuki, T.; Iwamoto, M. Stiffness and vibration characteristics of SMA/ER3 composites with shape memory alloy short fibers. *Compos. Struct.* **2007**, *79*, 501–507. [CrossRef]
33. Kurczyk, S.; Pawelczyk, M. Fuzzy Control for Semi-Active Vehicle Suspension. *J. Low Freq. Noise Vib. Act. Control* **2013**, *32*, 217–226. [CrossRef]
34. Williams, K.; Chiu, G.; Bernhard, R. Adaptive-Passive Absorbers Using Shape-Memory Alloys. *J. Sound Vib.* **2002**, *249*, 835–848. [CrossRef]
35. Williams, K.A.; Chiu, G.T.-C.; Bernhard, R.J. Dynamic modelling of a shape memory alloy adaptive tuned vibration absorber. *J. Sound Vib.* **2005**, *280*, 211–234. [CrossRef]
36. Williams, K.A.; Chiu, G.T.-C.; Bernhard, R.J. Nonlinear control of a shape memory alloy adaptive tuned vibration absorber. *J. Sound Vib.* **2005**, *288*, 1131–1155. [CrossRef]
37. Piedboeuf, M.C.; Gauvin, R.; Thomas, M. Damping Behaviour of Shape Memory Alloys: Strain Amplitude, Frequency and Temperature Effects. *J. Sound Vib.* **1998**, *214*, 885–901. [CrossRef]
38. Toker, G.P.; Saedi, S.; Acar, E.; Ozbulut, O.E.; Karaca, H.E. Loading frequency and temperature-dependent damping capacity of NiTiHfPd shape memory alloy. *Mech. Mater.* **2020**, *150*, 103565. [CrossRef]
39. Karakalas, A.A.; Machairas, T.T.; Lagoudas, D.C.; Saravanos, D.A. Quantification of Shape Memory Alloy Damping Capabilities Through the Prediction of Inherent Behavioral Aspects. *Shape Mem. Superelasticity* **2021**, *7*, 7–29. [CrossRef]

Article

Investigations of Effects of Intermetallic Compound on the Mechanical Properties and Shape Memory Effect of Ti–Au–Ta Biomaterials

Wan-Ting Chiu *, Kota Fuchiwaki †, Akira Umise, Masaki Tahara, Tomonari Inamura and Hideki Hosoda

Institute of Innovative Research (IIR), Tokyo Institute of Technology, 4259 Nagatsuta-cho, Midori-ku, Yokohama 226-8503, Japan; kfuchiwaki@tenaris.com (K.F.); umise.a.aa@m.titech.ac.jp (A.U.); tahara.m.aa@m.titech.ac.jp (M.T.); inamura.t.aa@m.titech.ac.jp (T.I.); hosoda.h.aa@m.titech.ac.jp (H.H.)
* Correspondence: chiu.w.aa@m.titech.ac.jp
† Now: Tenaris NKKTubes; Then: Tokyo Institute of Technology.

Abstract: Owing to the world population aging, biomedical materials, such as shape memory alloys (SMAs) have attracted much attention. The biocompatible Ti–Au–Ta SMAs, which also possess high X-ray contrast for the applications like guidewire utilized in surgery, were studied in this work. The alloys were successfully prepared by physical metallurgy techniques and the phase constituents, microstructures, chemical compositions, shape memory effect (SME), and superelasticity (SE) of the Ti–Au–Ta SMAs were also examined. The functionalities, such as SME, were revealed by the introduction of the third element Ta; in addition, obvious improvements of the alloy performances of the ternary Ti–Au–Ta alloys were confirmed while compared with that of the binary Ti–Au alloy. The Ti_3Au intermetallic compound was both found crystallographically and metallographically in the Ti–4 at.% Au–30 at.% Ta alloy. The strength of the alloy was promoted by the precipitates of the Ti_3Au intermetallic compound. The effects of the Ti_3Au precipitates on the mechanical properties, SME, and SE were also investigated in this work. Slight shape recovery was found in the Ti–4 at.% Au–20 at.% Ta alloy during unloading of an externally applied stress.

Keywords: biomedical materials; intermetallic compound; shape memory alloys; shape memory effect; Ti–Au–Ta

Citation: Chiu, W.-T.; Fuchiwaki, K.; Umise, A.; Tahara, M.; Inamura, T.; Hosoda, H. Investigations of Effects of Intermetallic Compound on the Mechanical Properties and Shape Memory Effect of Ti–Au–Ta Biomaterials. *Materials* **2021**, *14*, 5810. https://doi.org/10.3390/ma14195810

Academic Editor: Salvatore Saputo

Received: 10 September 2021
Accepted: 30 September 2021
Published: 4 October 2021

1. Introduction

Shape memory alloys (SMAs) have attracted much attention in the biomedical and biomaterials communities due to their functionalities, such as shape memory effect (SME) and superelastic (SE) behavior, which could be manipulated by controlling their operation temperature and externally applied stress [1–3]. Since the Ni–Ti (Nitinol) alloy was utilized as SMA stent, a commercial breakthrough was made in 1990s; thereafter, SMAs have been developed and widely accepted by the public [4]. However, a high potential of Ni hypersensitivity to the human body in the Nitinol impedes its applicability [5,6]; therefore, a substitution possessing biocompatibility for the Nitinol is urgently needed. The Ti–based alloys in their β–phase have been reported to perform good SME and SE [7,8]; in addition, Ti is known as a biocompatible metal [9,10]. Therefore, biocompatible β–Ti–based SMAs were chosen in this study for the applications in biomedical materials.

For manipulation of the performance of the β–Ti alloy, gold (Au) was selected in this study for the following reasons. First, from the biological point of view, Au is also known as a metal with excellent biocompatibility [11]. Second, Au meets the essential requirement of the high X-ray contrast [12] for the applications in the medical devices, such as a guidewire for surgery. Third, from the chemical point of view, Au possesses excellent corrosion resistance in the human body fluid [13,14]. Based on the aforementioned prerequisites, Au is a potential candidate to enhance the properties of the β–Ti SMAs towards biomedical applications; therefore, the binary Ti–Au-based alloy was chosen in this work. It is necessary to mention that the near–eutectoid Ti–4 at.% Au-based alloy was designated in this study since the relatively low temperature at the eutectoid reaction facilitates the preparation of the SMAs.

Regardless of the above–mentioned promising characteristics of the binary Ti–Au–based alloy, Plichta et al. reported that the massive transformation took place during cooling in the near–eutectoid Ti–4 at.% Au alloy, resulting in the formation of the α–massive martensite [15,16]. Here, the massive transformation indicates the phase transformation of $\beta \rightarrow \alpha_m$ (where α_m suggests the α–massive martensite). The α_m phase exhibits neither SME nor SE; thus, for revealing the functionality, such as SME and SE, a β–stabilizer, tantalum (Ta), was introduced into the near–eutectoid Ti–4 at.% Au–based alloy to stabilize the parent β–phase at operation temperature (i.e., body temperature).

Tantalum, which is known as a β–stabilizer, was chosen in this study for the manipulation of the functionalities of the Ti–Au–based SMA by tuning its martensite transformation start temperature (M_s) [8]. According to the binary Ti–Ta phase diagram, regardless of the addition amount of Ta to Ti, a solid solution (i.e., isomorphous) is formed through the entire concentration range [17]. The aforementioned α–massive martensite, which performs neither SME nor SE, could be suppressed by the Ta addition and could be transformed into the functional parent β–phase and/or α''–martensite phase [18–20]. Besides the purpose for revealing the functionalities, Ta, which is also a heavy element, could also contribute high X–ray contrast for the medical devices, such as guidewires [21,22]. Furthermore, concerning the biocompatibility to the human body, the Ta element, which is also considered as a promising candidate [23–25], shows high biocompatibility. Based on the aforementioned advantages of the Ta element, it was thus chosen as the third element for the manipulation of the Ti–Au–based SMAs in this work towards the applications of biomaterials.

For the binary Ti–Au SMAs, most of the researches focused on the near–$Ti_{50}Au_{50}$ alloys for the applications of high temperature SMAs [26–28]. In the case of the near–eutectoid Ti–Au alloy (i.e., Au concentration at approximately 4.2 at.% (or 15.3 wt.% Au)), the literature is relatively limited. As previously mentioned, Plichta et al. studied the α–massive martensite generated from the massive transformation around the near–eutectoid Ti–Au alloys [15,16]. Kikuchi et al. screened the elastic moduli of the Ti–Au alloys and other alloys from 5 wt.% to 30 wt.% addition amounts for the applications of biomedical materials, such as dental materials [29]. Concerning the binary Ti–Ta SMAs, Buenconsejo et al. claimed that the Ti–Ta–based alloys could be a good candidate for the applications of SMAs [30]. Ikeda et al. studied the effect of Ta addition concentration on phase constitution and the behaviors of the binary Ti–Ta alloys after aging treatment [31]. Some of the high–order Ti–Ta–based alloys have also been studied [32–34]. However, the ternary Ti–Au–Ta alloys, which are potential SMAs for the applications of the medical devices and biomaterials have not been investigated. This study thus investigated the unexplored effects of the Ta addition on the phase constituents, phase transformation, mechanical properties, shape memory effect, and superelasticity of the binary Ti–Au–based SMAs.

In this study, it was found that with the addition of the Ta element to the near–eutectoid Ti–Au alloy, the functionless α–massive martensite was successfully transformed into the functional β–parent phase. The intermetallic compound Ti_3Au was formed while Ta addition amount was concentrated up to 30 at.%. Having the precipitation of the intermetallic Ti_3Au compound, positive effect to the strength of the alloys were achieved. This work is a preliminary study of the Ti–Au–Ta–based SMAs and this ternary system is now still studying by our research group. More results concerning the Ti–Au–Ta–based alloys and their high–order systems will be published in the future.

2. Materials and Methods

Titanium sponges (Ti, purity = 98%), Gold plates (Au, purity > 99.9%), and Tantalum flakes (Ta, purity > 99.9%) were used as the starting materials for the preparations of the Ti–Au–Ta ingots. The purification of the Ti sponges could be found in our previous study [35].

Alloys with compositions of the Ti–4 at.% Au–20 at.% Ta and the Ti–4 at.% Au–30 at.% Ta in a total weight of eight grams were fabricated by conventional physical metallurgy techniques. In this article, these alloys are abbreviated to Ti–4Au–20Ta and Ti–4Au–30Ta, respectively, unless otherwise mentioned. The constituent elements were manufactured under an Ar–1 vol.% H_2 atmosphere by an arc–melting system, which is equipped with a non–consumable W electrode. The ingots were re–melted seven times and were flipped upside down prior to each re–melting to affirm the chemical homogeneity of the composition in the entire ingot. The arc–melted ingots are denoted as "as–melted" specimens.

Followed by an arc–melting, the as–melted ingots were cleaned, wrapped by Ti films, vacuum sealed into quartz tubes, and homogenized at 1273 K for 7.2 ks under pure Ar atmosphere for further composition homogenization after arc–melting process. The homogenized ingots were then quenched into iced–water for the termination of the homogenization treatment and the homogenized specimens were identified by "as–HT" specimens.

The as–HT specimens were then cold–rolled into thin films at room temperature (RT; i.e., 295 $\pm$ 3 K) via the cold–rolling machine equipped with a double–deck roll. The alloys were cold–rolled until a reduction in thickness of 98% was achieved. For that alloy, which cannot be cold rolled up to reduction in thickness of 98%, process–annealing at 1273 K for 1.8 ks was conducted followed by the identical cold–rolling procedure at RT until the reduction in thickness of 98% was obtained. On the other hand, for that alloy, which cannot be cold rolled owing to its low cold workability, was hot–rolled at 1173 K until the reduction in thickness of 98%. The thinned down specimens were named as "as–rolled" alloys. Specimens with specific shapes and dimensions were cut off from the as–rolled alloys for the following analysis and measurements.

The specimens with certain shapes, which were prepared from the as–rolled alloys, were then cleaned, wrapped into Ti films, vacuum sealed into quartz tubes, and solution–treated (ST) at 1273 K for 1.8 ks under the atmosphere of pure Ar. Afterward, specimens were quenched into iced–water for the termination of the solution–treatment. The solution–treated specimens were thus named as "ST" specimens. An illustration for the aforementioned thermomechanical processes in detail could be found in our previous articles [35,36].

The ST specimens were cleaned, grinded by emery papers with different meshes, and polished by the 1 μm diamond paste to obtain fine surface finish prior to the analysis of phase identification. A θ–2θ X–ray diffraction (XRD) measurement was carried out at RT under ambient for the determination of phase constituents. Tube current and voltage were at 40 mA and 45 kV while the Cu K_α radiation (λ = 1.5405 Å) was used as the X–ray source. The scan range was from 20° to 120° with a scan rate at approximately 2°/min. A standard Si plate was used for the correction of the external system error.

Microstructure observations and chemical composition analysis were carried out by the field–emission scanning electron microscopy (FE–SEM) along with an energy dispersive X–ray spectrometry (EDS) at RT at an electron beam voltage of 30 kV. Similar to the X–ray diffraction measurements, prior to the microstructure observation and chemical composition analysis, the specimens were cleaned, mechanically grinded by emery papers, and polished by diamond paste with a diameter of 1 μm to obtain fine surface finish for the analysis.

In addition to the aforementioned fundamental characterizations, examinations of the functionalities, such as shape memory effect, were also carried out by the test of bending deformation followed by a heating process by a cigar lighter, which reached temperatures higher than 1000 K. Specimens with a dimension of 0.2 mm (thickness) × 2 mm (width) × 20 mm (width), were obtained from the above–mentioned ST specimens. The specimens were firstly subjected a 5% surface strain, which was made by bending the specimens with a round bar with a diameter of 6 mm. The surface strain made to the specimens could be calculated by following the equation of $\varepsilon = [h/(2r)]$, where ε is surface strain, h is thickness of the specimen, and r is curvature radius, respectively. The shape recovery rates of the specimens were determined by recognizing and calculating the shape difference in the optical photos, which were taken at the (i) after bending deformation and (ii) upon heating stages. The bending tests and calculations for shape recovery rates were carried out for three times for ensuring the accuracy and also for calculations of average shape recovery rates. An image processing software, ImageJ, was used for certifying the average shape recovery rates.

For the evaluations of the mechanical properties, two different tensile tests were carried out by a universal testing machine of Shimadzu AG1kN Autograph. Specimens, which were in a dimension of 0.2 mm (thickness) × 2 mm (width) × 10 mm (width), possessed a dog–bone structure. The tensile tests were carried out at RT under ambient in two different manners: (1) the continuous tensile test and (2) the cyclic loading–unloading tensile tests. A video extensometer, which was attached to the universal testing machine, was used for the accurate measurement of shape deformation strain for both tensile tests. In the case of the continuous tensile test, the specimens were subjected to a continuous tensile stress until their fractures. On the other hand, in the case of the cyclic loading–unloading tensile tests, specimens were repeatedly subjected to a 1% strain per cycle until the 10th cycle or the fractures of the specimens. All the specimens were subjected to a strain rate of 8.3×10^{-4} s^{-1} and the stress loading direction was parallel to the rolling direction (RD) of the alloys.

3. Results and Discussion

3.1. Cold Workability

In the case of the Ti–4Au–20Ta specimen, it could be cold rolled up at RT to the reduction in thickness of 70% in the first cold–rolling. Followed by the first cold–rolling, process–annealing was carried out at 1273 K for 1.8 ks. Second, cold–rolling was thereafter conducted until the reduction in thickness reached 98%. The final thickness of the specimen after cold–rolling was at approximately 0.2 mm. On the other hand, in the case of the Ti–4Au–30Ta specimen, it was found that the alloy, which showed low workability, cracked at the first pass of the cold–rolling process. Hot–rolling was therefore executed at 1173 K instead for thinning down the specimen until the reduction in thickness reached 98%. The differences in the cold workability of these two alloys are explained by the phase constituents and microstructure observations in both Sections 3.2 and 3.3.

3.2. Phase Identification

The X–ray diffraction patterns of the (a) Ti–4Au–20Ta alloy and the (b) Ti–4Au–30Ta alloy at RT under ambient are shown in Figure 1. Put simply, the parent β–phase was found in the (a) Ti–4Au–20Ta alloy indicating a single β–phase of this alloy. On the other hand, in addition to the parent β–phase, a compound with a crystal structure of A15 was also found in the (b) Ti–4Au–30Ta alloy. According to the literature [37,38], the compound possessing A15 crystal structure corresponds to the intermetallic compound of Ti_3Au. More analysis concerning the Ti_3Au with an A15 crystal structure are shown in Section 3.3 via microstructure observations and chemical composition analysis.

It was reported that in the binary Ti–4Au specimen the apparent phase of the α''–martensite or the α–massive martensite (α_m), which were generated from the martensitic transformation or the massive transformation, were found [15,16]. Different from the binary Ti-4Au specimen, it is obvious that the parent β–phase was successfully stabilized via the introduction of the Ta element as the third element. These results observed in this study were also in a good agreement with literature [39,40].

On the other hand, according to the literature, the intermetallic compound of Ti_3Au, which is a brittle phase and is often found in the Ti–Au-based alloys [35,38,41,42], could deteriorate the cold workability of the specimens. Therefore, in Section 3.1, the specimen of the Ti–4Au–30Ta alloy, which could not be cold–rolled at RT, could be attributed to the embrittlement brought from the Ti_3Au precipitates. Hence, the cold workability agreed well with the phase constituents observed by the X–ray diffraction patterns (Figure 1) and the literatures [35,38,41,42]. More details of the Ti_3Au precipitates concerning its microstructure are shown in Section 3.3.

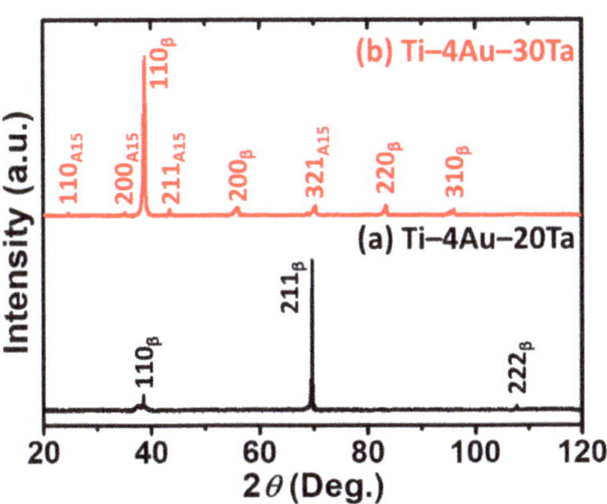

Figure 1. The X-ray diffraction patterns of the (**a**) Ti–4Au–20Ta alloy and the (**b**) Ti–4Au–30Ta alloy at RT under ambient.

3.3. Microstructure Observations and Elemental Mappings

Figure 2a,b show the SEM images of the Ti–4Au–20Ta alloy and the Ti–4Au–30Ta alloy, respectively. In addition, the elemental mapping results of the (c) Ti, (d) Au, and (e) Ta elements in the Ti–4Au–30Ta alloy are also shown in Figure 2. The observed region of the elemental mappings is surrounded by the dashed lines in the SEM image of the Ti–4Au–30Ta alloy (Figure 2b).

In the SEM images, only single phase was observed in the Ti–4Au–20Ta alloy (Figure 2a) and this result is in accordance with its X-ray diffraction pattern (Figure 1a), where merely the β-parent phase was found. On the other hand, it is clear that some precipitates in bright contrast were discerned in the Ti–4Au–30Ta alloy (Figure 2b). The size of the oval-shaped precipitates was approximately 1–3 μm and the volume fraction of the precipitates was about 5%. It is also worth mentioning that the microstructure of the precipitates was found very similar with those in the literature [38]. More details concerning the precipitates in the Ti–4Au–30Ta alloy are explained by the elemental mapping results in the following.

The elemental mapping results of the precipitates were further carried out and are shown in Figure 2c–e. Bright contrast suggests high concentration while dark contrast indicates low concentration. It was found that, in the case of Ti element, there was no obvious concentration difference between the matrix and the precipitates (Figure 2c). On the other hand, the precipitates were high in Au (Figure 2d) and low in Ta (Figure 2e). Given that the Ti_3Au phase was determined by the X-ray diffraction pattern of the Ti–4Au–30Ta alloy (Figure 1b) and the literature [35,38,41,42], it thus could be concluded that the precipitates in bright contrast could be corresponded to the Ti_3Au phase. In addition, it was reported that the Ti_3Au phase is a line compound [37]; therefore, the solubility of Ta was extremely low (i.e., dark contrast) to the Ti_3Au intermetallic compound. A good agreement was reached among the X-ray diffraction observations (Figure 1), SEM images (Figure 2b), elemental mapping results (Figure 2c–e), and also the literature [35,37,38,41,42].

Based on the aforementioned results, the addition of Ta as the third element promoted the precipitation of Ti_3Au intermetallic compound since the overall chemical composition in terms of the binary Ti–Au system shifted to the Au-rich side [37] while Ta was introduced into the binary Ti–Au system. Therefore, obvious precipitates of the oval-shaped Ti_3Au intermetallic compound with a bright contrast were observed in the Ti–4Au–30Ta alloy (Figure 1b–e).

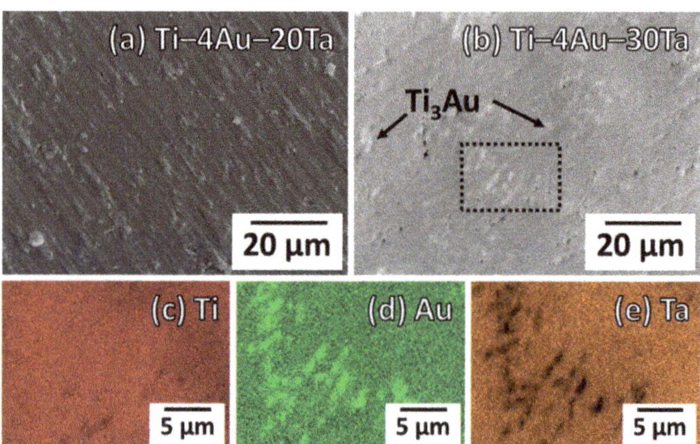

Figure 2. SEM images of the (**a**) Ti–4Au–20Ta alloy and the (**b**) Ti–4Au–30Ta alloy. Elemental mapping results of (**c**) Ti, (**d**) Au, and (**e**) Ta elementals of the (**b**) Ti–4Au–30Ta alloy. The elemental mapping analyzed regime of the (**b**) Ti–4Au–30Ta alloy is surrounded by a dashed square.

3.4. Bending Test

Results of the bending tests are shown in Figure 3. In Figure 3, column (i) indicates the specimens after bending deformation (stage i) while column (ii) suggests the shape recovery of the specimens upon heating (stage ii). Judging from the optical images, an obvious shape recovery was found in the Ti–4Au–20Ta alloy (Figure 3a); on the contrary, the shape recovery rate of the Ti–4Au–30Ta alloy (Figure 3b) was relatively faint. The shape recovery rates of (a) the Ti–4Au–20Ta alloy and (b) the Ti–4Au–30Ta alloy were determined to be 10% and 30%, respectively. Since the apparent phase of the (a) Ti–4Au–20Ta alloy was determined to be the single parent β–phase via the X-ray diffraction observation (Figure 1a) and the microstructure observation (Figure 2a), it thus could be deduced that the shape recovery behavior of the Ti–4Au–20Ta alloy was brought by the stress–induced martensite transformation (SIMT) during bending deformation (stage i) and followed by its backward transformation upon heating (stage ii).

Specimen	(i) Deformed	(ii) Heated	Recovery rate
(a) Ti–4Au–20Ta	0 1 2 3 cm	0 1 2 3 cm	30%
(b) Ti–4Au–30Ta	0 1 2 3 cm	0 1 2 3 cm	10%

Figure 3. Bending tests of the (**a**) Ti–4Au–20Ta alloy and the (**b**) Ti–4Au–30Ta alloy for the examinations of the shape memory effect and shape recovery rates. Stage (**i**) corresponds to after bending deformation while stage (**ii**) indicates shape recovery upon heating process.

Similar to the (a) Ti–4Au–20Ta alloy, the apparent phase of the (b) Ti–4Au–30Ta alloy was also the parent β–phase before deformation; however, precipitates of Ti_3Au compound was also found in the Ti–4Au–30Ta alloy (Figures 1b and 2b–e). The deteriorated shape recovery rate of the (b) Ti–4Au–30Ta alloy could be attributed to the suppression of the phase transformation by the Ti_3Au precipitates during bending deformation and shape recovery. These results are in accordance with those in the literature [38]. In addition to the inhibition of the phase transformation by the Ti_3Au precipitates, the relatively low shape recovery rates (i.e., 10% and 30%) of these two alloys could also be ascribed to the plastic deformation, which was introduced into the alloys during bending deformation.

3.5. Continuous Tensile Tests

Stress–strain (SS; σ–ε) curves of the (a) Ti–4Au–20Ta alloy and the (b) Ti–4Au–30Ta alloy by the continuous tensile tests are shown in Figure 4. The cross symbols at the end of the SS curves suggest the fractures of the specimens. The strain until fracture of the (a) Ti–4Au–20Ta alloy was at approximately 17%, which was almost twice larger than that of the (b) Ti–4Au–30Ta alloy (i.e., deformation strain at approximately 9%), showed high elongation. The different ductility of these two alloys could be elaborated through the following reasons.

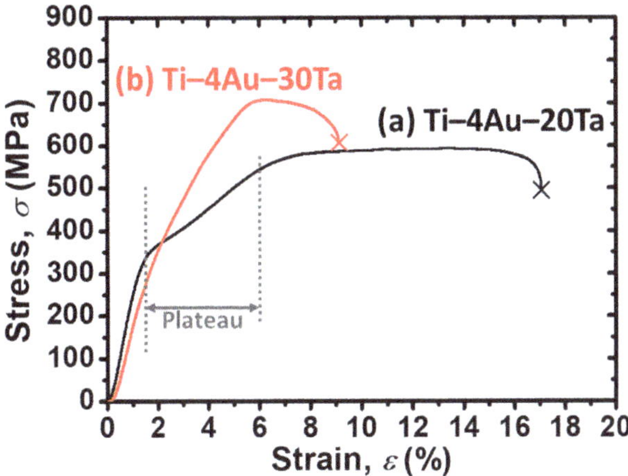

Figure 4. Stress–strain (SS) curves of the (**a**) Ti–4Au–20Ta alloy and the (**b**) Ti–4Au–30Ta alloy via the continuous tensile tests. The cross symbols at the end of the curves suggest fractures of the specimens. A stress plateau in the (**a**) Ti–4Au–20Ta alloy was indicated by the vertical dashed lines and the horizontal solid arrows.

Firstly, it was worth noticing that in the (a) Ti–4Au–20Ta alloy, there is a clear stress plateau, which is indicated by the vertical dashed lines and the solid horizontal line with arrows. Given that the apparent phase of the (a) Ti–4Au–20Ta alloy is the parent β–phase at the testing temperature (i.e., RT) (Figure 1a), the stress plateau could be attributed to the stress-induced martensitic transformation (SIMT) [38,41,43]. The first yielding could be attributed to the onset of the SIMT, while the second yielding could be ascribed to the commencement of the plastic deformation [38,41,43]. On the other hand, almost no stress plateau was found in the case of the (b) Ti–4Au–30Ta alloy. Therefore, the relatively high elongation of the (a) Ti–4Au–20Ta alloy could be ascribed to the SIMT. It is also worth mentioning that the SIMT found in the Ti–4Au–20Ta alloy in Figure 4a corresponds well with the result of the bending test in Section 3.4. While the Ti–4Au–30Ta alloy, which had a lower shape recovery rate upon heating (Figure 3b), showed almost no stress plateau in Figure 4b.

Secondly, the Ti_3Au intermetallic compound, which was merely found in the (b) Ti–4Au–30Ta alloy (Figures 1b and 2b–e), is known as a brittle phase [44]. The ductility of the (b) Ti–4Au–30Ta alloy was thus deteriorated by the formation of the Ti_3Au precipitates. On the contrary, the (a) Ti–4Au–20Ta alloy showed relatively high deformation strain before fracture since no brittle Ti_3Au precipitates were found in this alloy (Figures 1a and 2a).

To sum up, the relatively high elongation of the (a) Ti–4Au–20Ta alloy could be attributed to the occurrence of SIMT and lack of Ti_3Au precipitates. On the other hand, the relatively low elongation of the (b) Ti–4Au–30Ta alloy was due to the Ti_3Au precipitates and almost free of SIMT during deformation.

However, reverse to the ductility, comparatively high strength was observed in the (b) Ti–4Au–30Ta alloy than that of the (a) Ti–4Au–20Ta alloy in the continuous tensile tests (Figure 4). It was found that the ultimate tensile strength (UTS) was at approximately 737 MPa for the (b) Ti–4Au–30Ta alloy and at approximately 600 MPa for the (a) Ti–4Au–20Ta alloy. This could be simply explained by the precipitation–hardening effect brought from the precipitates of the Ti_3Au compound and also the solution–hardening effect brought by the Ta addition as the third element.

3.6. Functional Mapping

A functional mapping, which reveals (a) yielding stress (black squares), (b) ultimate tensile strength (UTS) (red circles), and (c) fracture strain (blue triangles) as a function of Ta concentration in the alloys, is shown as Figure 5. It is necessary to mention that for the purposes of completeness and comparison, the binary Ti–4Au alloy was cited from our preliminary research [35] and plotted in Figure 5. It is obvious that both the (a) yielding stress and the (b) UTS increased monotonically with Ta addition concentration. Similar to the explanation in Section 3.5, this enhancement in strength could be attributed to the solid–solution strengthening effect of the introduction of Ta element and the precipitation hardening effect caused by the Ti_3Au precipitates. Slip of dislocation (i.e., plastic deformation) was inhibited by the introduction of the Ta impurity and the Ti_3Au precipitates. Opposite to the strength, fracture strain (Figure 5c) of the specimens decreased with the introduction of Ta concentration due to the precipitation of the brittle Ti_3Au intermetallic compound.

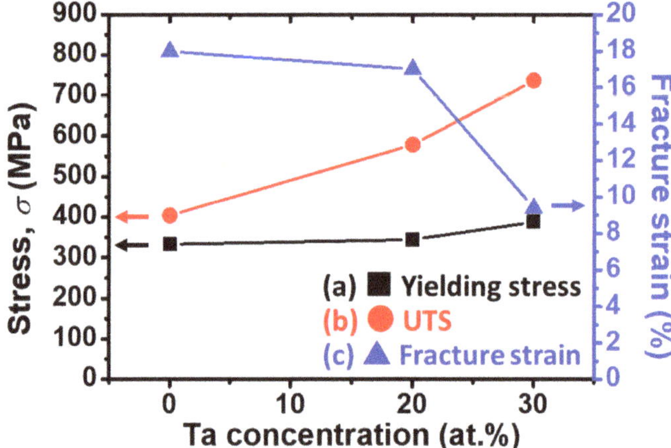

Figure 5. Functional mapping of (**a**) yielding stress (black squares), (**b**) ultimate tensile strength (UTS) (red circles), and (**c**) fracture strain (blue triangles) as a function of the Ta addition concentration in the alloys. (Left *y*–axis: stress (MPa); Right *y*–axis: fracture strain (%)) (Note that the results of the binary Ti–4Au alloy was cited from our preliminary research [35]).

3.7. Cyclic Loading–Unloading Tensile Tests

The cyclic loading–unloading tests of the (a) Ti–4Au–20Ta alloy and the (b) Ti–4Au–30Ta alloy are shown in Figure 6. Overall, it was found that the (a) Ti–4Au–20Ta alloy which exhibited slight pseudoelasticity (PE), showed slight shape recovery during unloading. While the (b) Ti–4Au–30Ta alloy showed almost no shape recovery during unloading in the cyclic loading–unloading tests at RT. Therefore, it could be deduced that the austenite transformation start temperature (A_s) is higher than the operation temperature (i.e., RT). In addition, the vanished shape recovery in the (b) Ti–4Au–30Ta alloy could be ascribed to the inhibition of the Ti_3Au precipitates. On the other hand, the slight shape recovery in the (a) Ti–4Au–20Ta alloy could be attributed to the pseudoelasticity via the reorientation of martensitic variants, which is commonly observed in the martensite phase [45–47]. Moreover, similar to the results from the continuous tensile tests (Figure 4) in Section 3.5, the (b) Ti–4Au–30Ta alloy showed improved strength than that of the (a) Ti–4Au–20Ta alloy.

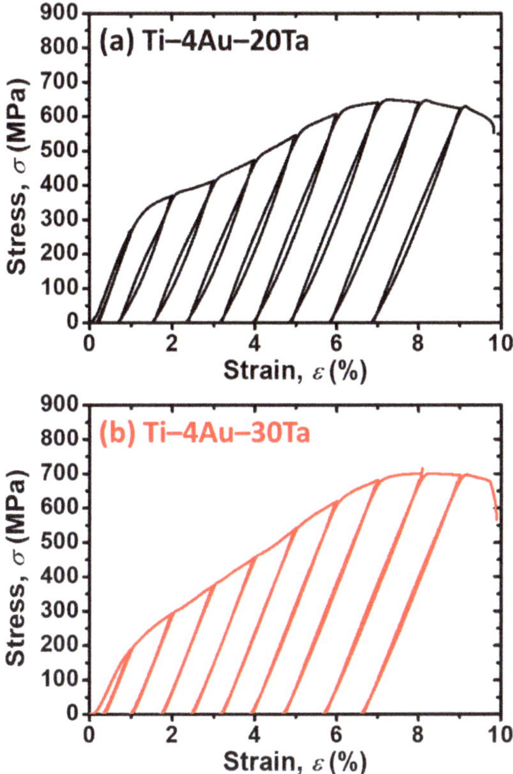

Figure 6. Cyclic loading–unloading tensile tests of the (**a**) Ti–4Au–20Ta alloy and the (**b**) Ti–4Au–30Ta alloy at RT under ambient.

For further analysis of the pseudoelasticity, every cycle in the Ti–4Au–20Ta alloy was detached and the stress for SIMT (i.e., first-yielding stress), which is shown in Figure 7a, was read from every cycle of the cyclic loading–unloading tensile test (Figure 6a). During the cyclic tensile test, dislocations, which increased the internal stress, was introduced into the specimen. The martensite phase was stabilized by the elevated internal stress; as a result, the external stress for the SIMT (σ_{SIMT}) was reduced accordingly (Figure 7a). This phenomenon was in good agreement with that in the literature [41].

The ninth cycle in the cyclic loading–unloading tensile test of the Ti–4Au–20Ta specimen is also shown in Figure 7b. It is obvious that during unloading of externally applied stress, there was a slight shape recovery, which is at approximately 0.5%, was brought from the pseudoelasticity. The shape recovery originated from the pseudoelasticity was determined by reading strain difference between the overall recovery strain after fully unloading and the strain produced by the elastic shape recovery. Further explanations for the specific terms are shown and explained by the illustration in Figure 7d.

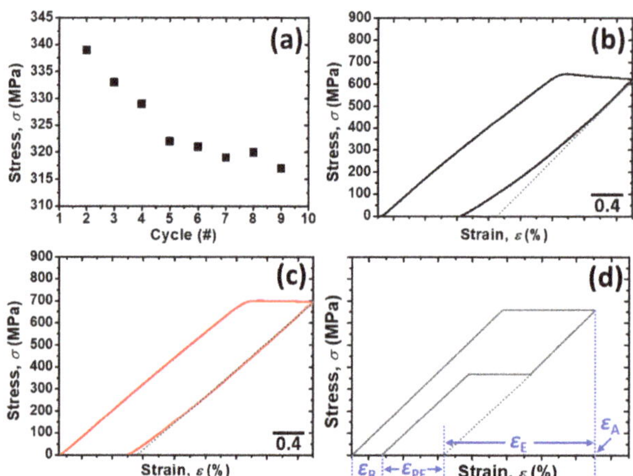

Figure 7. (**a**) Stress for the first yielding stress (i.e., stress for SIMT) of the Ti–4Au–20Ta alloy, (**b**) the ninth cycle in the cyclic loading–unloading tensile test of the Ti–4Au–20Ta alloy, (**c**) the ninth cycle in the cyclic loading–unloading tensile test of the Ti–4Au–30Ta alloy, and (**d**) the illustration for the explanations of the terms in the SS curve. Where ε_A indicates the total applied strain, ε_E suggests the elastic shape recovery strain, ε_{PE} corresponds to shape recovery strain brought by pseudoelasticity, and ε_R represents the remaining strain after unloading.

On the other hand, almost no shape recovery was found in the Ti–4Au–30Ta alloy (Figure 7c) during unloading of the externally applied stress. The aforementioned Ti–4Au–20Ta alloy and Ti–4Au–30Ta alloy results are in good accordance with those in the bending tests (Figure 3) and the continuous tensile tests (Figure 4). An illustration for the explanations of the specific terms are shown in Figure 7d. In Figure 7d, ε_A indicates the total applied strain, ε_E suggests the elastic shape recovery strain, ε_{PE} corresponds to shape recovery strain brought by pseudoelasticity, and ε_R represents the remaining strain after unloading.

This study worked on the Ti–4Au–20Ta alloy and Ti–4Au–30Ta alloy for the preliminary screening of the Ti–Au–Ta–based shape memory alloys. Some of the Ti–Au–based studies have already been carried out by our research group [35,36,38,41,43] while some of the Ti–Au–based studies are now under investigations and preparations. Further results concerning the Ti–Au–based shape memory alloys will be published in the future by our research group.

4. Conclusions

For the biomedical applications in this work, the Ti–4Au–20Ta and the Ti–4Au–30Ta alloys were prepared by physical metallurgy techniques and their phase constituents, composition analysis, mechanical behaviors, and functional properties, such as shape memory effect (SME) and pseudoelasticity (PE), have been studied. The important findings are listed in the following points.

1. The single parent β–phase was found in the Ti–4Au–20Ta alloy while the parent β–phase along with the precipitates of the Ti_3Au intermetallic compound in a volume fraction of approximate 5% was observed in the Ti–4Au–30Ta alloy.
2. The Ti–4Au–20Ta alloy showed about 30% shape recovery upon heating in the bending tests while only 10% shape recovery was found in the Ti–4Au–30Ta alloy due to the inhibition of phase transformation between austenite and martensite from the Ti_3Au precipitates.
3. An obvious two–stage yielding, which corresponds to the stress for the stress–induced martensite transformation (SIMT) and the stress for plastic deformation, was found in

the Ti–4Au–20Ta alloy. On the other hand, no obvious two–stage yielding was found in the Ti–4Au–30Ta alloy due to the inhibition of SIMT from the Ti_3Au intermetallic compound.

4. In the continuous tensile test, the strength of the Ti–4Au–30Ta alloy was higher than that of the Ti–4Au–20Ta alloy due to the truth of the solid–solution hardening effect from the Ta addition as the third element and the precipitation hardening effect from the Ti_3Au intermetallic compound. On the contrary, better elongation was found in the Ti–4Au–20Ta alloy than that of the Ti–4Au–30Ta alloy owing to the SIMT and no precipitation of Ti_3Au compound.
5. Slight pseudoelasticity at approximately 0.5% was found in the Ti–4Au–20Ta alloy showing slight shape recovery after unloading of externally applied stress while almost no shape recovery was found in the Ti–4Au–30Ta alloy due to the inhibition of Ti_3Au precipitates. These results are in accordance with the bending tests and the continuous tensile tests.

Author Contributions: Conceptualization, T.I. and H.H.; data curation, W.-T.C. and K.F.; formal analysis, W.-T.C., K.F., M.T. and H.H.; funding acquisition, M.T. and H.H.; investigation, W.-T.C., K.F., M.T., T.I. and H.H.; methodology, T.I. and H.H.; project administration, M.T. and H.H.; resources, M.T., T.I. and H.H.; software, W.-T.C. and K.F.; supervision, W.-T.C., Akira Umise, M.T., T.I. and H.H.; validation, W.-T.C., M.T., T.I. and H.H.; visualization, W.-T.C.; writing–original draft, W.-T.C.; writing–review & editing, K.F., M.T. and H.H. All authors have read and agreed to the published version of the manuscript.

Funding: This work is supported by the Tokyo Tech Fund (ASUNARO Grant) and Japan Society for the Promotion of Science (JSPS) (KAKENHI 19H02417 and KAKENHI 20K20544).

Institutional Review Board Statement: Not applicable.

Informed Consent Statement: Not applicable.

Data Availability Statement: All data contained within the article.

Conflicts of Interest: The authors declare no conflict of interest.

References

1. Jani, J.M.; Leary, M.; Subic, A.; Gibson, M.A. A review of shape memory alloy research, applications and opportunities. *Mater. Des.* **2014**, *56*, 1078–1113. [CrossRef]
2. Yamauchi, K.; Ohkata, I.; Tsuchiya, K.; Miyazaki, S. *Shape Memory and Superelastic Alloys: Applications and Technologies*; Elsevier: Amsterdam, The Netherlands, 2011.
3. Morgan, N.B. Medical shape memory alloy applications—the market and its products. *Mater. Sci. Eng. A* **2004**, *378*, 16–23. [CrossRef]
4. Fang, C.; Wang, W. *Shape Memory Alloys for Seismic Resilience*; Springer: New York, NY, USA, 2020.
5. Magnusson, B.; Bergman, M.; Bergman, B.; Söremark, R. Nickel allergy and nickel-containing dental alloys. *Eur. J. Oral Sci.* **1982**, *90*, 163–167. [CrossRef] [PubMed]
6. Christensen, O.B.; Möoller, H. External and internal exposure to the antigen in the hand eczema of nickel allergy. *Contact Derm.* **1975**, *1*, 136–141. [CrossRef] [PubMed]
7. Li, Y.; Yang, C.; Zhao, H.; Qu, S.; Li, X.; Li, Y. New Developments of Ti-Based Alloys for Biomedical Applications. *Materials* **2014**, *7*, 1709–1800. [CrossRef]
8. Biesiekierski, A.; Wang, J.; Gepreel, M.A.-H.; Wen, C. A new look at biomedical Ti-based shape memory alloys. *Acta Biomater.* **2012**, *8*, 1661–1669. [CrossRef]
9. Eisenbarth, E.; Velten, D.; Müller, M.; Thull, R.; Breme, J. Biocompatibility of β-stabilizing elements of titanium alloys. *Biomaterials* **2004**, *25*, 5705–5713. [CrossRef]
10. Ratner, B.D. A Perspective on Titanium Biocompatibility. In *Titanium in Medicine*; Engineering Materials; Springer: Berlin/Heidelberg, Germany, 2001.
11. Wataha, J.C. Biocompatibility of dental casting alloys: A review. *J. Prosthet. Dent.* **2000**, *83*, 223–234. [CrossRef]
12. Hainfeld, J.F.; Slatkin, D.N.; Focella, T.M.; Smilowitz, H.M. Gold nanoparticles: A new X-ray contrast agent. *Br. J. Radiol.* **2006**, *79*, 248–253. [CrossRef]
13. Goodman, P. Current and future uses of gold in electronics. *Gold Bull.* **2002**, *35*, 21–26. [CrossRef]
14. Hanawa, T. In vivo metallic biomaterials and surface modification. *Mater. Sci. Eng. A* **1999**, *267*, 260–266. [CrossRef]

15. Plichta, M.R.; Williams, J.C.; Aaronson, H.I. On the existence of the $\beta \rightarrow \alpha_m$ transformation in the alloy systems Ti–Ag, Ti–Au, and Ti–Si. *Metall. Trans. A* **1977**, *8*, 1885–1892. [CrossRef]
16. Plichta, M.R.; Perepezko, J.H.; Aaronson, H.I.; Lange, W.F. Part I Nucleation kinetics of the $\beta \rightarrow \alpha_m$ transformation in Ti–Ag and Ti–Au alloys. *Acta Metall.* **1980**, *28*, 1031–1040. [CrossRef]
17. Murray, J.L. *Phase Diagrams of Binary Titanium Alloys*; ASM International: Materials Park, OH, USA, 1987.
18. Soro, N.; Attar, H.; Brodie, E.; Veidt, M.; Molotnikov, A.; Dargusch, M.S. Evaluation of the mechanical compatibility of additively manufactured porous Ti–25Ta alloy for load-bearing implant applications. *J. Mech. Behav. Biomed. Mater.* **2019**, *97*, 149–158. [CrossRef]
19. Meisner, L.L.; Markov, A.B.; Rotshtein, V.P.; Ozur, G.E.; Meisner, S.N.; Yakovlev, E.V.; Semin, V.O.; Mironov, Y.P.; Poletika, T.M.; Girsova, S.L.; et al. Microstructural characterization of Ti-Ta-based surface alloy fabricated on TiNi SMA by additive pulsed electron-beam melting of film/substrate system. *J. Alloys Compd.* **2018**, *730*, 376–385. [CrossRef]
20. Liu, Y.; Li, K.; Wu, H.; Song, M.; Wang, W.; Li, N.; Tang, H. Synthesis of Ti–Ta alloys with dual structure by incomplete diffusion between elemental powders. *J. Mech. Behav. Biomed. Mater.* **2015**, *51*, 302–312. [CrossRef]
21. He, W.; Ai, K.; Lu, L. Nanoparticulate X-ray CT contrast agents. *Sci. China Chem.* **2015**, *58*, 753–760. [CrossRef]
22. Jin, Y.; Li, Y.; Ma, X.; Zha, Z.; Shi, L.; Tian, J.; Dai, Z. Encapsulating tantalum oxide into polypyrrole nanoparticles for X-ray CT/photoacoustic bimodal imaging-guided photothermal ablation of cancer. *Biomaterials* **2014**, *35*, 5795–5804. [CrossRef] [PubMed]
23. Yang, C.; Li, J.; Zhu, C.; Zhang, Q.; Yu, J.; Wang, J.; Wang, Q.; Tang, J.; Zhou, H.; Shen, H. Advanced antibacterial activity of biocompatible tantalum nanofilm via enhanced local innate immunity. *Acta Biomater.* **2019**, *89*, 403–418. [CrossRef]
24. Matsuno, H.; Yokoyama, A.; Watari, F.; Uo, M.; Kawasaki, T. Biocompatibility and osteogenesis of refractory metal implants, titanium, hafnium, niobium, tantalum and rhenium. *Biomaterials* **2001**, *22*, 1253–1262. [CrossRef]
25. Niinomi, M. Mechanical properties of biomedical titanium alloys. *Mater. Sci. Eng. A* **1998**, *243*, 231–236. [CrossRef]
26. Zarinejad, M.; Wada, K.; Pahlevani, F.; Katal, R.; Rimaz, S. Martensite and reverse transformation temperatures of TiAu-based and TiIr-based intermetallics. *J. Alloys Compd.* **2021**, *870*, 159399. [CrossRef]
27. Casalena, L.; Bigelow, G.S.; Gao, Y.; Benafan, O.; Noebe, R.D.; Wang, Y.; Mills, M.J. Mechanical behavior and microstructural analysis of NiTi-40Au shape memory alloys exhibiting work output above 400 °C. *Intermetallics* **2017**, *86*, 33–44. [CrossRef]
28. Wu, S.K.; Wayman, C.M. Martensitic transformations and the shape memory effect in $Ti_{50}Ni_{10}Au_{40}$ and $Ti_{50}Au_{50}$ alloys. *Metallography* **1987**, *20*, 359–376. [CrossRef]
29. Kikuchi, M.; Takahashi, M.; Okuno, O. Elastic moduli of cast Ti–Au, Ti–Ag, and Ti–Cu alloys. *Dent. Mater.* **2006**, *22*, 641–646. [CrossRef]
30. Buenconsejo, P.J.S.; Kim, H.Y.; Hosoda, H.; Miyazaki, S. Shape memory behavior of Ti–Ta and its potential as a high-temperature shape memory alloy. *Acta Mater.* **2009**, *57*, 1068–1077. [CrossRef]
31. Ikeda, M.; Komatsu, S.-Y.; Nakamura, Y. The Effect of Ta Content on Phase Constitution and Aging Behavior of Ti-Ta Binary Alloys. *Mater. Trans.* **2002**, *43*, 2984–2990. [CrossRef]
32. Kim, H.Y.; Fukushima, T.; Buenconsejo, P.J.S.; Nam, T.-H.; Miyazaki, S. Martensitic transformation and shape memory properties of Ti–Ta–Sn high temperature shape memory alloys. *Mater. Sci. Eng. A* **2011**, *528*, 7238–7246. [CrossRef]
33. Bertrand, E.; Castany, P.; Yang, Y.; Menou, E.; Gloriant, T. Deformation twinning in the full-α'' martensitic Ti–25Ta–20Nb shape memory alloy. *Acta Mater.* **2016**, *105*, 94–103. [CrossRef]
34. Buenconsejo, P.J.S.; Kim, H.Y.; Miyazaki, S. Effect of ternary alloying elements on the shape memory behavior of Ti–Ta alloys. *Acta Mater.* **2009**, *57*, 2509–2515. [CrossRef]
35. Chiu, W.-T.; Ishigaki, T.; Nohira, N.; Umise, A.; Tahara, M.; Hosoda, H. Effect of 3d transition metal additions on the phase constituent, mechanical properties, and shape memory effect of near–eutectoid Ti–4Au biomedical alloys. *J. Alloys Compd.* **2021**, *857*, 157599. [CrossRef]
36. Chiu, W.-T.; Wakabayashi, K.; Umise, A.; Tahara, M.; Hosoda, H. Enhancement of Mechanical Properties and Shape Memory Effect of Ti–Cr–based Alloys via Au and Cu Modifications. *J. Mech. Behav. Biomed. Mater.* **2021**, *123*, 104707. [CrossRef]
37. Murray, J.L. The Au–Ti (Gold–Titanium) system. *Bull. Alloy Phase Diagr.* **1983**, *4*, 278–283. [CrossRef]
38. Chiu, W.-T.; Ishigaki, T.; Nohira, N.; Umise, A.; Tahara, M.; Inamura, T.; Hosoda, H. Influence of the precipitates on the shape memory effect and superelasticity of the near–eutectoid Ti–Au–Fe alloy towards biomaterial applications. *Intermetallics* **2021**, *133*, 107180. [CrossRef]
39. Kim, H.Y.; Hashimoto, S.; Kim, J.I.; Inamura, T.; Hosoda, H.; Miyazaki, S. Effect of Ta addition on shape memory behavior of Ti–22Nb alloy. *Mater. Sci. Eng. A* **2006**, *417*, 120–128. [CrossRef]
40. Ijaz, M.F.; Zhukova, Y.; Konopatsky, A.; Dubinskiy, S.; Korobkova, A.; Pustov, Y.; Brailovski, V.; Prokoshkin, S. Effect of Ta addition on the electrochemical behavior and functional fatigue life of metastable Ti-Zr-Nb based alloy for indwelling implant applications. *J. Alloys Compd.* **2018**, *748*, 51–56. [CrossRef]
41. Chiu, W.-T.; Ishigaki, T.; Nohira, N.; Umise, A.; Tahara, M.; Inamura, T.; Hosoda, H. Effect of Cr additions on the phase constituent, mechanical properties, and shape memory effect of near–eutectoid Ti–4Au towards the biomaterial applications. *J. Alloys Compd.* **2021**, *867*, 159037. [CrossRef]
42. Karimi, A.; Cattin, C. Intermetallic β-Ti_3Au hard thin films prepared by magnetron sputtering. *Thin Solid Films* **2018**, *646*, 1–11. [CrossRef]

43. Chiu, W.-T.; Ishigaki, T.; Nohira, N.; Umise, A.; Tahara, M.; Inamura, T.; Hosoda, H. Evaluations of mechanical properties and shape memory behaviors of the aging–treated Ti–Au–Mo alloys. *Mater. Chem. Phys.* **2021**, *269*, 124775. [CrossRef]
44. Xin, Y.; Han, K.; Svanidze, E.; Besara, T.; Siegrist, T.; Morosan, E. Microstructure of hard biocompatible $Ti_{1-x}Au_x$ alloys. *Mater. Charact.* **2019**, *149*, 133–142. [CrossRef]
45. Bhadeshia, H.K.D.H.; Wayman, C.M. Phase Transformations: Nondiffusive. In *Physical Metallurgy*, 5th ed.; Elsevier: Amsterdam, The Netherlands, 2014; pp. 1021–1072.
46. Popov, P.; Lagoudas, D.C. A 3-D constitutive model for shape memory alloys incorporating pseudoelasticity and detwinning of self-accommodated martensite. *Int. J. Plast.* **2007**, *23*, 1679–1720. [CrossRef]
47. Otsuka, K.; Wayman, C.M. *Shape Memory Materials*; Cambridge University Press: Cambridge, UK, 1999.

 materials

MDPI

Article

Strain Rate Effect upon Mechanical Behaviour of Hydrogen-Charged Cycled NiTi Shape Memory Alloy

Fehmi Gamaoun [1,2]

[1] Department of Mechanical Engineering, College of Engineering, King Khalid University, Abha 61421, Saudi Arabia; fgamaoun@kku.edu.sa

[2] Laboratory of Mechanics of Sousse, National Engineering School of Sousse, University of Sousse, Sousse 4054, Tunisia

Abstract: The rate dependence of thermo-mechanical responses of superelastic NiTi with different imposed strain rates after cycling from 1 to 50 cycles under applied $10^{-5}s^{-1}$, $10^{-4}s^{-1}$ and $10^{-3}s^{-1}$ strain rates, immersion for 3 h and ageing has been investigated. The loaded and unloaded as-received NiTi alloy under an imposed strain of 7.1% have shown an increase in the residual deformation at zero stress with an increase in strain rates. It has been found that after 13 cycles and hydrogen charging, the amount of absorbed hydrogen (291 mass ppm) was sufficient to cause the embrittlement of the tensile loaded NiTi alloy with $10^{-5}s^{-1}$. However, no premature fracture has been detected for the imposed strain rates of $10^{-4}s^{-1}$ and $10^{-3}s^{-1}$. Nevertheless, after 18 cycles and immersion for 3 h, the fracture has occurred in the plateau of the austenite martensite transformation during loading with $10^{-4}s^{-1}$. Despite the higher quantity of absorbed hydrogen, the loaded specimen with a higher imposed strain rate of $10^{-3}s^{-1}$ has kept its superelasticity behaviour, even after 20 cycles. We attribute such a behaviour to the interaction between the travelling distance during the growth of the martensitic domains while introducing the martensite phase and the amount of diffused hydrogen.

Keywords: shape memory alloys; superelastic; hydrogen; cyclic loading; martensite variants

Citation: Gamaoun, F. Strain Rate Effect upon Mechanical Behaviour of Hydrogen-Charged Cycled NiTi Shape Memory Alloy. *Materials* **2021**, *14*, 4772. https://doi.org/10.3390/ma14164772

Academic Editor: Carmine Maletta

Received: 6 July 2021
Accepted: 18 August 2021
Published: 23 August 2021

1. Introduction

NiTi Shape Memory Alloys (SMAs) have been employed on a large scale within various engineering applications such as Micro-Electro-Mechanical Systems (MEMS), aerospace structures [1,2], and biomedical applications [3,4]. The success of these applications hinges on their capability of having an excellent combination of several properties, including superelasticity, corrosion resistance, and biocompatibility. Moreover, due to its larger flexibility, the equiatomic superelastic NiTi archwire turns to be one integral part of orthodontic treatment because of exerting a relatively constant force in the range of tooth movement. However, the NiTi alloy is found to be very sensitive to the hydrogen diffusion caused by adding fluoride to prophylactic agents and toothpaste in the oral cavity [5–7]. In fact, the diffusion of hydrogen is a main environmental cause for premature fractures of the superelastic NiTi alloys [8,9].

We have shown, after immersion, that the NiTi alloy fractures in a brittle manner for different strain rates [10,11] and for varied immersion time [12,13]. In fact, at room temperature and after immersion in a 0.9% NaCl aqueous solution, while utilizing a varied current density of 5, 10, and 20 A/m^2 from 2 to 24 h, embrittlement takes place after ageing at room temperature rather than immediately after hydrogen charging. This behaviour suggests that when hydrogen diffuses to an entire alloy, not concentrated just close to the surface, it can most likely obstruct the stress-induced martensite phase (B19 phase). Ogawa et al. [14] showed by Thermal Desorption Analysis (TDA); i.e., in case the diffused hydrogen exceeded 50–200 mass ppm, that the mechanical SMA characteristics could change, hence the appearance of premature failure. The superelastic NiTi archwire embrittlement, after hydrogen-absorption, was greatly linked to the transformation of stress-induced austenite-martensite [15]. This behaviour was attributed to the accumulation of hydrogen atoms, preferentially within regions that contain continual deficiencies, because of being trapped by retained martensite plates and dislocations, forming an obstacle which would cause fracture when the martensite domains were forced to pass across these boundaries [11]. In several studies [16,17], permanent defects could be obtained by cyclic deformation. In [18], the authors indicated that the pseudo-elastic hysteresis loops would decrease as the number of cycles increased and the strain hardening slope grew. B. Kockar et al. [19] mentioned that the main reason for the cyclic instability of the NiTi SMAs' pseudoelastic response was the local plastic accommodation of incompatibility between the B2 parent phase and the B19 phase during transformation. The authors in [20] indicated that during thermomechanical cycling, the variant interfaces would move backward and forward, while transformation-induced defects such as dislocations would build up and obstruct any interface mobility in the subsequent cycles. As a result, accumulations of retained martensite domains and local residual stresses would eventually affect the macroscopic superelastic behaviour of the NiTi alloy. In addition, we note that the effect of the strain rate on the cyclic deformations of superelastic NiTi alloys was studied by some experimental connotations [21–23]. It was found that rate dependence mainly resulted from the interaction between internal heat production (due to the mechanic dissipation and transformation latent heat) as well as the heat exchange with the external environment. Moreover, during cycling, the higher the strain rate the bigger the number of martensite domains in the austenite-martensite transformation.

Therefore, as the NiTi archwires are subjected to varied mechanical cycling before insertion in the oral cavity, this study aims to determine the hydrogen effect on tensile tests of the superelastic NiTi alloy under different imposed strain rates. The effect of the residual deformation after cycling with $10^{-5}s^{-1}$, $10^{-4}s^{-1}$ and $10^{-3}s^{-1}$ before immersion and tensile testing will be discussed.

2. Experimental Procedure

A commercial superelastic Ni-Ti (50.9 (at.%) Ni–49.1 (at.%) Ti, from Forestdent Co.), is used in this study. We obtained the heat treatment material's transformation temperature from the differential scanning calorimetry test at a scan rate of 10 °C/min. The start and finish temperatures of the martensite transformation (M_s and M_f) are equal to −8 and−26 °C, respectively; and the start and finish temperatures of the austenite transformation (A_s and A_f) are equal to −11 °C and −4 °C, respectively. The initial Young's modulus of the austenite and martensite phases were, respectively, 24 GPa and 12 GPa. During the mechanical tensile tests, we used rectangular specimens that had a gage length of 20 mm and across-sectional dimension of 0.64×0.43 mm^2.

Cyclic tensile tests, from 1 to 50, and a uniaxial tensile test at room temperature were carried out on a standard Instron screw-driven mechanical testing 5566-type machine (Figure 1a). With the aim of studying the loading rate influence up on the NiTi archwire's superelastic behaviour, we utilized the$10^{-5}s^{-1}$, $10^{-4}s^{-1}$ and $10^{-3}s^{-1}$ strain rates. The imposed strain of 7.1% was performed for the applied cyclic tensile test. In order to run the test, 30 mm was cut from the commercial superelastic NiTi archwire (Figure 1b) and clamped between the bottom and top flat grips of the Instron machine (5 mm distance) (Figure 1c). The strain was measured as across-head overall displacement in addition to a 20 mm initial gauge length. A brand-new specimen was used after each monotonic or cyclic test.

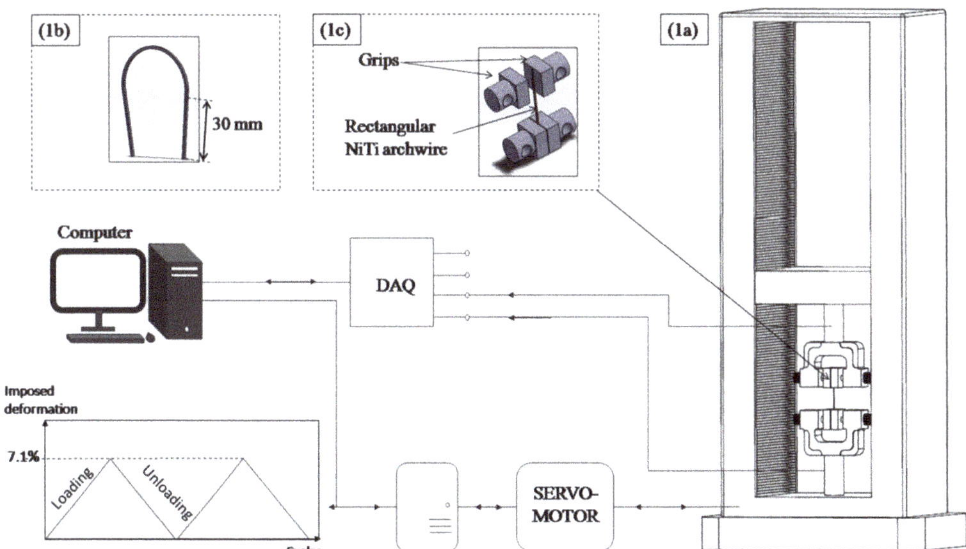

Figure 1. Loading protocol of the monotonic and cyclic loading: (**a**) Schematic representation of the Instron tensile machine, (**b**) used part of the as-received alloy, and (**c**) clamping method of the NiTi archwire.

Specimens' hydrogen charging was done through the immersion in a 0.9% NaCl solution. This was also performed for 3 h at room temperature at a 10 A/m^2 current density. A Direct-Current (DC) power supply was used for immersion, as shown in Figure 2. Circular platinum and the NiTi archwire were based on the 0.9% NaCl solution as the anode and the cathode, respectively. The current density controlled the generated gaseous hydrogen on the surface of the sample. Before immersion, samples were polished with a No. 600-grit SiC paper. After that, it was ultrasonically washed for 5 min in acetone. To homogeneously distribute the diffused hydrogen after immersion, we would age the specimens for one day in air at room temperature [24,25].

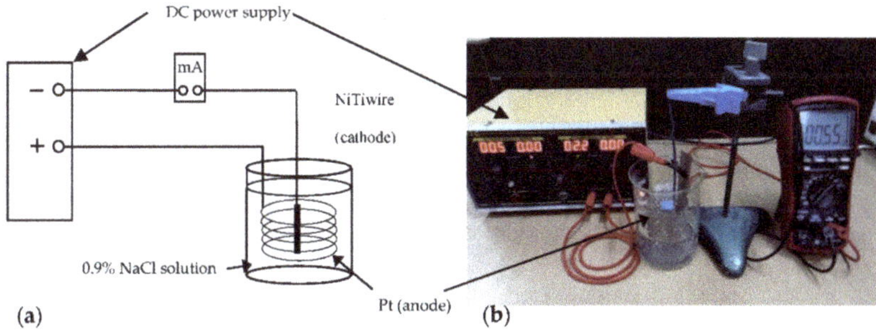

Figure 2. (**a**) Scheme of hydrogen charging (**b**) setup by electrolysis.

After immersion for 3 h and ultrasonic cleaning with acetone for 5 min, the amount of desorbed hydrogen was measured by TDA through the use of a gas chromatograph. This latter was actually calibrated under 10^{-6} Pa of vacuum with a standard mixture of argon gases and hydrogen. Moreover, we employ the high-purity argon as a carrier gas. TDA was performed on a 20 mm NiTi alloy total length. The sampling time of the carrier gas to the gas chromatograph was 5-min intervals at a heating rate of 100 °C/h. All TDA examinations were carried out after 24 h hydrogen charging. The total amount of hydrogen within the specimen was determined to be the sum of desorbed hydrogen within peaks of a hydrogen desorption rate [26,27]. At least two analyses have been done for each cyclic condition, and the experimental errors of the absorbed-hydrogen amount have been estimated between 5% and 10%.

3. Results and Discussion

The non-charged and charged specimens' tensile stress–stain curves for 3 h with the current density of 10 A/m^2 besides annealing for 24 hat room temperature are shown in Figure 3. For strain rates of a non-charged hydrogen specimen between 10^{-5} and 10^{-3}s^{-1}, the elastic loading of an austenite parent phase is observed, which is followed by "a plateau" during the austenite-martensite transformation. After that, there follows the elastic deformation of a martensite phase until fracture. It can be seen that the martensite start stress goes up from about 430 MPa to 460 as well as 490 MPa for a strain rate of 10^{-5}s^{-1} to 10^{-4}s^{-1} and 10^{-3}s^{-1}, respectively. On the other hand, for the 10^{-3}s^{-1} strain rate, the strain hardening effect of the austenite-martensite transformation goes up.

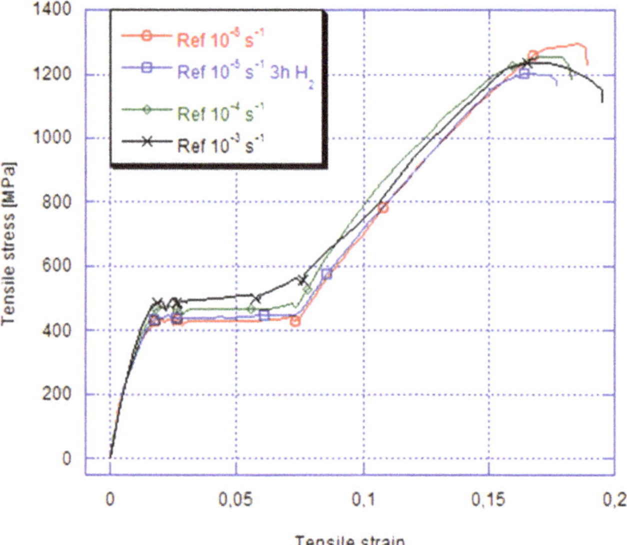

Figure 3. Typical engineering stress–strain curves of superelastic Ni-Ti alloy immersed for 3 h and aged for 24 h at different strain rates of 10^{-5}s^{-1}, 10^{-4}s^{-1} and 10^{-5}s^{-1} and as-received at 10^{-5}s^{-1}.

The strain rate dependency appears in a clear way during the austenite martensite transformation with the growth of the "plateau" slope. After that, it increases in the martensite start stress. Such a rise could be interpreted by the material's viscoelastic behaviour, and this is because of the friction between the martensite interface variants as well as the strain rate dependency that is actually caused by a thermomechanical NiTibehaviour due to the fact that the austenite-martensite transformation was exothermic [28,29]. Indeed, when the strain rate goes up, an internal heat production is as fast as the strain rate and the latent heat of the specimen is confined in the alloy. Consequently, a quantity of heat cannot be transferred outside of the NiTi alloy since the conduct time is shorter. This enhancement in temperature leads to a rise in the necessary applied load for inducing the martensite phase (Figure 4). Thus, both the amount of the required stress inducing the martensite and the slope of the austenite-martensite transformation will increase [17,30,31].

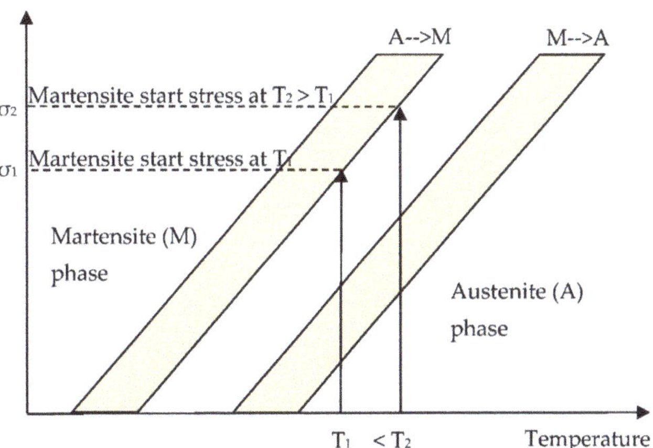

Figure 4. Simplified NiTi phase diagram showing the increase in the martensite starting stress from σ_1 to σ_2 when the temperature goes up from T_1 to T_2, respectively.

After 4 h of immersion with a 10 A/m^2 current density, Figure 3 illustrates a typical tensile curve for the applied 10^{-5}s^{-1} strain rate. This result shows that the material has conserved its superelastic behaviour and that there has not been any fracture detected. However, compared to the non-immersed alloy in the 0.9% NaCl solution, the NiTi archwire indicates arise in the martensite start stress of about 10 MPa. This phenomenon is attributed to the solid-solution hardening during hydrogen charging [18–32].

Figure 5a depicts the tested specimen's stress–strain curves for 50 cycles at the lowest 10^{-5}s^{-1} strain rate under an imposed strain of 7.1%. This latter represents the end of the plateau of the austenite-martensite transformation of the as-received NiTi archwire (Figure 3). The result highlights that the slope of the austenite-martensite transformation increases in a progressive manner during cycling. On the other hand, the martensite start transformation decreases with the growth of the loading and unloading cycles. This martensite start stress decreases from approximately 430 MPa at cycle 1 to about 310 MPa at cycle20. We can attribute this reduction to the introduction of dislocations and to the residual martensite phase during cycling, which is the main cause for establishing the residual stress [20,33]. Therefore, this residual stress assists the stress-induced transformation during the next loading by enabling martensite bands to activate at a lower macroscopic stress. Cyclic loading and unloading with imposed 10^{-4}s^{-1} and 10^{-3}s^{-1} strain rates are represented by Figure 5b,c, respectively. It is clear that the slope of the austenite martensite transformation plateau increases steadily with the number of cycles, besides raising the imposed strain rate. In addition, after 10^{-4}s^{-1} and 10^{-3}s^{-1} of imposed strain rates, the aforementioned yield stress falls approximately by 80 MPa and 110 MPa, respectively, after 20 cycles of loading and unloading. These results show that the internal residual stress becomes more significant with the increase in the imposed strain rate.

Figure 6 depicts the residual strain evolution as a function of the number of cycles for the three various strain rates. This residual strain is significant in the first cycles and goes up progressively according to the number of cycles. Thus, after 20 cycles, for the imposed strain rate of 10^{-3}s^{-1}, the residual deformation at zero stress is almost double of the value obtained with 10^{-5}s^{-1}. According to these results, we can deduce that there is an interaction between the decrease in the martensite start stress and the increment or increase in the residual strain at zero stress, and that there is an imposed strain rate during cycling. The interaction between the critical stress for introducing the martensite and the increase in the residual strain at zero stress was studied previously [20]. In fact, during thermo-mechanical cycling, the phase boundaries propagate forward and backward [34]. Throughout this reversible transformation, defects such as dislocations will form and act as an obstacle for the interface movement for the next cycles. Nevertheless, Chu K et al. [35] showed that during the fatigue tests of NiTi micropillars, the nanosized residual martensite domains, which were in fact blocked by the dislocations, could decrease in a considerable way the overall transformation stress throughout a generated remanent stress and the direct variant increase, and that was without overcoming the martensite nucleation obstacle. Similarly, Kan Q. et al. [36] studied the effect of the strain rates upon loading or unloading the SMA NiTi alloy. They found that the residual strain was accumulated by raising the number of cycles and strain rates. Indeed, after 20 cycles, the residual strain was about 2.04% for a 3.3 $10^{-4}s^{-1}$ strain rate and it reached 4.2% with 3.3 $10^{-2}s^{-1}$. In addition, they found that the NiTi SMAs' super-elasticity degradation after cycling was caused by interacting between the dislocation slipping and the martensitic transformation. Moreover, they concluded that the super-elastic NiTi SMAs' dislocation slipping could be generated through a high local stress close to the austenite-martensite interfaces. This was in general during the repeated martensitic transformation as well as during its reverse [37–39]. Consequently, during the cyclic deformation, the dislocation density and then the internal stress of NiTi SMAs went up rapidly.

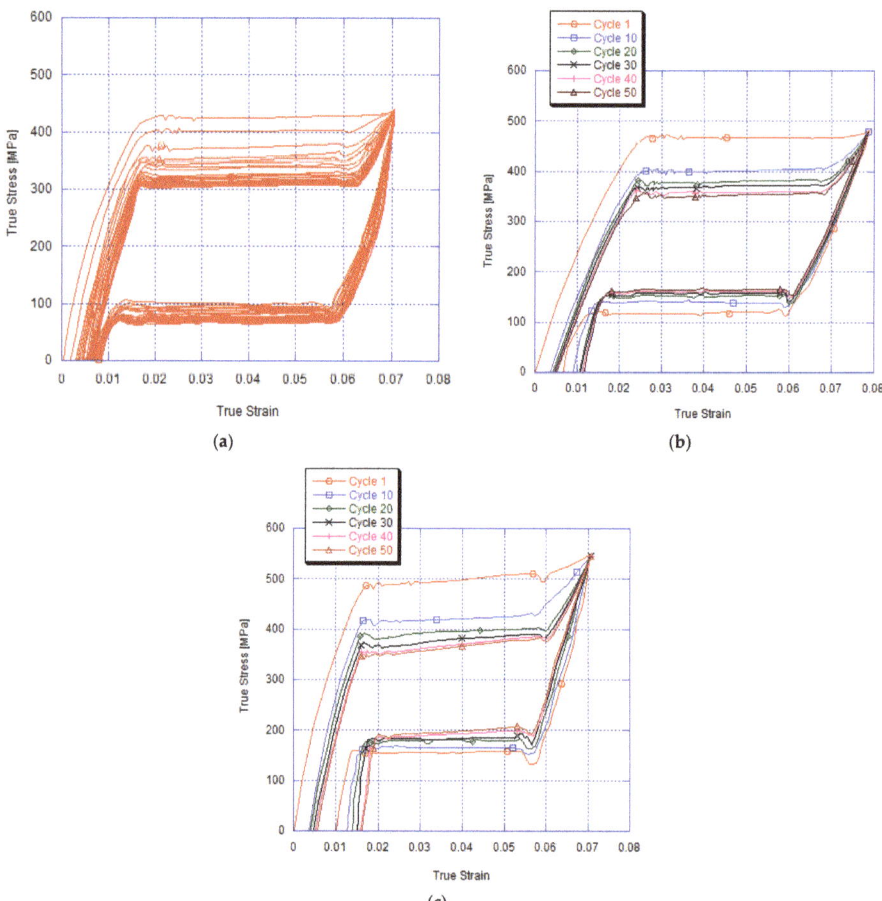

Figure 5. Typical strain cycling curve after 50 cycles for deformed specimen until 7.1%, at imposed strain rate of (**a**) $10^{-5}s^{-1}$, (**b**) $10^{-4}s^{-1}$ and (**c**) $10^{-3}s^{-1}$, showing reduction in phase-transformation yield stress and increase in residual strain.

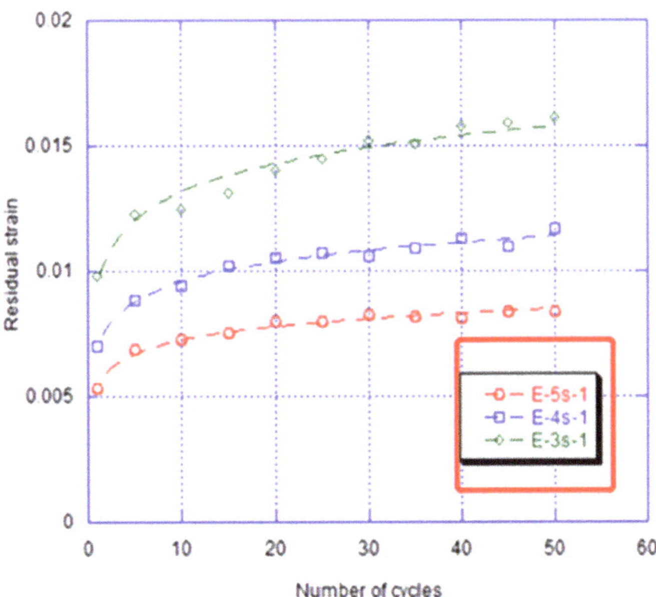

Figure 6. Evolution of residual strain at zero stress as a function of number of cycles for cyclic deformed specimen with 7.1% strain after different imposed strain rates of $10^{-5}s^{-1}$, $10^{-4}s^{-1}$ and $10^{-3}s^{-1}$ (represents the fitting curve).

In our case, a decrease in the martensite start stress and an increase in the residual strain at zero stress can be explained by the rise in the dislocation density and then the internal stress, which acts as an obstacle for the growth of the martensite bands and the nucleation of new variants. In addition, during the reverse transformation, some of the martensite bands can be pinned by dislocations and remain untransformed after unloading [36,40]. During the NiTi specimen's cyclic deformation, the temperature variation is based upon interacting between the heat conduct and the internal heat production. This can greatly depend on the strain rate. In the condition that the strain rate is low, the convected and conducted heat transfer is quicker than the heat production, and this is because of the latent heat and inelastic dissipation. When a strain rate is high, the inelastic-dissipation and latentheat production is quicker than the heat transfer. As the critical stress of the introducing martensite increases linearly with the rising test temperature [24], the required driving force of dislocation slipping will become higher, and then the density of dislocation will build up [34]. As a result, the increase in the residual stress with the strain rates is due to the higher density of dislocation. Additionally, the bigger the number of cycles, the more accentuated the internal stress and the retained martensite by dislocation will be. This is done by raising the dislocation density after the reverse transformation.

The amounts of absorbed hydrogen by the NiTi superelastic alloy obtained by TDA after 3 h of immersion of hydrogen charging within the 0.9% NaCl solution under an imposed current density of 10 A/m^2 at room temperature are illustrated in Figure 7. This quantity of diffused hydrogen is calculated by subtracting the amount of pre-dissolved hydrogen in the as-received NiTi archwire (8 mass ppm). Before mechanical testing and after 3 h of immersion, the as-received specimen shows approximately 96 mass ppm. The maximum peak of desorption is obtained at about 244 °C. Moreover, a shoulder of the desorption peak appears at about 142 °C. The TDA after loading and unloading from 1 to 20 cycles with $10^{-5}s^{-1}$ and $10^{-4}s^{-1}$ of the imposed strain rate and hydrogen charging is represented in Figure 7a,b, respectively. We can see that after one cycle and immersion, the amount of absorbed hydrogen is about 187 mass ppm with the lower strain rate, and it grows up to 301 mass ppm after cycling with an imposed deformation of $10^{-4}s^{-1}$. Besides, the diffused hydrogen into the NiTi archwire increases progressively with the number of cycles of loading and unloading at the imposed strain of 7.1%. After 20 cycles and 3 h of electrolytic charging in 0.9% NaCl solution for the lower strain rate, this value is about 319 mass ppm, and it becomes 456 mass ppm at $10^{-4}s^{-1}$. We can notice that for both strain rates, the primary peak temperature of hydrogen is in the rage of 270–300 °C, which is quite higher than the as-received and charged by hydrogen specimen. According to Yokoyama et al. [41], this behaviour is due to the fact that the hydrogen states can vary from a parent phase to a martensite one. Additionally, it is remarkable that the shoulder of the desorption peaks, between 145 and 175 °C, is higher than the as-received and charged specimen. Concerning the imposed $10^{-3}s^{-1}$ strain rate, the quantity of absorbed hydrogen after one cycle and immersion can be about 411 mass ppm (Figure 7c). This value is almost double of the one obtained with an imposed strain rate of $10^{-5}s^{-1}$ and four times higher than the as-received and hydrogen-charged alloy. In addition, after 20 cycles, we notice that the quantity of hydrogen is about 587 mass ppm, which is almost double of the measured value with the lower strain rate obtained after the same number of cycles. Similar to the results obtained with $10^{-5}s^{-1}$ and $10^{-4}s^{-1}$, the maximum absorbed peak remains in the same range of temperature, from 270 to 300 °C, hence the similarity between hydrogen states and/or trapping. However, a shoulder of the desorption peak is affected by the imposed strain rate. Indeed, this peak is increased and appears in the range 200–230 °C. T. Shimada et al. [42] indicated that the shoulder of the hydrogen desorption peak of a superelastic NiTi alloy was lower than that for the specimen charged with an applied stress of 600 MPa. The authors concluded that this phenomenon was due to the formation of the martensite phase under the applied stress where the diffusion distance and hydrogen trapping sites were different from the parent phase. On the other hand, an increase in the shoulder would be in fact detected for the NiTi alloy, which was charged by hydrogen in the martensite phase presence and under a static applied strain [43]. The authors mentioned that the diffused hydrogen tended to form unstable states. After a dynamic cyclic tensile test, carried out under the stress plateau region, the unstable hydrogen states would accumulate, leading to a hydrogen desorption peak shoulder [8]. The latter would increase by raising the strain rate as well as the number of deformation cycles.

The evolution of the mean value of the amount of absorbed hydrogen as a function of cycles number at different imposed strain rates, i.e., $10^{-3}s^{-1}$, $10^{-4}s^{-1}$, and $10^{-5}s^{-1}$, is depicted in Figure 8, respectively. It is seen that there is a direct relation between the amount of absorbed hydrogen, the imposed strain rate, and the number of cycles. Moreover, this behaviour is in coherence with the residual strain evolution at zero stress as a function of the number of cycles observed in Figure 6. In fact, as explained previously, the increase in the residual deformation with the strain rate and the number of cycles is caused by the rise in the internal stress due to the density of dislocations and to the pinned martensite by dislocations after unloading. The dislocations and the retained martensite represent a preferential site for hydrogen diffusion rather than the parent phase. In fact, the diffusion of hydrogen contributes to the stress relaxation of the infinite dislocation strain at the core. Additionally, the amount and diffusion rate of hydrogen are higher in the martensite phase in comparison with those in the austenite phase [41]. Consequently, we can deduce that the amount of absorbed hydrogen goes up with the increase in the residual strain at zero stress, since the density of the dislocation and the quantity of the retained martensite grow after each cycle.

Figure 9a shows the NiTi archwire's stress strain behaviour for the three quasi-static strain rates after 13 cycles with a strain rate of $10^{-5}s^{-1}$ and immersion for 3 h at room temperature. It can be seen that for a lower strain rate of $10^{-5}s^{-1}$, the NiTi alloy is embrittled by hydrogen charging. Indeed, the failure stress is detected in the plateau of the martensite start stress; whereas, the tested $10^{-4}s^{-1}$ and $10^{-3}s^{-1}$ NiTi archwires, for the same charging and ageing conditions, present a superelastic behaviour. We notice only an increase in the critical stress for martensite start stress by 50 MPa and 70 MPa, respectively, compared to the cycled specimen for 13 cycles without immersion. It is important to mention that the tested specimen with different strain rates after 13 cycles with $10^{-4}s^{-1}$ and $10^{-3}s^{-1}$ and immersion does not show any embrittlement during the tensile tests with different strain rates. The TDA after 13 cycles and immersion with the current density of 10 A/m^2 at different imposed strain rates is represented in Figure 9b. Despite the important amount of absorbed hydrogen of 411 and 543 mass ppm, it is clear that the mechanical behaviour of the NiTi alloy is not remarkably affected during the tensile tests with$10^{-4}s^{-1}$ and $10^{-3}s^{-1}$. After 18 cycles of loading and unloading with $10^{-3}s^{-1}$ of the imposed strain rate and hydrogen charging and aging for one day, the tensile test with $10^{-5}s^{-1}$ and $10^{-4}s^{-1}$ shows an embrittlement in the Luder-like plateau of the austenite martensite transformation, as depicted in Figure 10a. Nevertheless, the tensile tested NiTi archwire with a higher imposed strain rate of $10^{-3}s^{-1}$ (corresponding to the shorter time) conserves its superelasticity. This specimen presents a higher martensite start stress and in addition to the lower total strain compared to the non-charged ones. Figure 10a illustrates the TDA after 18 cycles at $10^{-3}s^{-1}$ and hydrogen charging in comparison with loaded specimens at $10^{-5}s^{-1}$ after 13 cycles, where the first embrittlement has happened. It can be seen that the amount of absorbed hydrogen after 18 cycles at $10^{-3}s^{-1}$ (584 mass ppm) is almost twice the one cycled with $10^{-5}s^{-1}$ (291 mass ppm), and no fracture has been detected yet (Figure 10b). Table 1 summarizes the results obtained forall performed experimental conditions. It is important to mention that the experimental results obtained after 20 cycles are similar to the ones given with 18 cycles. Accordingly, the hydrogen quantity may not be the only principal factor while considering in particular the studied superelastic NiTi alloy's hydrogen embrittlement mechanism.

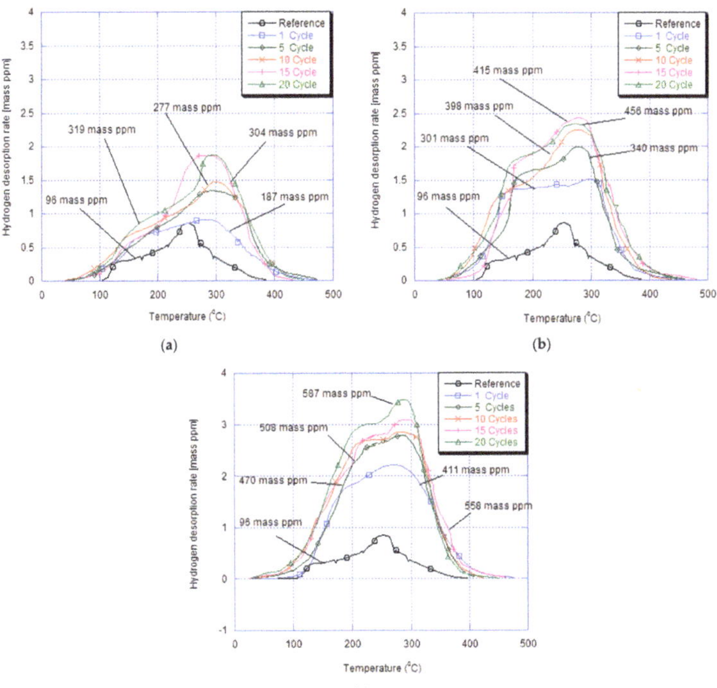

Figure 7. Hydrogen thermal desorption curves for specimens immersed for 3 h and aged for 24 h at room temperature after 13 cycles with imposed deformation of (**a**) $10^{-5}s^{-1}$, (**b**) $10^{-4}s^{-1}$ and (**c**) $10^{-3}s^{-1}$, showing an increase in the amount of absorbed hydrogen with stain rates.

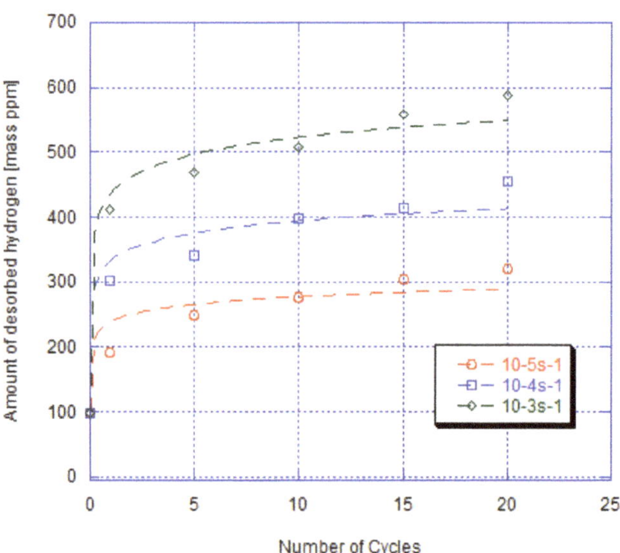

Figure 8. Amount of absorbed hydrogen vs. number of cycles of loaded and unloaded specimens at imposed $10^{-5}s^{-1}$, $10^{-4}s^{-1}$, and $10^{-3}s^{-1}$ strain rates and hydrogen charging for 3 h (represents the fitting curve).

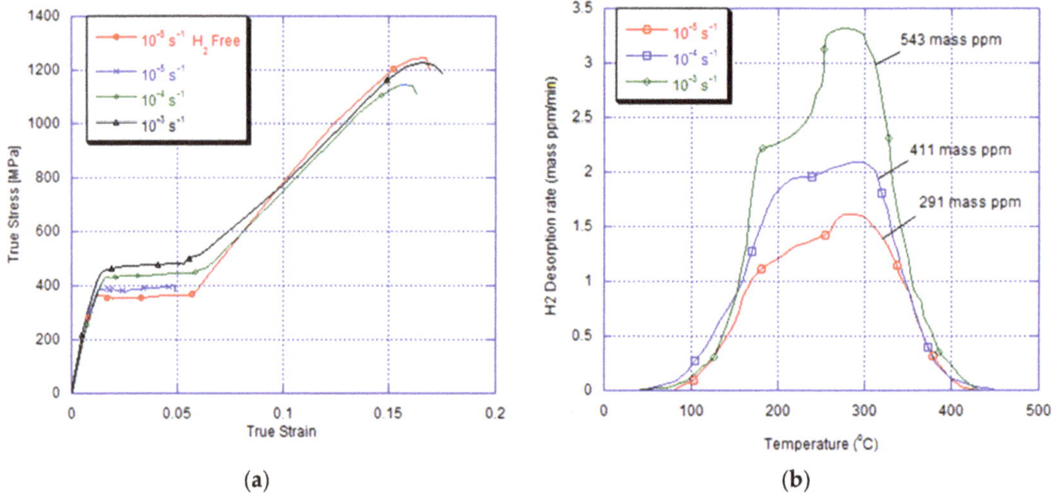

Figure 9. (**a**) Tensile curve at $10^{-5}s^{-1}$ showing fracture after 13 cycles with imposed strain rate of $10^{-5}s^{-1}$, and (**b**) amount of absorbed hydrogen after the same number of cycles at $10^{-5}s^{-1}$, $10^{-4}s^{-1}$, and $10^{-3}s^{-1}$.

The strain rate response of the superelastic NiTi alloy can be explained based on a thermo-mechanical phenomenon [44,45]. For a low applied strain rate ($10^{-5}s^{-1}$ and $10^{-4}s^{-1}$), once the first variants are generated, the austenite-martensite interfaces move slowly enough in order to enable the latent transformation to be transferred quickly outside. Temperature change in the specimen is eliminated and the transformation is considered under an isothermal process. Consequently, the flat Luder-like plateau of the austenite martensite transformation results from the growth of a few number of martensite bands, not the nucleation of new martensite variants. In a high strain rate of $10^{-3}s^{-1}$, the latent heat cannot be timely convected to the environment since the martensite variants are forced to propagate at a somewhat higher velocity. For the domain-front growth, the domain-fronts' localized self-heating raises the stress [46], which becomes more important than the untransformed parent zone's domain-nucleation stress. Therefore, during the phase transformation, the variation in the average sample temperature is significant. The slope of the stress–strain curve goes up and new domains nucleate to fulfill the applied strain, with the continuously changing temperature.

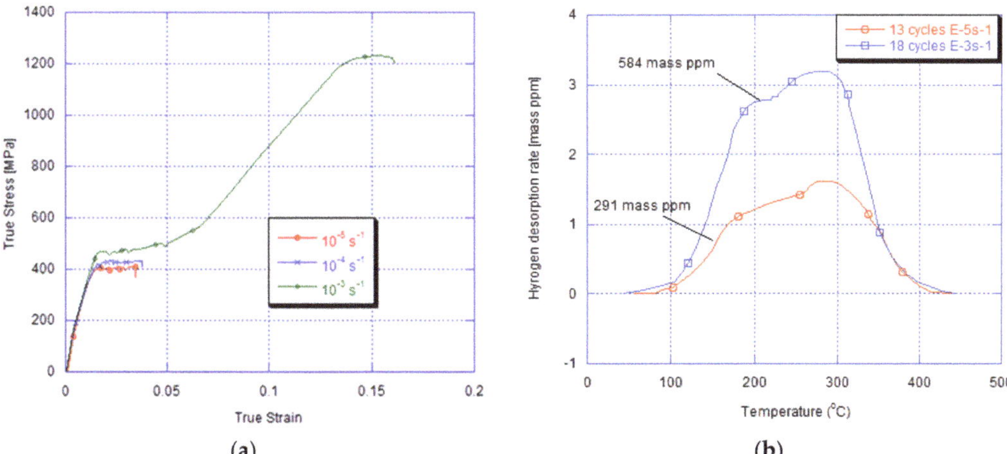

Figure 10. (**a**) Tensile curves obtained after 18 cycles and immersion for 3 h showing the embrittlement of loaded specimen with imposed strain rates of $10^{-5}s^{-1}$ and $10^{-4}s^{-1}$ and superelastic behaviour of loaded specimens with $10^{-3}s^{-1}$, and (**b**) comparison between critical amount of absorbed hydrogen causing fracture at $10^{-5}s^{-1}$ and quantity of absorbed hydrogen after 18 cycles with $10^{-3}s^{-1}$ and immersion.

Table 1. Summary of conducted tensile tests at different strain rates after immersion and cycling with $10^{-5}s^{-1}$, $10^{-4}s^{-1}$ and $10^{-3}s^{-1}$.

Nbre of Cycles / Cyclic Strain Rate	13 Cycles			18 Cycles		
	Absorbed Hydrogen (Mass ppm)	**Tensile Strain Rate**	**Fracture**	**Absorbed Hydrogen (Mass ppm)**	**Tensile Strain Rate**	**Fracture**
$10^{-5}s^{-1}$	291	$10^{-5}s^{-1}$	F	313	$10^{-5}s^{-1}$	F
		$10^{-4}s^{-1}$			$10^{-4}s^{-1}$	
		$10^{-3}s^{-1}$			$10^{-3}s^{-1}$	
$10^{-4}s^{-1}$	411	$10^{-5}s^{-1}$	F	436	$10^{-5}s^{-1}$	F
		$10^{-4}s^{-1}$			$10^{-4}s^{-1}$	
		$10^{-3}s^{-1}$			$10^{-3}s^{-1}$	
$10^{-3}s^{-1}$	543	$10^{-5}s^{-1}$	F	584	$10^{-5}s^{-1}$	F
		$10^{-4}s^{-1}$			$10^{-4}s^{-1}$	F
		$10^{-3}s^{-1}$			$10^{-3}s^{-1}$	

F: Fracture.

The maximum number of martensite variants n_{max} (or the minimum domain spacing) increases as a function of the strain rate. In fact, the maximal number of martensite domains n_{max} and as well as the strain rate $\dot{\varepsilon}$ canbe fitted into a power-law relationship [47,48]:

$$n_{max} = c\,(\dot{\varepsilon})^{x}, \tag{1}$$

where the fitting coefficient x is an exponent factor that is around 0.5, and c is a fitting coefficient.

Hao Yu et al. [49] showed that during a high strain rate, the precipitation of Ni_4Ti_3 is found to increase the NiTi yield strength, which is caused by the impediment of Ni_4Ti_3 precipitation to dislocation motion and leading to an increase in the martensite phase nucleation at the interface between the precipitation and the B2 matrix. This indicates that during the tensile tests with the same applied strain rate, the number of martensite variants, obtained in the plateau of austenite martensite transformation, should increase with the imposed strain rate during cycling.

The sensitivity to the strain rate after cycling at different strain rates and after hydrogen charging for 3 h and ageing at room temperature appears to be as follows. After 13 cycles with a $10^{-5}s^{-1}$ strain rate of, the 291 mass ppm of absorbed hydrogen seems to represent the critical amount to cause fracture during the austenite martensite plateau. Indeed, during tensile loading with $10^{-5}s^{-1}$, a few martensite variants tend to grow during the austenite-martensite transformation plateau, since the latent transformation heat is transferred rapidly to ambient media. However, this growth is hindered by the trapped hydrogen into the retained martensite bands and the local dislocation stress field near the austenite-martensite interfaces and this actually happens during the repeated martensitic transformation as well as during its reverse. Thus, the growth of martensite bands cannot overcome the obstacle induced by hydrogen. It is important to point out that during tensile loading with $10^{-5}s^{-1}$, the number of martensite variants growing in the plateau for austenite martensite transformation obtained after 13 cycles with an imposed strain rate of $10^{-4}s^{-1}$ and $10^{-3}s^{-1}$ should be more important than after cycling with $10^{-5}s^{-1}$, since we have more local stress near the austenite-martensite interfaces [50]. However, an embrittlement occurs. Thus, it is assumed that the amount of absorbed hydrogen of 543 mass ppm is sufficient to act as an obstacle for the short travelling distance of martensite bands to cause fracture. This points out that there is a competition between the number of martensite variants and the amount of absorbed hydrogen to cause fracture.

With an imposed $10^{-3}s^{-1}$ strain rate, after 18 cycles, a premature fracture is detected for a loaded tensile NiTi alloy with both $10^{-5}s^{-1}$ and $10^{-4}s^{-1}$ strain rates. The corresponding amount of absorbed hydrogen, after 3 h of immersion, is about 584 mass ppm, which is considered as the critical value causing earlier fracture during the monotonic tensile test with a $10^{-4}s^{-1}$ strain rate. The delay of embrittlement with $10^{-4}s^{-1}$ can be related to a competition between the number of martensite bands in the austenite martensite plateau and the quantity of absorbed hydrogen. In fact, after13 cycles, the number of nucleated variants is higher than $10^{-5}s^{-1}$. Consequently, the amount of absorbed hydrogen is not enough to act as an obstacle for the variant growth caused by $10^{-4}s^{-1}$, since the travelled distance is less important than the one obtained with $10^{-5}s^{-1}$. However, after 18 cycles, despite the increase in nucleated bands, and due to (i) the multiplication of local stress near the austenite-martensite interfaces after cycling with $10^{-3}s^{-1}$ and (ii) the higher applied strain rate during tensile test, the quantity of absorbed hydrogen seems to be sufficient to act as an obstacle for varied growth in the quasi-flat plateau. This explanation seems to be in coherence with the obtained superelastic behaviour after 18 cycles and the tensile test with a higher imposed strain rate. In this case, the time for transferring the latent heat is highly restricted, causing the nucleation of a large number of martensite bands with small domain spacing between them. As a result, the travelled distance of these martensite bands will be very limited and the amount of absorbed hydrogen will have no effect.

4. Conclusions

We have investigated in this study the effects of the strain rate response after hydrogen charging for 3 h and ageing for one day on pre-cycled superelastic NiTi archwires. Cycling from 1 to 50 loading and unloading at $10^{-5}s^{-1}$, $10^{-4}s^{-1}$, and $10^{-3}s^{-1}$ has been carried out to generate a gradual rise in the residual strain at zero stress. Moreover, cycling, which is characterized by an increase in the density of dislocation and retained martensite after unloading, represents a preferential site for hydrogen diffusion during immersion. After hydrogen charging with 191 mass ppm of absorbed hydrogen, an embrittlement occurs for the low applied tensile strain rate. In addition, this fracture happens only with a higher quantity of absorbed hydrogen (584 mass ppm) with a $10^{-4}s^{-1}$ imposed strain rate, and no embrittlement will be detected for the higher strain rate of $10^{-3}s^{-1}$. We attribute this behaviour to the interaction between the travelled distance during the increase in martensite bands and the amount of absorbed hydrogen during the step of austenite-martensite transformation.

Funding: The author would like to extend his appreciation to the Deanship of Scientific Research at King Khalid University, Saudi Arabia for finding this work through the Reseach Group Program under grant no. RGP.1/14/42.

Institutional Review Board Statement: Not applicable.

Informed Consent Statement: Not applicable.

Data Availability Statement: The data presented in this study are available on request from the corresponding author.

Conflicts of Interest: The author declare no conflict of interest.

References

1. Kang, G.; Kan, Q.; Yu, C.; Song, D.; Liu, Y. Whole-life transformation ratchetting and fatigue of super-elastic NiTi Alloy under uniaxial stress-controlled cyclic loading. *Mater. Sci. Eng.* **2012**, *2*, 28–234. [CrossRef]
2. Hartl, D.J.; Lagoudas, D.C. Aerospace applications of shape memory alloys, Proceedings of the Institution of Mechanical Engineers. *Part G J. Aerosp. Eng.* **2007**, *221*, 535–552.
3. AFathallah, A.; THassine, T.; FGamaoun, F.; Wali, M. Three-Dimensional Coupling Between Orthodontic Bone Remodeling and Superelastic 4 Behavior of a NiTi Wire applied for initial alignment. *J. Orofac. Orthop.* **2021**, *82*, 2.
4. Elkhal Letaief, W.; Fathallah, A.; Hassine, T.; Gamaoun, T. Finite-Element Analysis of Hydrogen Effects on Superelastic NiTi Shape Memory Alloys: Orthodontic Application. *J. Intell. Mater. Syst. Struct.* **2018**, *29*, 3188–3198. [CrossRef]
5. Sarraj, R.; Elkhal Letaief, W.; Hassine, T.; Gamaoun, F. Modeling of rate dependency of mechanical behavior of superelastic NiTi alloy under cyclic loading. *Int. J. Adv. Manuf. Technol.* **2019**, *9–12*, 2715–2724. [CrossRef]
6. Sarraj, R.; Elkhal Letaief, W.; Hassine, T.; Gamaoun, F.; El Ouni, M.H. Modeling of Hydrogen Difusion Towards a NiTi Arch Wire Under Cyclic Loading. *Met. Mater. Int.* **2021**, *27*, 413–424. [CrossRef]
7. Elkhal Letaief, W.; Hassine, T.; Gamaoun, F.; Sarraj, R.; Kahla, N.B. Coupled Diffusion-Mechanical Model of NiTi Alloys Accounting for Hydrogen Diffusion and Ageing. *Int. J. Appl. Mech.* **2020**, *12*, 4. [CrossRef]
8. Yokoyama, K.; Tomita, M.; Sakai, J. Hydrogen embrittlement behavior induced by dynamic martensite transformation of Ni-Tisuperelastic alloy. *Acta Mater.* **2009**, *57*, 1875–1885. [CrossRef]
9. Sarraj, R.; Hassine, T.; Gamaoun, F. Mechanical behavior of NiTi arc wires under pseudoelastic cycling and cathodically hydrogen charging. *Mater. Res. Express* **2018**, *1*, 015704. [CrossRef]
10. Elkhal Letaief, W.; Hassine, T.; Gamaoun, T. Rate Dependency During Relaxation of Superelastic Orthodontic NiTi Alloys After Hydrogen Charging. *Shape Mem. Superelasticity* **2016**, *2*, 121–127. [CrossRef]
11. Gamaoun, F.; Hassine, T.; Bouraoui, T. Strain rate response of a Ni-Ti shape memory alloy after hydrogen charging. *Philos. Mag. Lett.* **2014**, *94*, 30–36. [CrossRef]
12. Gamaoun, F.; Ltaief, M.; Bouraoui, T.; Ben Zineb, T. Effect of hydrogen on the tensile strength of aged Ni-Ti superelastic alloy. *J. Intell. Mater. Syst. Struct.* **2011**, *22*, 2053–2059. [CrossRef]
13. Elkhal Letaief, W.; Hassine, T.; Gamaoun, T. In situ stress relaxation mechanism of a superelastic NiTi shape memory alloy under hydrogen charging. *Philos. Mag. Lett.* **2017**, *97*, 50–57. [CrossRef]
14. Ogawa, T.; Yokoyama, K.; Asaoka, K.; Sakai, J. Effects of moisture and dissolved oxygen in methanol and ethanol solutions containing hydrochloric acid on hydrogen absorption and desorption behaviors of Ni-Ti superelastic alloy. *Mater. Sci. Eng. A* **2005**, *422*, 218–226. [CrossRef]

15. Yokoyama, K.; Eguchi, T.; Asaoka, K.; Nagumo, M. Effect of constituent phase of Ni-Ti shape memory alloy on susceptibility to hydrogen embrittlement. *Mater. Sci. Eng. A* **2004**, *374*, 177–183. [CrossRef]
16. Gamaoun, F.; Skhiri, I.; Bouraoui, T. Effect of the Residual Deformation on the Mechanical Behavior of the Ni-Ti Alloy Charged by Hydrogen. *Adv. Mater. Res.* **2011**, *324*, 181–184. [CrossRef]
17. Sarraj, R.; Kessentini, A.; Hassine, T.; Algahtani, A.; Gamaoun, F. Hydrogen Effect on the Cyclic Behavior of the SuperelasticNi-TiArchwire. *Metals* **2019**, *9*, 316. [CrossRef]
18. Gamaoun, F.; Skhiri, I.; Bouraoui, T.; Ben Zineb, T. Hydrogen effect on the austenite–martensite transformation of the cycled Ni-Ti alloy. *J. Intell. Mater. Syst. Struct.* **2014**, *25*, 980–988. [CrossRef]
19. Kockar, B.; Karaman, I.; Kim, J.I.; Chumlyakov, Y.I.; Sharp, J.; Yu, C.-J.M. Thermomechanical cyclic response of an ultrafine-grained NiTi shape memory alloy. *Acta Mater.* **2008**, *56*, 3630–3646. [CrossRef]
20. Miyazaki, S.; Imai, T.; Igo, Y.; Otsuka, K. Effect of Cyclic Deformation on the Pseudoelasticity Characteristics of Ti-Ni Alloys. *Metall. Trans. A* **1986**, *17A*, 115–120.
21. Nemat-Nasser, S.; Guo, W.G. Superelastic and cyclic response of NiTi SMA at various strain rates and temperatures. *Mech. Mater.* **2006**, *38*, 463–474. [CrossRef]
22. Morin, C.; Moumni, Z.; Zaki, W. Thermomechanical coupling in shape memory alloys under cyclic loadings: Experimental analysis and constitutive modeling. *Int. J. Plast.* **2011**, *27*, 1959–1980. [CrossRef]
23. Morin, C.; Moumni, Z.; Zaki, W. A constitutive model for shape memory alloys accounting for thermomechanical coupling. *Int. J. Plast.* **2014**, *27*, 748–767. [CrossRef]
24. Gamaoun, F.; Hassine, T. Ageing Effect and Rate Dependency of a NiTi Shape Memory Alloy after Hydrogen Charging. *J. Alloy. Compd.* **2014**, *615*, 680–683. [CrossRef]
25. Elkhal Letaief, W.; Hassine, T.; Gamaoun, T. A coupled model between hydrogen diffusion and mechanical behavior of superelastic NiTi alloys. *Smart Mater. Struct.* **2017**, *26*, 075001. [CrossRef]
26. Takai, K.; Watanuki, R. Hydrogen in trapping states innocuous to environmental degradation of high-strength steels. *ISIJ Int.* **2003**, *43*, 520–526. [CrossRef]
27. Merzlikin, S.V.; Borodin, S.; Vogel, D.; Rohwerder, M. Ultra high vacuum high precision low background setup with temperature control for thermal desorption mass spectroscopy (TDA-MS) of hydrogen in metals. *Talanta* **2015**, *136*, 108–113. [CrossRef]
28. Grandi, D.; Maraldi, M.; Molari, L. A macroscale phase-field model for shape memory alloys with non-isothermal effects: Influence of strain rate and environmental conditions on the mechanical response. *Acta Mater.* **2012**, *60*, 179–191. [CrossRef]
29. Chao, Y.; Guozheng, K.; Qianhua, K.; Yilin, Z. Rate-dependent cyclic deformation of super-elastic NiTi shape memory alloy: Thermo-mechanical coupled and physical mechanism-based constitutive model. *Int. J. Plast.* **2015**, *72*, 60–90.
30. Shaw, J.A. Simulations of localized thermo-mechanical behavior in a NiTi shape memory alloy. *Int. J. Plast.* **2000**, *16*, 541–562. [CrossRef]
31. Schlosser, P.; Louche, H.; Favier, D.; Orgéas, L. Image processing to estimate the heat sources related to phase transformations during tensile tests of NiTi tubes. *Strain* **2007**, *43*, 260–271. [CrossRef]
32. Elkhal Letaief, W.; Hassine, T.; Gamaoun, T. Tensile behaviour of superelastic NiTi alloys charged with hydrogen under applied strain. *Mater. Sci. Technol.* **2017**, *33*, 1533–1538. [CrossRef]
33. Tang, W.; Sandstrom, R. Analysis of the influence of cycling on NiTi shape memory alloy properties. *Mater. Des.* **1993**, *14*, 103–113. [CrossRef]
34. Xie, X.; Kan, Q.; Kang, G.; Lu, F.; Chen, K. Observation on rate-dependent cyclic transformation domain of super-elastic NiTi shape memory alloy. *Mater. Sci. Eng. A* **2016**, *671*, 32–47. [CrossRef]
35. Chu, K.; Sun, Q. Reducing functional fatigue, transition stress and hysteresis of NiTi micropillars by one-step overstressed plastic deformation. *Scr. Mater.* **2021**, *201*, 113958. [CrossRef]
36. Kan, Q.; Yu, C.; Kang, G.; Li, J.; Yan, W. Experimental observations on rate-dependent cyclic deformation of super-elastic NiTi shapememory alloy. *Mech. Mater.* **2016**, *97*, 48–58. [CrossRef]
37. Hamilton, R.F.; Sehitoglu, H.; Chumlyakov, Y.; Maier, H.J. Stress dependence of the hysteresis in single crystal NiTi alloys. *Acta Mater.* **2004**, *52*, 3383–3402. [CrossRef]
38. Norfleet, D.M.; Sarosi, P.M.; Manchiraju, S.; Wagner, M.X.; Uchic, M.D.; Anderson, P.M.; Mills, M.J. Transformation-induced plasticity during pseudoelastic deformation in Ni-Ti microcrystals. *Acta Mater.* **2009**, *57*, 3549–3561. [CrossRef]
39. Simon, T.; Kröger, A.; Somsen, C.; Dlouhy, A.; Eggeler, G. On the multiplication of dislocations during martensitic transformations in NiTi. *Acta Mater.* **2010**, *58*, 1850–1860. [CrossRef]
40. Yawny, A.; Sade, M.; Eggeler, G. Pseudoelastic cycling of ultra-fine grained NiTi shape-memory wires. *Z. Met.* **2005**, *96*, 608–618. [CrossRef]
41. Yokoyama, K.; Ogawa, T.; Takashima, K.; Asaoka, K.; Sakai, J. Hydrogen embrittlement of Ni-Tisuperelastic alloy aged at room temperature after hydrogen charging. *Mater. Sci. Eng. A* **2007**, *466*, 106–113. [CrossRef]
42. Shimada, K.; Yokoyama, K.; Sakai, J. Improved fracture properties of Ni-Tisuperelastic alloy under sustained tensile load in physiologic al saline solution containing hydrogen peroxide by hydrogen charging. *J. Alloy. Compd.* **2018**, *752*, 1–7. [CrossRef]
43. Yokoyama, K.; Tomita, M.; Asaoka, K.; Sakai, J. Hydrogen absorption and thermal desorption in NiTi superelasticsubjected to tensile straining test with hydrogen charging. *Scr. Mater.* **2007**, *57*, 393–396. [CrossRef]

44. Zhang, X.; Feng, P.; He, Y.; Yu, T.; Sun, Q. Experimental study on rate dependence of macroscopic domain and stress hysteresis in NiTi shape memory alloys trips. *Int. J. Mech. Sci.* **2010**, *52*, 1660–1670. [CrossRef]
45. Elibol, C.; Wagner, M.F.-X. Strain rate effects on the localization of the stress-induced martensitic transformation in pseudoelastic NiTi under uniaxial tension, compression and compression–shear. *Mater. Sci. Eng. A* **2015**, *643*, 194–202. [CrossRef]
46. Leo, P.H.; Shield, T.W.; Bruno, O.P. Transient heat transfer effects on the pseudoelasticbehavior of shape-memory wires. *Acta Metall. Mater.* **1993**, *41*, 2477–2485. [CrossRef]
47. He, Y.J.; Sun, Q.P. Rate-dependent domain spacing in a stretched NiTi strip. *Int. J. Solids Struct.* **2010**, *47*, 2775–2783. [CrossRef]
48. Iadicola, M.A.; Shaw, J.A. Rate and thermal sensitivities of unstable transformation behavior in a shape memory alloy. *Int. J. Plast.* **2004**, *20*, 577–605. [CrossRef]
49. Yu, H.; Qiu, Y.; Marcus, L. Young, Influence of Ni_4Ti_3 precipitate on pseudoelasticity of austenitic NiTi shape memory alloys deformed at high strain rate. *Mater. Sci. Eng. A* **2021**, *804*, 140753. [CrossRef]
50. Fan, G.; Otsuka, K.; Ren, X.; Yin, F. Twofold role of dislocations in the relaxation behavior of Ti–Ni martensite. *Acta Mater.* **2008**, *56*, 632–641. [CrossRef]

 *materials* 

Communication

On the Decrease in Transformation Stress in a Bicrystal Cu-Al-Mn Shape-Memory Alloy during Cyclic Compressive Deformation

Tung-Huan Su [1], Nian-Hu Lu [2], Chih-Hsuan Chen [2,3,*] and Chuin-Shan Chen [1,3,*]

1 Department of Civil Engineering, Civil Engineering Department Building, National Taiwan University, Room 205, No.1, Sec. 4, Roosevelt Road, Taipei 10617, Taiwan; caedbwind95@caece.net
2 Department of Mechanical Engineering, National Taiwan University, No. 1, Sec. 4, Roosevelt Road, Taipei 10617, Taiwan; f06522712@ntu.edu.tw
3 Department of Materials Science and Engineering, National Taiwan University, No. 1, Sec. 4, Roosevelt Road, Taipei 10617, Taiwan
* Correspondence: chchen23@ntu.edu.tw (C.-H.C.); dchen@ntu.edu.tw (C.-S.C.)

Abstract: The evolution of the inhomogeneous distribution of the transformation stress (σ_s) and strain fields with an increasing number of cycles in two differently orientated grains is investigated for the first time using a combined technique of digital image correlation and data-driven identification. The theoretical transformation strains (ε_T) of these two grains with crystal orientations $[5\ 3\ 26]_\beta$ and $[6\ 5\ 11]_\beta$ along the loading direction are 10.1% and 7.1%, respectively. The grain with lower ε_T has a higher σ_s initially and a faster decrease in σ_s compared with the grain with higher ε_T. The results show that the grains with higher σ_s might trigger more dislocations during the martensite transformation, and thus result in greater residual strain and a larger decrease in σ_s during subsequent cycles. Grain boundary kinking in bicrystal induces an additional decrease in transformation stress. We conclude that a grain with crystal orientation that has high transformation strain and low transformation stress (with respect to loading direction) will exhibit stable transformation stress, and thus lead to higher functional performance in Cu-based shape memory alloys.

Keywords: full-field stress and strain measurements; shape memory alloys; digital image correlation; data-driven identification; superelasticity; functional fatigue

Citation: Su, T.-H.; Lu, N.-H.; Chen, C.-H.; Chen, C.-S. On the Decrease in Transformation Stress in a Bicrystal Cu-Al-Mn Shape-Memory Alloy during Cyclic Compressive Deformation. *Materials* **2021**, *14*, 4439. https://doi.org/10.3390/ma14164439

Academic Editor: Francesco Iacoviello

Received: 9 July 2021
Accepted: 5 August 2021
Published: 8 August 2021

1. Introduction

Superelastic shape-memory alloys (SMAs) are functional materials capable of sustaining a large recoverable deformation strain as a result of a stress-induced martensitic transformation (MT) between austenite and martensite. Among SMAs, Cu-Al-Mn SMAs possess superior features such as low cost, high cold workability, and large transformation strain compared with TiNi-based SMAs [1]. They are considered to be suitable candidate materials for a variety of applications ranging from civil engineering to the space industry, in which the SMAs are subjected to cyclic loading. However, the issue of SMA fatigue and fracture is challenging because fatigue problems in Cu-Al-based SMAs are mainly attributed to the constraints of grain boundaries during MT and its resulting plastic deformation [2]. Several studies have extensively investigated the prevention of intergranular fracture problems caused by high elastic anisotropy in the Cu-Al-Mn alloys with the aim of enhancing the superelasticity of polycrystalline Cu-based shape memory alloys [3–5]. Therefore, treatments of the microstructure designed to increase grain size, such as the introduction of texture and the reduction of triple junctions, significantly improve the functional performance of Cu-Al-Mn SMAs [6–8].

Recently, Cu-Al-Mn single crystals with excellent superelasticity have been fabricated using abnormal grain growth (AGG) induced by a cyclic heat treatment [9,10]. These AGG methods also enable the preparation of large bicrystal samples for mechanical tests, providing insight into the inhomogeneous MT phenomenon and the elastocaloric effect of the sample under compression [11]. The experimental results [11] demonstrate the generation of microcracks at the grain and twin boundaries of the bicrystal Cu-Al-Mn sample under cyclic compression. Although compressive deformation is preferred as a deformation mode to delay the fatigue fracture of metallic materials [12,13], much less is known about the compressive fatigue behavior of Cu-Al-Mn SMAs during cyclic phase transformation.

Based on the above-mentioned motivations and the knowledge of large differences in superelasticity properties of the bicrystal Cu-Al-Mn sample [11], it was expected that such differences in superelasticity properties would cause functional instability of the Cu-based bicrsytal sample during cyclic compressive deformation. As Cu-Al-Mn SMAs are regarded as potential candidates of functional materials, the correlations between strain field and transformation stress, and the evolutions of stress and strain distributions during cyclic superelastic deformation, are critical factors for the compressive fatigue behavior of Cu-Al-Mn SMAs. In this study, we investigated the cyclic compressive behavior of the superelasticity of macro-scale Cu-Al-Mn bicrystals using the digital image correlation (DIC) technique and the data-driven identification (DDI) method. Both methods are used to determine the distributions of transformation stress and strain in the bicrystal and near the grain boundary. Based on the full-field measurement results, the correlations between the decrease in transformation stress, accumulation of residual strain, and martensite transformation are determined.

2. Materials and Methods

Figure 1 illustrates the methods employed to characterize the cyclic behavior of the superelasticity of bicrystal Cu-Al-Mn SMAs. The strain and stress fields of a Cu-Al-Mn shape-memory bicrystal were measured using the DIC technique and DDI method, respectively. In this work, we used the same Cu-Al-Mn bicrystal sample prepared in our previous study [11], with dimensions of 8 mm × 4.2 mm × 4.2 mm. Please note that the specimen was subjected to five compression–unloading tests with a maximum global deformation strain from 1% to 5% in 1% increments for each test (see Figure 1 in the previous study [11]). It was found that some plastic deformation occurred when the global deformation strain was higher than 4%. In this work, we further conducted twenty compression–unloading tests using the same Cu-Al-Mn bicrystal sample, which has already undergone the five compression–unloading tests mentioned above. The grain boundary within the sample is indicated by a dashed line in Figure 1. The crystal orientations of both grains were determined via electron backscatter diffraction (EBSD, Oxford Instruments, Abingdon, UK), as shown in the inset of Figure 2a. The thermal analyses of the sample were conducted in a differential scanning calorimeter (DSC, DSC 25, TA Instrument, New Castle, DE, USA) with cooling and heating rates of 10 °C/min. The microstructures were observed by transmission electron microscope (TEM, FEI Tecnai™ G2 F30, Hillsboro, OR, USA) operated at 300 kV. The sample was mechanically ground to a thickness of about 70 μm and then electropolished at −40 °C using HNO_3 and CH_3OH (2:8 in volume). The preparation procedures for the bicrystal Cu-Al-Mn sample were detailed in the literature [11].

The cyclic compression–unloading test was performed under the strain-controlled mode using a universal tester with a 50 kN load cell (AG-IS 50 KN, Shimadzu, Japan). The strain rate used in the compression–unloading test was about 2.4×10^{-3} s^{-1} such that experiments can be considered as quasi-static. Each compression cycle took about 120 s. A speckle pattern was applied on the observed surface (i.e., area of interest, AOI) of the specimen using black and white sprays. The pattern was used for in situ strain tracing and ex post strain field analysis. The deformation strain of the specimen was measured with a virtual strain gauge by optical DIC (VIC-Gauge 3D, Correlated Solutions, Irmo, SC, USA). Three deformation strains (i.e., global gauge strain ε_g as shown in Figure 1 and the local strain gauges at the top and bottom grains ε_t and ε_b as shown in inset of Figure 2b) were measured using the virtual strain gauge technique. Notably, regardless of the residual strain, a 5% strain (relative to each unloaded state) was applied to the sample during each compression cycle. During the compression test, images of the deformed sample were taken at a rate of 5 Hz using two cameras. Around 600 snapshots were taken for each compression cycle. Because the imaging rate (5 s^{-1}) is higher than the strain rate (2.4 × 10^{-3} s^{-1}), the deformation behavior of the material can be captured. These snapshots were analyzed ex post in the VIC 3D 8 software to obtain the strain distribution at the surface of the specimen, as shown in the full-field strain measurement in Figure 1.

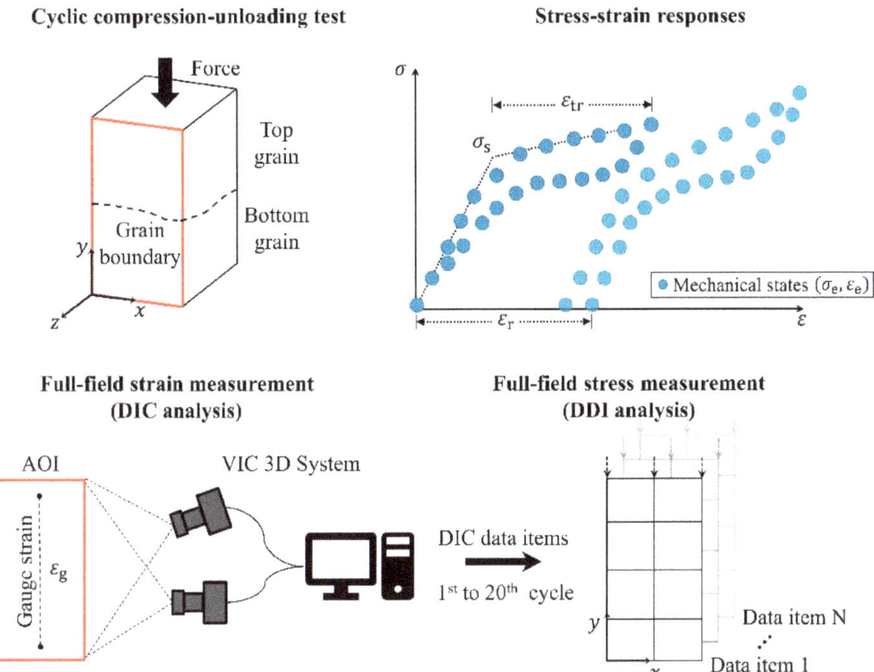

Figure 1. The digital image correlation (DIC) technique and data-driven identification (DDI) method were employed to measure the strain and stress distributions, respectively, at the surface of the specimen to characterize the cyclic behavior of the superelasticity of the bicrystal Cu-Al-Mn SMAs. The cyclic compression–unloading test was performed under the strain-controlled mode. The strain fields in the area of interest (AOI) can be obtained using the DIC technique. Based on the experimentally determined strain fields, the stress fields in the AOI can be computed using the DDI method. Finally, three parameters (i.e., transformation stress (σ_s), residual strain (ε_r), and transformation strain (ε_{tr})) can be computed from the stress–strain responses.

Based upon the measured results (i.e., DIC data items) of the strain fields and the applied loads from the full-field strain measurement, the equilibrated stress distribution at the surface of the specimen can be numerically computed using the DDI method [14–16] under a finite-element framework without the use of constitutive modeling, as illustrated in the full-field stress measurement in Figure 1. For a given set of compression–unloading cycle data, the DDI method uses the governing equations (i.e., stress equilibrium equations) to compute the mechanical stresses at the surface of the specimen. Then, the method is used to identify a database of material states that samples the mechanical stress–strain pairs of material to satisfy the criterion of minimum distance between mechanical stress–strain pairs and material database. Mathematically, this process can be formulated as a constrained minimization problem, which was originally proposed in [14]. The DDI method was validated with synthetic data for linear and non-linear elasticity [14] and was further applied experimentally using real experimental data (i.e., DIC measurements) for elastomer sheet [15] and Cu-Al-Mn shape memory alloy single crystal [16]. In the DDI method, the only governing equations used to determine the stress components are the stress equilibrium equations. Thus, the bias introduced by the choice and the calibration of a constitutive model was removed. Further details about the full-field stress and strain measurements can be found in the literature [15,16].

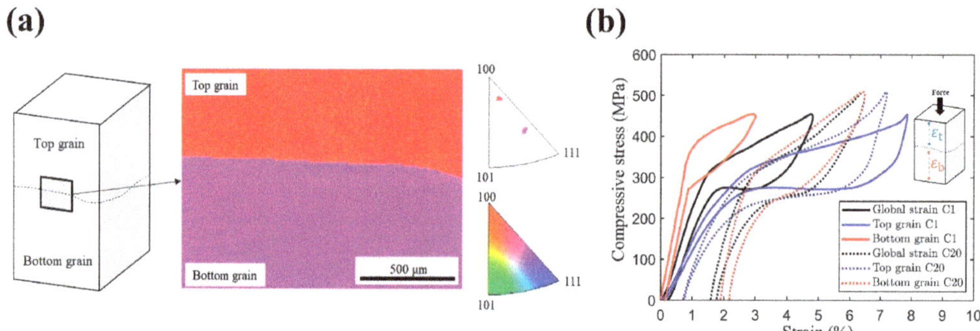

Figure 2. (**a**) Geometry of the bicrystal Cu-Al-Mn SMA. The loading directions of the top and bottom grains are shown in the inverse pole figure. (**b**) Average stress–strain curves of the top grain (ε_t), bottom grain (ε_b), and the entire specimen (ε_g). The bicrystal sample was loaded to a gauge strain (ε_g) of 5% during cyclic deformation. Local virtual strain gauges ε_t and ε_b were used to measure the average strains in the top and bottom grains, respectively (inset of (**b**)).

The mechanically admissible stress–strain pairs, obtained from the full-field stress and strain measurements shown in Figure 1, are stored at each element and are considered to represent the mechanical state $(\sigma_e, \varepsilon_e)$, as shown in the stress–strain responses in Figure 1. Using a linear regression analysis of these mechanical states in both the elastic and plateau regions, the distribution of MT stress σ_s can be obtained from the intersection of the two linear stress–strain curves, as illustrated by the stress–strain responses shown in Figure 1. The residual strain ε_r and transformation strain ε_{tr} can be obtained readily, as shown in the stress–strain responses of Figure 1.

3. Results

Figure 2a shows the crystal orientations of the top and bottom grains along the loading direction (LD), as determined using EBSD, which were along $[5326]_\beta$ and $[6511]_\beta$, respectively. The theoretical transformation strain (ε_T) of the transition from the β phase to 6M martensite during compression was calculated based on the Wechsler–Liebermann–Read theory [8,11,17]. The values of ε_T were 10.1% and 7.1% for the top and bottom grains, respectively. Note that the compressive transformation strain of these grains significantly differed in the loading direction.

Figure 2b shows the compressive stress–strain curves of both grains when a gauge strain (ε_g) of 5% was applied, which covered the entire specimen during deformation, as illustrated in Figure 1. Local virtual strain gauges ε_t and ε_b were used to measure the average strains in the top and bottom grains, respectively (inset of Figure 2b). It can be seen that the stress-induced martensitic transformation (SIMT) of the top grain (blue lines) occurred at transformation stresses of 306 MPa and 297 MPa for the first and twentieth cycles, respectively. By contrast, for the bottom grain (red lines), the transformation stresses of the first and twentieth compression cycles were 398 MPa and 292 MPa, respectively. Both grains exhibited different cyclic behaviors on average, including the decrease in transformation stress, accumulation of irrecoverable strain, and the transformation strain (see Table 1). As shown in Table 1, the properties (i.e., σ_s, ε_r, and ε_{tr}) of the total curve are between those of the top and bottom grains, which were reported in [11]. Please note that the total stress–strain curve is denoted as "Average" in first column of Table 1. According to the results of residual strain ε_r (sixth column of Table 1), it was found that the plastic deformation of the entire specimen was mainly contributed from the bottom grain. In short, combinations of crystal orientation in bicrystal SMAs will result in varied mechanical properties of the entire specimen.

Table 1. Loading direction (LD), theoretical transformation strain (ε_T), transformation stress (σ_s), residual strain (ε_r), and transformation strain (ε_{tr}) of the top grain, bottom grain, and the entire specimen (average) for the first and twentieth compression cycles. Three parameters (i.e., σ_s, and ε_{tr}) were computed from the stress–strain curves of the top grain, bottom grain, and the entire specimen (average) shown in Figure 2b.

	Loading Direction	Theoretical Transformation Strain (%)	Number of Cycles	Transformation Stress (MPa)	Residual Strain (%)	Transformation Strain (%)
Top	[5 3 26]	10.1	1st	306	0.12	5.7
			20th	297	0.77	5.1
Bottom	[6 5 11]	7.1	1st	398	0.14	2.0
			20th	292	2.17	3.9
Average	–	–	1st	313	0.24	3.5
			20th	290	1.78	4.0

Figure 3a shows the evolution of the axial strain field ε_{yy} while loading toward and unloading away from a gauge strain ε_g of 5% during the first, tenth, and twentieth compression–unloading cycles. The transformation stress fields are shown in Figure 3b, which illustrates the distribution of transformation stress in the specimen. As shown in the first compression cycle, the top grain underwent most of the deformation during the loading process. By contrast, the bottom grain began its partial MT after an ε_g of 3%. The difference in transformation behavior between the top grain and the bottom grain can be ascribed to differences in the MT stresses required to trigger MT, as shown in Figure 3b. The transformation stress of the top grain was approximately 325 MPa, which was less than that of the bottom grain (approximately 400 MPa), indicating that the top grain was more likely to begin MT earlier until the loading force was high enough to initiate MT in the bottom grain. The transformation stresses of both grains near the grain boundary (indicated by white dashed lines) were smaller than those further away from the grain boundary, as shown in Figure 3b, indicating that the stress state around the grain boundary promoted MT at a lower stress level.

Before the beginning of the tenth compression cycle, some regions in the bottom grain had residual strain. At the tenth deformation, the top grain experienced less deformation relative to the first cycle, whereas the bottom grain began to exhibit increased deformation. As shown in Figure 3b, at the tenth cycle, the transformation stress of the upper part of the bottom grain (near the grain boundary) decreased, bringing the values closer to those of the top grain. This decrease in transformation stress resulted in an increase in regions in which the MT could be triggered in the bottom grain, leading to increased transformation strain in the bottom grain.

At the beginning of the twentieth compression cycle, the band of residual strain at the bottom grain extended, and more residual strain remained. Furthermore, during the loading process, the upper part of the bottom grain showed a level of transformation stress closer to that of the top grain, as shown in Figure 3b. According to Figure 3b, the decrease in transformation stress in the bottom grain was initiated around the grain boundary and then propagated to the lower part of the bottom grain.

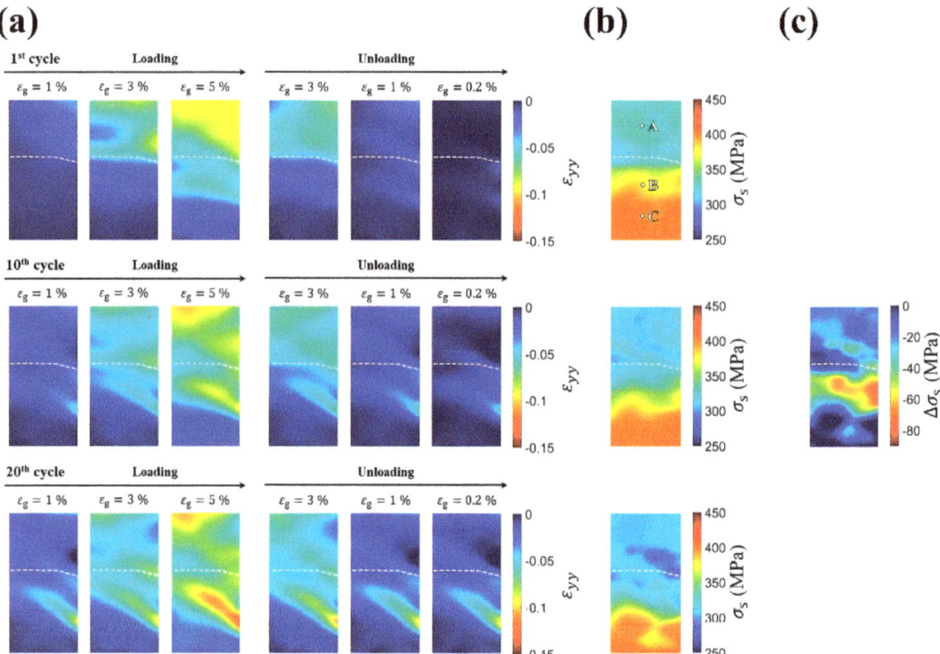

Figure 3. Distribution of (**a**) axial strain fields ε_{yy} during loading toward and unloading away from the gauge strain ε_g of 5% and (**b**) transformation stress fields σ_s in the bicrystal Cu-Al-Mn SMA sample for selected compression–unloading cycles: C1, C10, and C20. Points A, B, and C are probing points for recording the local axial stress–strain responses $(\sigma_{yy}, \varepsilon_{yy})$ as shown in Figure 4a. (**c**) Transformation stress difference $\Delta\sigma_s$, which is the difference in transformation stress between cycles 1 and 20, shown in the plot in Figure 3b.

By comparing the evolution of the strain distributions during the cyclic deformation, it can be deduced that the region exhibiting a decrease in transformation stress was highly correlated with the region undergoing MT. In the first compression test, the MT band in the bottom grain was clearly identified. During cyclic deformation, the MT in the bottom grain mainly originated from this band, and the residual strain in this region accumulated. This band, which was associated with accumulated permanent deformation, also experienced a more severe decrease in transformation stress than that of the top grain, as shown in Figure 3c, which reveals the decrement in transformation stress after twenty cycles. Because the transformation stress in the top and bottom grains (near the grain boundary) became similar after cyclic deformation, a concurrent MT occurred in the later cycles in these grains, leading to a significantly different deformation behavior from that of the first cycle.

To further investigate the relationships between transformation stress (σ_s) and residual strain (ε_r), the local axial stress–stain responses (σ_{yy}, ε_{yy}) at probing points A, B, and C (Figure 3b) are shown in Figure 4a. The evolutions of the transformation stresses and the accumulations of residual strains at these points are shown in Figure 4b,c. Point A was set in the top grain, and points B and C in the bottom grain were placed in the regions that underwent full MT and partial MT, respectively. At point A in the top grain, a stable transformation stress was observed after twenty compression cycles. By contrast, varied mechanical responses in terms of transformation stress and accumulation of irrecoverable strain were observed for points B and C in the bottom grain. The transformation stress and residual strain at these three points were quantified and are presented in Figure 4b,c, respectively. Figure 4b shows a comparison of different grain orientations (points A and B). The transformation stress at point A was observed to have slight decreasing behavior, while at point B, the decrease in transformation stress was more significant (i.e., from 400 MPa to

311 MPa). The results also show that, for points having the same grain orientation (points B and C), point B, which undergoes more MT (i.e., higher transformation strain, ε_{tr}), exhibits faster decreasing behavior than point C, as determined from the stress–strain curve in Figure 4a. In other words, in a single grain, a region that underwent more MT (i.e., high ε_{tr}) experienced a greater decrease in transformation stress. Notably, after twenty cycles, the transformation stresses at points A and B became nearly equal, as shown in Figure 4b, which resulted in more MT in the bottom grain.

Considering the loading cycles, a region that underwent more MT in the bottom grain also caused a faster accumulation of unrecoverable strain, as shown in Figure 4c. An ε_r of 4% at point B was observed after twenty compression cycles, roughly four times greater when compared with the accumulation at point C. For point A, an ε_r of 0.5% was observed, which is the minimum strain among these points owing to its lower transformation stress. In different grains, the grain requiring a higher stress to induce MT (the bottom grain) showed a larger residual strain and faster decrease in transformation stress. In addition, in the same grain (the bottom grain), the regions with more MT accumulated more residual strain and exhibited a clear decrease in transformation stress.

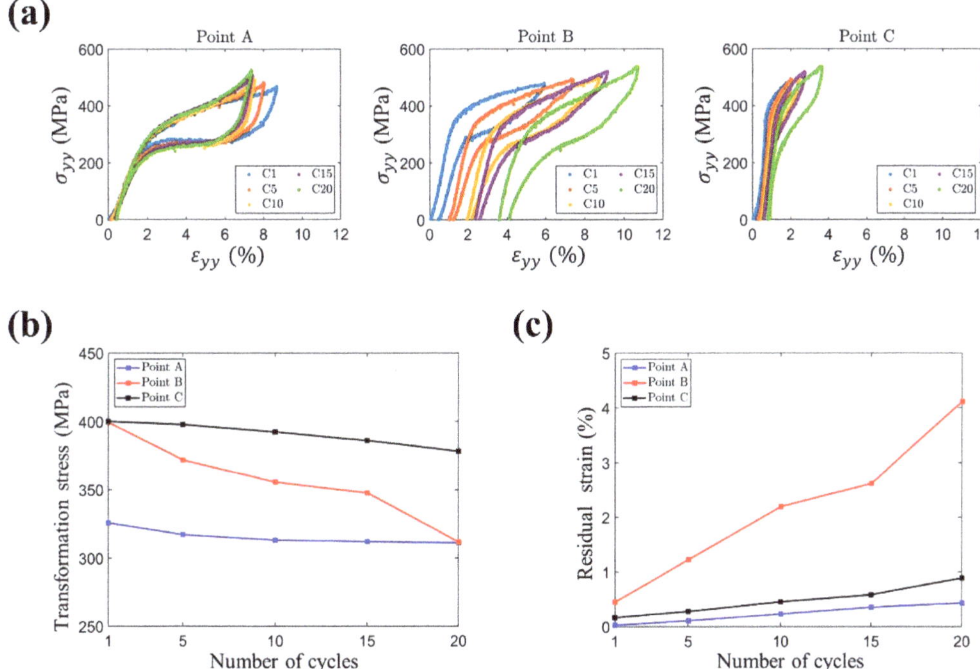

Figure 4. (**a**) Local axial stress–strain responses (σ_{yy}, ε_{yy}) recorded by the probing points (according to Figure 3b) along the axial centerline for several selected compression–unloading cycles (C1, C5, C10, and C20). The evolution of the (**b**) transformation stress σ_s and (**c**) residual strain ε_r with respect to the number of cycles. These values are computed from the local axial stress–strain responses shown in (**a**).

4. Discussion

The difference in transformation stress between the two crystal grains is attributed to the difference of their crystal orientations. According to [18], the habit planes of Cu-Al-Mn martensite are {0.16 −0.72 −0.68} and the shear directions are <0.14 −0.65 0.74>. With these transformation systems, the maximum Schmid factor of the two grains with orientations $[5\,3\,26]_\beta$ (top grain) and $[6\,5\,11]_\beta$ (bottom grain) is determined to be 0.49 and 0.32, respectively. As the Schmid factor of the bottom grain is smaller than that of the top grain, larger stress is needed to trigger MT in the bottom grain. Therefore, the transformation stress of the bottom grain (398 MPa) is larger than that of the top grain (306 MPa), as shown in Figure 2b and Table 1.

The previous results (Figure 3) show that the decrease in transformation stress in a bicrystal Cu-Al-Mn SMA under a strain-controlled cyclic compression–unloading test was affected by the accumulation of residual strain. These macroscopic residual strains are mainly a result of dislocation slips in the austenite phase [19–24] and accumulated residual martensite phase owing to an incomplete reverse MT [19,25,26]. These dislocation slips, which can be triggered separately during forward and reverse MTs [27], are fostered by localized stress fields between austenite–martensite interfaces during forward and reverse MTs [20,28–30]. Such dislocation slips and residual martensite can also cause mesoscopic residual stress fields within the specimen [31]. Therefore, the mesoscopic residual stress field, which is of the same type as the applied stress, assists in the nucleation of martensite variants [19], and thus leads to a significant reduction in the macroscopic transformation stress required to trigger MT during subsequent cycles [32–36].

In order to provide clear evidence, the bottom grain was cut from the bicrystal sample for thermal analysis. As shown in Figure 5a, after 20 cyclic deformations, the first heat curve shows that the reverse MT occurred at about 140.1 °C. The sample was then cooled to −140 °C (Step 2), and a forward MT was identified at −87.7 °C. During the second heating (Step 3), the reverse transformation occurred at −69.7 °C, instead of the 140.1 °C in the first heating curve. This feature indicated that the martensite was stabilized during the cyclic compressions. The stabilized martensite needed a higher temperature to transform back to austenite, as shown in the first heating curve. In the successive cooling and heating, the martensite was thermally induced and thus was not stabilized, causing the reverse transformation temperature to be restored to its normal value (−69.7 °C). The feature of martensite being stabilized after deformation was also reported in other Cu-based [37] and TiNi-based SMAs [38].

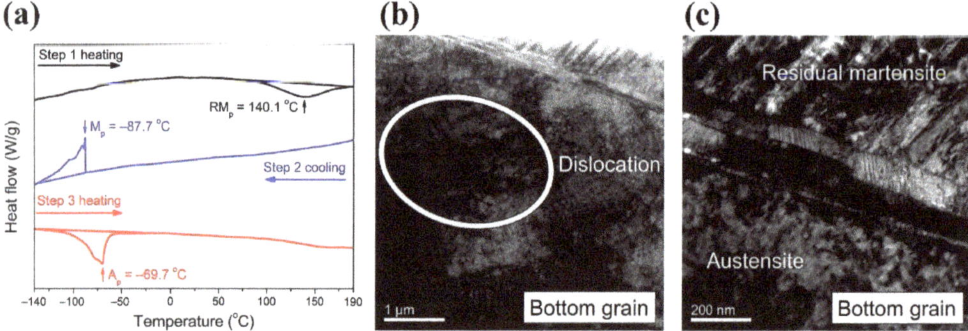

Figure 5. (**a**) Thermal analysis of the bottom grain after 20 compression cycles. (**b**,**c**) TEM bright field images of the bottom grain, which show the formation of dislocations and residual martensite after cyclic compression, respectively.

TEM analyses were performed on the bottom grain after cyclic compression. Figure 5b shows the TEM bright-field image of the bottom grain, in which entangled dislocations can be observed. Figure 5c shows large amounts of residual martensite in the bottom grain, indicating that martensite was stabilized at room temperature by dislocations after cyclic deformation. The TEM observations confirmed that the residual strain is caused by the dislocation and residual martensite formed during cyclic deformation.

As shown in Figure 2b, the higher transformation stress in the bottom grain generates more dislocations in the austenite phase and more residual martensite during the forward and reverse MT. Both mechanisms lead to plasticity formed mainly in the bottom grain. Both the dislocation and stabilized martensite can cause mesoscopic residual stress fields within the specimen. Such residual stress assisted the nucleation of martensite variants and thus led to a significant reduction in the macroscopic transformation stress required to trigger MT during subsequent cycles. On the other hand, the top grain exhibited smaller residual strain owing to its smaller transformation stress, and thus its transformation stress remained stable. In addition, in the bottom grain, the decreasing behavior varied significantly between regions undergoing different levels of transformation strain (points B and C). As shown in Figure 3c or Figure 4b, point B experienced more MT and associated residual strain compared with point C. Consequently, point B exhibited a larger decrease in transformation stress than point C. These results support the hypothesis that the mesoscopic stress field caused by dislocations or residual martensite assisted MT.

In this study, the horizontal strain fields ε_{xx} at the surface of the specimen were examined to investigate the deformation in the vicinity of the grain boundary. Figure 6a shows the evolution of ε_{xx} during loading toward and unloading away from a gauge strain ε_g of 5% for the first, tenth, and twentieth compression–unloading cycles. As shown in the unloaded state (ε_g = 0.2%) of the first compression cycle, the bottom grain exhibits an accumulation of residual ε_{xx} near the grain boundary (highlighted in the red rectangle), while the remaining part of the grain boundary shows relatively less residual ε_{xx}. In the tenth and twentieth cycles, this accumulation of residual ε_{xx} in the red rectangle kept increasing and expanding toward the remaining part of the grain boundary. Note that, in the right half-part of the grain boundary, the bottom grain accumulated more residual ε_{xx} than the top grain. These differences in the accumulation of residual ε_{xx} or strain incompatibility in the vicinity of the grain boundary in a bicrystal reveal the relative movement between the top grain and bottom grain.

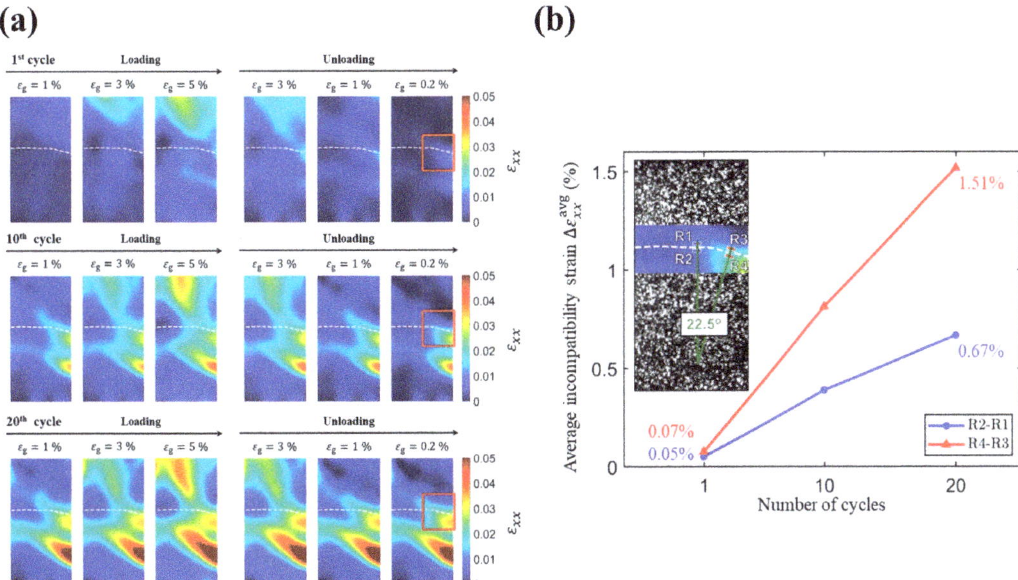

Figure 6. (**a**) Distribution of horizontal strain fields ε_{xx} during loading toward and unloading away from the gauge strain ε_g of 5% in the bicrystal Cu-Al-Mn SMA sample for selected compression–unloading cycles: C1, C10, and C20. (**b**) The evolution of average incompatibility strain $\Delta\varepsilon_{xx}^{avg}$ in the regions (R2-R1 and R4-R3) with respect to the number of cycles. These values are computed from the strain fields multiplied by transformation matrix based on the angle between loading direction and normal direction of the grain boundary (inset of (**b**)).

To further investigate incompatibility conditions of the grain boundary, we quantified the average incompatibility strain $\Delta\varepsilon_{xx}^{avg}$ near the grain boundary. The definition of $\Delta\varepsilon_{xx}^{avg}$ is the average ε_{xx} in the selected region of the bottom grain minus the average ε_{xx} in the selected region of the top grain. As shown in inset of Figure 6b, we selected four regions near the grain boundary (R1, R2, R3, and R4) and paired the regions (i.e., R2–R1 and R4–R3) to compute their evolution of $\Delta\varepsilon_{xx}^{avg}$ regarding the selected compression–unloading cycles, C1, C10, and C20. In Figure 6b, the average incompatibility strain $\Delta\varepsilon_{xx}^{avg}$ in the region R4–R3 increased from 0.07% to 1.51%, while the $\Delta\varepsilon_{xx}^{avg}$ in the region R2–R1 increased from 0.05% to 0.67%. These results show that the incompatibility of the grain boundary in the region R4–R3 is more severe than that in the region R2–R1. Such a difference between these two regions can be ascribed to the angle between the loading direction and normal direction of the grain boundary. As can be seen in the inset of Figure 6b, the grain boundary in the region R4-R3 deviates from the horizontal plane by approximately 22.5 degrees. This variation in the direction of grain boundary will introduce the major axial deformation ε_{yy} as an extra component of tangential movement to the inclined grain boundary (see the right half-part of the grain boundary during the loading process in Figure 3a). Hence, the regions near the inclined grain boundary will experience not only larger axial deformation (Figure 3a), but also larger relative movement (Figure 6a) compared with the regions near the flat grain boundary. Consequently, a significant difference in average incompatibility strain $\Delta\varepsilon_{xx}^{avg}$ between the two regions (i.e., R2–R1 and R4–R3) occurs, as shown in Figure 6b. Additionally, as reported in the previous work [11], the top and bottom grains underwent outward and inward out-of-plane deformations during compression–unloading cycles. This out-of-plane motion was considered for causing generation of microcracks. In this study, we found that the incompatibility along the x direction of the bicrystal Cu-Al-Mn sample under cyclic compression could be another mechanism for formation of microcrack at the grain boundary. Thus, even though compressive deformation is considered a preferred deformation mode to delay fatigue fracture of metallic materials [12,13], the out-of-plane motion and relative deformation near the grain boundary along the x direction in a bicrystal Cu-Al-Mn sample may cause cracking and even fracture in Cu-Al-Mn SMAs.

With the aid of the full-field stress and strain measurements (i.e., DIC and DDI techniques) for revealing strain and transformation stress fields, it was found that grain boundary kinking plays an important role in the decrease in transformation stress near the grain boundary in the bicrystal Cu-Al-Mn sample. As can be seen in Figure 3a, during the loading process, because the right half-part of the grain boundary underwent more MT than the other part of the grain boundary, more residual strain remained near the right half-part of the grain boundary at the end of the cycle (ε_g = 0.2% in Figure 3a). As mentioned above, residual strain causes the decrease in transformation stress. Hence, this additional residual strain near the kink grain boundary will induce an extra decrease in transformation stress. As can be seen in Figure 3c, the transformation stress near the right half-part of the grain boundary degraded faster than other part of the grain boundary. Thus, we concluded that the grain boundary kinking in bicrystal induces an additional decrease in transformation stress.

Furthermore, during the strain-controlled cyclic loading (5% strain for each cycle), the transformation stress of the bottom grain gradually decreased to the values close to those of the top grain (Figure 3b,c). Hence, the bottom grain gradually had more MT, and thus a larger ε_{tr} than in its first cycle, as shown in Figure 3a. This explains why the average ε_{tr} in the bottom grain increased with the increasing numbers of deformation cycles, as shown in Figure 2b and Table 1. By contrast, because the bottom grain contributed more deformation, the MT and the associated strain contributed by the top grain decreased when the deformation cycles increased, as shown in Figure 2b or Figure 3a. Hence, the average ε_{tr} in the top grain decreased with the increasing number of deformation cycles, as shown in Table 1. Therefore, the initially inhomogeneous deformation behavior in the bicrystal sample became slightly more homogeneous after twenty cyclic compressive deformations, as shown in Figure 3a.

The full-field stress and strain measurements provide a promising technique for measuring stress–strain responses in SMAs. This new method offers not only insights into cyclic superelastic deformation, but also the compressive fatigue behavior of Cu-Al-Mn

SMAs. With this method, several future research directions could be considered, including the influence of misorientation between two grains on its mechanical properties, the effect of grain boundary geometry on the grain boundary strength [39–41], and the shape memory recoverability between two grains in a bicrystal sample. Furthermore, a direct connection between macroscopic shape memory effects and stress and strain states at the materials' grains could also be elucidated in the future.

5. Conclusions

In summary, this study investigated the distribution of transformation stress and strain fields in a bicrystal Cu-Al-Mn sample under cyclic compression. The decrease in transformation stress in both grains correlated with the accumulation of residual strain. The accumulation of residual strain depends on factors such as grain orientation along the loading directions, transformation strain, and grain boundaries. These experimental results and analyses demonstrated that dislocation slip and residual martensite were triggered more easily when a higher transformation stress was required to trigger MT, thus resulting in greater residual strain and a larger decrease in transformation stress. The decreasing behavior at the grain boundary was related to strain incompatibility and the angle between loading direction and the normal direction of grain boundary. Consequently, microstructures with low-angle grain boundaries or single crystals, which exhibit more homogenous deformation behaviors and less restrictions from grain boundaries, will demonstrate higher functional stability and thus longer lifetimes during their cyclic service lives.

Author Contributions: Conceptualization, T.-H.S., C.-H.C. and C.-S.C.; data curation, T.-H.S.; formal analysis, T.-H.S. and N.-H.L.; investigation, T.-H.S. and N.-H.L.; methodology, T.-H.S. and N.-H.L.; software, T.-H.S. and N.-H.L.; visualization, T.-H.S. and N.-H.L.; resources, N.-H.L.; writing—original draft, T.-H.S., N.-H.L. and C.-H.C.; writing—review and editing, T.-H.S., N.-H.L., C.-H.C. and C.-S.C.; funding acquisition, C.-H.C. and C.-S.C.; supervision, C.-H.C. and C.-S.C.; project administration, C.-S.C. All authors have read and agreed to the published version of the manuscript.

Funding: This work was supported by the Ministry of Science and Technology (MOST), Taiwan, under Grant 108-2221-E-002-005-MY3, and by the Young Scholar Fellowship program of MOST, under Grant 109-2636-E-002-031.

Institutional Review Board Statement: Not applicable.

Informed Consent Statement: Not applicable.

Data Availability Statement: The data that support the findings of this study are available from the corresponding author upon reasonable request.

Conflicts of Interest: The authors declare that they have no known competing financial interests or personal relationships that could have appeared to influence the work reported in this paper.

References

1. Sutou, Y.; Omori, T.; Kainuma, R.; Ishida, K. Ductile Cu-Al-Mn based shape memory alloys: General properties and applications. *Mater. Sci. Technol.* **2008**, *24*, 896–901. [CrossRef]
2. Sutou, Y.; Omori, T.; Kainuma, R.; Ishida, K. Grain size dependence of pseudoelasticity in polycrystalline Cu–Al–Mn-based shape memory sheets. *Acta Mater.* **2013**, *61*, 3842–3850. [CrossRef]
3. Liu, J.-L.; Huang, H.-Y.; Xie, J.-X. The roles of grain orientation and grain boundary characteristics in the enhanced superelasticity of $Cu_{71.8}Al_{17.8}Mn_{10.4}$ shape memory alloys. *Mater. Des.* **2014**, *64*, 427–433. [CrossRef]
4. Xie, J.X.; Liu, J.L.; Huang, H.Y. Structure design of high-performance Cu-based shape memory alloys. *Rare Met.* **2015**, *34*, 607–624. [CrossRef]
5. Babacan, N.; Ma, J.; Turkbas, O.S.; Karaman, I.; Kockar, B. The effects of cold rolling and the subsequent heat treatments on the shape memory and the superelasticity characteristics of $Cu_{73}Al_{16}Mn_{11}$ shape memory alloy. *Smart Mater. Struct.* **2018**, *27*, 015028. [CrossRef]
6. Sutou, Y.; Omori, T.; Yamauchi, K.; Ono, N.; Kainuma, R.; Ishida, K. Effect of grain size and texture on pseudoelasticity in Cu-Al-Mn-based shape memory wire. *Acta Mater.* **2005**, *53*, 4121–4133. [CrossRef]
7. Tuncer, N.; Schuh, C.A. Melt-cast microfibers of Cu-based shape memory alloy adopt a favorable texture for superelasticity. *Scr. Mater.* **2016**, *117*, 46–50. [CrossRef]

8. Sutou, Y.; Omori, T.; Kainuma, R.; Ishida, K.; Ono, N. Enhancement of superelasticity in Cu-Al-Mn-Ni shape-memory alloys by texture control. *Metall. Mater. Trans. A* **2002**, *33*, 2817–2824. [CrossRef]
9. Omori, T.; Kusama, T.; Kawata, S.; Ohnuma, I.; Sutou, Y.; Araki, Y.; Ishida, K.; Kainuma, R. Abnormal grain growth induced by cyclic heat treatment. *Science* **2013**, *341*, 1500–1502. [CrossRef]
10. Kusama, T.; Omori, T.; Saito, T.; Kise, S.; Tanaka, T.; Araki, Y.; Kainuma, R. Ultra-large single crystals by abnormal grain growth. *Nat. Commun.* **2017**, *8*, 354. [CrossRef]
11. Lu, N.-H.; Chen, C.-H. Inhomogeneous martensitic transformation behavior and elastocaloric effect in a bicrystal Cu-Al-Mn shape memory alloy. *Mater. Sci. Eng. A* **2021**, *800*, 140386. [CrossRef]
12. Wu, Y.; Ertekin, E.; Sehitoglu, H. Elastocaloric cooling capacity of shape memory alloys—Role of deformation temperatures, mechanical cycling, stress hysteresis and inhomogeneity of transformation. *Acta Mater.* **2017**, *135*, 158–176. [CrossRef]
13. Zhang, K.; Kang, G.; Sun, Q. High fatigue life and cooling efficiency of NiTi shape memory alloy under cyclic compression. *Scr. Mater.* **2019**, *159*, 62–67. [CrossRef]
14. Dalémat, M.; Coret, M.; Leygue, A.; Verron, E. Measuring stress field without constitutive equation. *Mech. Mater.* **2019**, *136*, 103087. [CrossRef]
15. Su, T.-H.; Lu, N.-H.; Chen, C.-H.; Chen, C.-S. Full-field stress and strain measurements revealing energy dissipation characteristics in martensitic band of Cu-Al-Mn shape memory alloy. *Mater. Today Commun.* **2020**, *24*, 101321. [CrossRef]
16. Leygue, A.; Coret, M.; Réthoré, J.; Stainier, L.; Verron, E. Data-based derivation of material response. *Comput. Method Appl. Mech. Eng.* **2018**, *331*, 184–196. [CrossRef]
17. Lieberman, D.S.; Wechsler, M.S.; Read, T.A. Cubic to orthorhombic diffusionless phase change—Experimental and theoretical studies of AuCd. *J. Appl. Phys.* **1955**, *26*, 473–484. [CrossRef]
18. Omori, T.; Kawata, S.; Kainuma, R. Orientation Dependence of Superelasticity and Stress Hysteresis in Cu–Al–Mn Alloy. *Mater. Trans.* **2020**, *61*, 55–60. [CrossRef]
19. Miyazaki, S.; Imai, T.; Igo, Y.; Otsuka, K. Effect of cyclic deformation on the pseudoelasticity characteristics of Ti-Ni alloys. *Metall. Trans. A* **1986**, *17*, 115–120. [CrossRef]
20. Norfleet, D.M.; Sarosi, P.M.; Manchiraju, S.; Wagner, M.F.X.; Uchic, M.D.; Anderson, P.M.; Mills, M.J. Transformation-induced plasticity during pseudoelastic deformation in Ni–Ti microcrystals. *Acta Mater.* **2009**, *57*, 3549–3561. [CrossRef]
21. Delville, R.; Malard, B.; Pilch, J.; Sittner, P.; Schryvers, D. Transmission electron microscopy investigation of dislocation slip during superelastic cycling of Ni–Ti wires. *Int. J. Plast.* **2011**, *27*, 282–297. [CrossRef]
22. Simon, T.; Kröger, A.; Somsen, C.; Dlouhy, A.; Eggeler, G. On the multiplication of dislocations during martensitic transformations in NiTi shape memory alloys. *Acta Mater.* **2010**, *58*, 1850–1860. [CrossRef]
23. Pfetzing-Micklich, J.; Ghisleni, R.; Simon, T.; Somsen, C.; Michler, J.; Eggeler, G. Orientation dependence of stress-induced phase transformation and dislocation plasticity in NiTi shape memory alloys on the micro scale. *Mater. Sci. Eng. A* **2012**, *538*, 265–271. [CrossRef]
24. Mohamadnejad, S.; Basti, A.; Ansari, R. Analyses of Dislocation Effects on Plastic Deformation. *Multiscale Sci. Eng.* **2020**, *2*, 69–89. [CrossRef]
25. Brinson, L.C.; Schmidt, I.; Lammering, R. Stress-induced transformation behavior of a polycrystalline NiTi shape memory alloy: Micro and macromechanical investigations via in situ optical microscopy. *J. Mech. Phys. Solids* **2004**, *52*, 1549–1571. [CrossRef]
26. Kato, H.; Ozu, T.; Hashimoto, S.; Miura, S. Cyclic stress–strain response of superelastic Cu–Al–Mn alloy single crystals. *Mater. Sci. Eng. A* **1999**, *264*, 245–253. [CrossRef]
27. Heller, L.; Seiner, H.; Šittner, P.; Sedlák, P.; Tyc, O.; Kadeřávek, L. On the plastic deformation accompanying cyclic martensitic transformation in thermomechanically loaded NiTi. *Int. J. Plast.* **2018**, *111*, 53–71. [CrossRef]
28. Kato, H.; Sasaki, K. Transformation-induced plasticity as the origin of serrated flow in an NiTi shape memory alloy. *Int. J. Plast.* **2013**, *50*, 37–48. [CrossRef]
29. Wang, J.; Sehitoglu, H.; Maier, H.J. Dislocation slip stress prediction in shape memory alloys. *Int. J. Plast.* **2014**, *54*, 247–266. [CrossRef]
30. Paranjape, H.M.; Bowers, M.L.; Mills, M.J.; Anderson, P.M. Mechanisms for phase transformation induced slip in shape memory alloy micro-crystals. *Acta Mater.* **2017**, *132*, 444–454. [CrossRef]
31. Yu, C.; Kang, G.; Kan, Q. A micromechanical constitutive model for anisotropic cyclic deformation of super-elastic NiTi shape memory alloy single crystals. *J. Mech. Phys. Solids* **2015**, *82*, 97–136. [CrossRef]
32. Moumni, Z.; Herpen, A.V.; Riberty, P. Fatigue analysis of shape memory alloys: Energy approach. *Smart Mater. Struct.* **2005**, *14*, S287–S292. [CrossRef]
33. Zaki, W.; Moumni, Z. A 3D model of the cyclic thermomechanical behavior of shape memory alloys. *J. Mech. Phys. Solids* **2007**, *55*, 2427–2454. [CrossRef]
34. Kan, Q.; Yu, C.; Kang, G.; Li, J.; Yan, W. Experimental observations on rate-dependent cyclic deformation of super-elastic NiTi shape memory alloy. *Mech. Mater.* **2016**, *97*, 48–58. [CrossRef]
35. Gu, X.; Zhang, W.; Zaki, W.; Moumni, Z. An extended thermomechanically coupled 3D rate-dependent model for pseudoelastic SMAs under cyclic loading. *Smart Mater. Struct.* **2017**, *26*, 095047. [CrossRef]
36. Wang, J.; Moumni, Z.; Zhang, W. A thermomechanically coupled finite-strain constitutive model for cyclic pseudoelasticity of polycrystalline shape memory alloys. *Int. J. Plast.* **2017**, *97*, 194–221. [CrossRef]

37. Picornell, C.; Pons, J.; Cesari, E. Stabilisation of martensite by applying compressive stress in Cu-Al-Ni single crystals. *Acta Mater.* **2001**, *49*, 4221–4230. [CrossRef]
38. Lin, H.C.; Wu, S.K. Determination of heat of transformation in a cold-rolled martensitic tini alloy. *Metall. Trans. A* **1993**, *24*, 293–299. [CrossRef]
39. Miyazaki, S.; Kawai, T.; Otsuka, K. Study of fracture in cualni shape memory bicrystals. *J. Phys. Colloq.* **1982**, *43*, C4-813. [CrossRef]
40. Creuziger, A.; Crone, W.C. Grain boundary fracture in CuAlNi shape memory alloys. *Mater. Sci. Eng. A* **2008**, *498*, 404–411. [CrossRef]
41. Takezawa, K.; Izumi, T.; Chiba, H.; Sato, S. Coherency of the transformation strain at the grain boundary and fracture in Cu-Zn-Al alloy. *J. Phys. Colloq.* **1982**, *43*, C4-819–C4-824. [CrossRef]

MDPI
St. Alban-Anlage 66
4052 Basel
Switzerland
Tel. +41 61 683 77 34
Fax +41 61 302 89 18
www.mdpi.com

Materials Editorial Office
E-mail: materials@mdpi.com
www.mdpi.com/journal/materials